T0269276

HEMOPHILIA AND VON WILLEBRAND DISEASE

HEMOPHILIA AND VON WILLEBRAND DISEASE

Factor VIII and Von Willebrand Factor

DAVID GREEN

ACADEMIC PRESS

An imprint of Elsevier

Academic Press is an imprint of Elsevier
125 London Wall, London EC2Y 5AS, United Kingdom
525 B Street, Suite 1650, San Diego, CA 92101, United States
50 Hampshire Street, 5th Floor, Cambridge, MA 02139, United States
The Boulevard, Langford Lane, Kidlington, Oxford OX5 1GB, United Kingdom

Notices
Knowledge and best practice in this field are constantly changing. As new research and experience broaden our understanding, changes in research methods, professional practices, or medical treatment may become necessary.

Practitioners and researchers must always rely on their own experience and knowledge in evaluating and using any information, methods, compounds, or experiments described herein. In using such information or methods they should be mindful of their own safety and the safety of others, including parties for whom they have a professional responsibility.

To the fullest extent of the law, neither the Publisher nor the authors, contributors, or editors, assume any liability for any injury and/or damage to persons or property as a matter of products liability, negligence or otherwise, or from any use or operation of any methods, products, instructions, or ideas contained in the material herein.

Library of Congress Cataloging-in-Publication Data
A catalog record for this book is available from the Library of Congress

British Library Cataloguing-in-Publication Data
A catalogue record for this book is available from the British Library

ISBN 978-0-12-812954-8

For information on all Academic Press publications
visit our website at https://www.elsevier.com/books-and-journals

  Working together
to grow libraries in
developing countries

www.elsevier.com • www.bookaid.org

Publisher: John Fedor
Acquisition Editor: Tari K. Broderick
Editorial Project Manager: Tracy Tufaga
Production Project Manager: Sreejith Viswanathan
Cover Designer: Miles Hitchen

Typeset by SPi Global, India

Contents

10. Von Willebrand Disease: Classification and Diagnosis

11. Treatment of Von Willebrand Disease

12. Acquired Von Willebrand Syndrome

13. Factor VIII and Thrombosis

14. Von Willebrand Factor and Thrombosis

15. Factor VIII/Von Willebrand Factor: The Janus of Coagulation

Author Biography

David Green, MD, PhD, is Professor Emeritus in Medicine, Division of Hematology/Oncology, at Northwestern University Feinberg School of Medicine in Chicago, Illinois. He received his medical degree from Jefferson Medical College in Philadelphia, Pennsylvania, and Doctorate in Biochemistry from Northwestern University, Evanston, Illinois. He is a clinician-investigator and author of more than 300 published scientific articles. His most recent book, *Linked by Blood: Hemophilia and AIDS*, describes the AIDS epidemic in the early 1980s that ravaged the hemophilia community and led to major changes in the collection and processing of blood and blood products. He is a Master of American College of Physicians and recipient of many other awards.

Preface

My first encounter with factor VIII occurred during my medical residency in the early 1960s. Trainees were given the opportunity to perform laboratory research, and I elected to work under a famed hematologist, Leandro M. Tocantins (1901–63) [1]. Doctor Tocantins was interested in the role of platelets and clotting factors in blood coagulation. During the course of his work, he observed that the clotting time of blood containing the anticoagulant, sodium citrate, became shorter with storage. He suggested that I investigate this phenomenon.

During the course of this work, I considered the possibility that activation of factor VIII (FVIII) might be responsible for the shortening of the clotting times. FVIII, a protein required for normal blood coagulation, is decreased or absent in people with classical hemophilia. A method for measuring FVIII had been developed by Biggs and Douglas in 1953 [2], but was technically very demanding, available in very few laboratories, and was beyond my capabilities at the time. Fortunately, two researchers in Doctor Tocantins' laboratory, Ruth Holburn and Margaret DeSipin, were very experienced in conducting clotting studies, and performed the FVIII measurements for me. Although the results of these assays did not enlighten my studies, I became intrigued by FVIII and resolved to learn more about this clotting protein.

During the course of my clinical training and in the decades since, my interest in FVIII has only grown more intense. Not only is the factor absolutely essential for the clotting of blood of nearly all vertebrate species, but it is also intricately related to the Von Willebrand Factor (VWF), another essential hemostatic protein. In recent years, the relationship between these two factors has finally been clarified, and the treatment of hemophilia and Von Willebrand Disease greatly improved. I believe that a book devoted to these two proteins and their disorders is long overdue.

David Green
Feinberg School of Medicine of Northwestern University,
Chicago, IL, United States

References

[1] Dameshek W. Leandro M Tocantins, M.D. Blood 1963;22:360–1.
[2] Biggs R, Douglas AS. The thromboplastin generation test. J Clin Pathol 1953;6:23–9.

Acknowledgments

The author thanks Anaa Zakarija, MD, for helpful discussions and contributing clinical material, and Sandy Harris, RN, for her long-term commitment to the care of people with bleeding disorders. He is grateful to his wife, Theodora, for her critical review of the manuscript, cogent advice, and unflagging support and encouragement.

Introduction

FVIII/VWF: THE JANUS OF HEMOSTASIS

Factor VIII (FVIII) and Von Willebrand Factor (VWF) are proteins that are deficient or defective in hemophilia A (classical hemophilia) and Von Willebrand Disease (VWD), respectively. They have an impressive history. FVIII first appeared at least 430 million years ago and has been discovered among the coagulation proteins of all jawed vertebrates. Similarly, VWF evolved in ancestral vertebrates some 500 million years ago and has been identified in the Atlantic hagfish (slimy eel), a fish of ancient origin. More recently, there are references to a disease resembling hemophilia in the 5th century Talmud, and VWD was first described in the medical literature in 1926. The relationship between FVIII and VWF baffled investigators for years because FVIII and VWF circulate in the blood bound together in a single complex. It was not until the early 1970s that FVIII was clearly distinguished from VWF. Furthermore, the two congenital bleeding disorders, hemophilia and VWD, have decreased levels of FVIII, leading to the misdiagnosis of hemophilia in some patients with VWD.

The FVIII/VWF complex can be symbolically represented by Janus, the double-faced Roman god, who was known as the "God of Beginnings" and "Gatekeeper," and decreed who and what might pass through his portals. In like fashion, VWF initiates hemostasis by mediating the binding of platelets to the injured vessel wall, and FVIII opens the door to coagulation by binding to the tenase complex, an essential step in thrombin generation. Because of the intimate association of FVIII and VWF in the physiology of hemostasis, it seems logical to include the two clotting proteins and their disorders in a single volume. This format enables clinicians to consult one source for the information they might need about the pathophysiology, diagnosis, and treatment of both hemophilia and VWD.

FVIII and VWF are synthesized by endothelial cells and packaged into a specific cell organelle, the Weibel-Palade Body (WPB). Upon stimulation by agents such as desmopressin (DDAVP), the WPB releases the FVIII/VWF complex into the circulation, which accounts for the therapeutic efficacy of desmopressin in selected patients with hemophilia and VWD. However, those with severe disease usually require clotting factor concentrates; in *Hemophilia and Von Willebrand Disease*, the reader will find a listing of commercial products approved for the treatment of hemophilia and

VWD. The pharmaceutical industry is actively engaged in engineering FVIII concentrates that have a longer half-life and are less immunogenic, because alloantibodies or inhibitors arise in up to 30% of hemophiliacs exposed to concentrates. This book describes current concepts of antibody formation as well as patient management with bypassing agents and the induction of FVIII tolerance. A new approach is the subcutaneous administration of drugs that enhance plasma procoagulant activity without requiring FVIII. These agents have been shown to decrease annual bleeding rates in hemophiliacs with inhibitors and are now in clinical trials.

The text explores the genetic landscape of hemophilia and VWD and provides detailed descriptions of the genes known to affect the synthesis, release, and clearance of FVIII and VWF. Hundreds of mutations in these genes result in clinical conditions that vary greatly in their presentation and severity. Gene therapy for serious congenital disorders is rapidly becoming a reality, and hemophilia is no exception; briefly summarized are recent data from pilot studies in patients with severe hemophilia A.

Several acquired diseases induce a hemorrhagic diathesis by attacking FVIII or VWF. Some of these disorders induce autoantibodies, while others affect the release or clearance of the clotting factors. The clinical features, laboratory diagnosis, and treatment of these disorders are described in detail in *Hemophilia and Von Willebrand Disease.* A high degree of clinical suspicion is required to recognize these conditions, but once the diagnosis is established, most can be successfully treated.

There is considerable evidence that excessive concentrations of FVIII and VWF fuel the formation of pathologic thrombi that obstruct normal blood flow. Increased levels of the factors are inherited in families with specific genetic polymorphisms or are acquired in patients with cancer, diabetes, hyperthyroidism, and other diseases. The book reviews epidemiologic and other evidence that high concentrations of FVIII and VWF contribute to cardiovascular disease, venous thromboembolism, and embolic phenomena in patients with atrial fibrillation.

Some common **misperceptions** about hemophilia and VWD are

- Affected members of The Royal Families of Europe had classical hemophilia A
- The female carriers of hemophilia rarely have bleeding problems
- People with hemophilia are spared from atherosclerosis and coronary artery disease
- FVIII alloantibodies (inhibitors) only affect individuals with severe hemophilia
- Reduced FVIII levels are restricted to people with hemophilia A
- Hemarthroses do not occur in individuals with VWD
- Desmopressin is used in the same dose for diabetes insipidus and hemophilia
- The levels of FVIII & VWF in people with mild hemophilia or VWD do not increase with aging

Some of these erroneous ideas might potentially harm patients and are fully discussed and corrected within the pages of *Hemophilia and Von Willebrand Disease*.

The science of hemostasis is rapidly expanding and biomedical professionals are being challenged to remain current with vast amounts of published information. This compact book will be useful for healthcare workers, geneticists, and members of the pharmaceutical industry. It includes contemporary data on FVIII and VWF that will assist researchers in hematology, pathology, cell biology, and those interested in drug discovery. For clinicians, the book describes the major features of hemophilia A and VWD, indicates the laboratory studies required for a comprehensive diagnosis, and includes a differential diagnosis. It provides descriptions of approved therapeutic concentrates, their indications, recommended doses, and potential adverse effects. Most of the chapters conclude with a list of topics for future investigation.

Although the treatment of hemostatic disorders has dramatically improved in recent decades, many patients have been left behind because of the high costs of products and services. The goals of Partners for Better Care are transparent and comprehensive healthcare, fair and equitable access to medicines, assuring patients' rights to dignified and culturally competent care, and stable and reasonable costs (HFA *Dateline Federation*, Spring 2016, p. 21). Their Patient Charter states that patients should have an active and formal voice in payment and delivery system reform; reasonable and timely access to providers within their network; information about covered services, providers, formularies, and out-of-pocket costs of insurance plans; access to medications, services, devices and other care without discrimination created by unreasonable tiering or excessive cost sharing; not subject to cumbersome preauthorization and renewal processes that restrict access to care and therapies; and timely access to a rapid and fair appeals process. Assisting patients in achieving these goals is a shared responsibility of all healthcare providers.

It is the hope of the author that this book will make a positive contribution to the knowledge of everyone involved in the care of people with bleeding disorders, and will result in an improved quality of life and better clinical outcome for our patients.

Historical Background of Blood Coagulation

It's such a wonderful trick, isn't it? You tilt a tube and the contents fall out and then you do it again and it doesn't fall out. It's so sudden, and it's so easy to see. - **Rosemary Biggs**

Blood spurts from a wound when a vessel is severed, and the formation of a clot at the site of the injury prevents further blood loss. Blood flow is not compromised because the clot covers the wound and does not extend very far into the injured vessel. As long as the clot remains in place, bleeding does not resume and eventually new tissue growth permanently seals the punctured vessel. This seemingly miraculous healing process has inspired many scientists to study how blood coagulates. The history of clotting discoveries has been ably summarized by Oscar Ratnoff [1]. He writes that as long ago as 1686, Malphighi described a network of white fibrous strands that remained after a blood clot was washed free of red cells. However, it was not until 1830 that Babington [2] noted that these strands, subsequently called fibrin, were not found in blood prior to clotting, and suggested that they arose from a precursor, subsequently named fibrinogen by Virchow in 1856 [3]. A major step forward was Schmidt's observation at the end of the 19th century that fibrinogen underwent conversion to fibrin through the action of an enzyme he called thrombin, and this occurred only when blood clotted [4]. Thrombin arose from a precursor, prothrombin, which was present in unclotted blood, but how prothrombin was converted to thrombin was unclear.

An important conceptual advance was the formulation of a blood coagulation scheme by Paul Morawitz in 1905 [5]. He proposed that it was a two-step process: first, injured tissues released a factor, termed thromboplastin, which converted prothrombin to thrombin, and second, thrombin

1

converted fibrinogen to fibrin. He also recognized that calcium was required for thromboplastin to be active. This theory accounted for how blood forms clots when tissues are injured, but did not explain how blood coagulates when drawn into a test tube, since no tissue thromboplastin is present in the tube. An even more relevant clinical problem is that fibrinogen, prothrombin, and calcium are all present in normal amounts in people with hemophilia, who nonetheless have a severe bleeding problem.

FACTOR VIII AND HEMOPHILIA A

The conundrum posed by hemophilia was addressed by Arthur Patek, Jr. and his coworkers in the 1930s. They observed that a substance was present in normal blood that shortened the clotting time of hemophilic blood [6,7]. On investigation, this substance had the characteristics of a protein—it contained nitrogen, it was too large to pass through a membrane filter, and it lost its ability to shorten the clotting time if it was heated. A precipitate formed if the putative clotting factor was mixed with acidified water, a characteristic of globular proteins. When the precipitate was dissolved in saline solution and infused intravenously into people with hemophilia, it shortened the clotting time of their blood, suggesting it might have a potential role in the treatment of this disorder. Patek and colleagues wrote: "The fact that a normal globulin substance reduces the clotting time *in vivo*, we believe changes the complexion of the disease from an abnormality that was immutably fixed to one that is amenable to change." This assertion might have been the first time that scientists suggested that an inherited disorder could be mitigated by a medical intervention.

Patek and coworkers called their protein antihemophilic globulin (AHG) because it accelerated the clotting time of hemophilic blood [7]. At the time of their discovery, hemophilia was defined as a bleeding disorder that affected boys and men; although their mothers and sisters could pass the trait on to some of their male children, the sons of hemophiliacs were normal. The defects in other bleeding disorders that affected women as well as men were also eventually characterized. As the factor deficient in each type of hemorrhagic disorder was elucidated, researchers named the involved proteins after the individuals in whom the condition was discovered, giving origin to names such as Christmas Factor and Hageman Factor. Stuart-Prower factor was named after the two patients with the same defect studied by different workers. Other investigators selected names that described how the factor functioned in coagulation; some of these unwieldy monikers were serum prothrombin conversion accelerator, plasma thromboplastic component, and proaccelerin. Each worker selected the name that best fit his or her experimental studies, and one influential researcher, Walter Seegers, created

TABLE 1.1 Synonyms Used for Factor VIII Prior to 1960

Antihemophilic globulin
Antihemophilic globulin A
Antihemophilic factor (AHF)
Plasma thromboplastic factor (PTF)
Plasma thromboplastic factor A
Thromboplastic plasma component (TPC)
Facteur antihemophilique A
Thromboplastinogen
Prothrombokinase
Platelet cofactor
Plasmokinin
Thrombokatilysin

Modified from Wright IS. Nomenclature of blood clotting factors. Can Med Assoc J 1959;80:659–61.

an entirely distinct set of names for the clotting factors he examined [8]. Eventually, there were several designations for each clotting protein; the appellations given to the protein deficient in classical hemophilia are shown in Table 1.1.

By the 1950s, it was apparent that the profusion of names was causing a great deal of confusion and hampering research. The International Society of Thrombosis and Haemostasis (ISTH) was founded in 1954; its mission was to advance the understanding, prevention, diagnosis, and treatment of thrombotic and bleeding disorders. As one of its first tasks, the ISTH organized a committee, the International Committee for the Standardization of the Nomenclature of Blood Clotting Factors, to bring order to the cacophony of clotting factors [9]. The Committee met between 1954 and 1959, and decided to take a rigorous approach to the identification and naming of clotting factors; it recommended that Roman numerals be assigned only to those factors that had data on the effects of pH, storage, absorbance, and inactivation by heating. In addition, there needed to be a clinical disorder associated with a deficiency or excess of the factor, and a reproducible method of assay. They adopted this rigorous approach because a presumed clotting activity had been previously designated factor VI, but its existence could not be confirmed by further study; this number was never assigned to any other factor. Table 1.2 lists the clotting factors that were assigned Roman numerals; it includes Von Willebrand Factor, which has not been assigned a Roman numeral.

TABLE 1.2   Clotting Factors and Their Physiologic Concentrations in Plasma

Clotting factor	Physiologic concentration (µg/mL)
Factor I: fibrinogen	2000–4000
Factor II: prothrombin	100–150
Factor III: tissue factor	–
Factor IV: calcium	90–105
Factor V: proaccelerin	5–10
Factor VII: proconvertin	0.5
Factor VIII: antihemophilic globulin	0.1–0.2
Von Willebrand Factor	10
Factor IX: Christmas factor	4–5
Factor X: Stuart-Prower factor	8–10
Factor XI: Plasma thromboplastin antecedent	5
Factor XII: Hageman factor	30
Factor XIII: Fibrin-stabilizing factor	10

Modified from Roberts HR, Monroe III DM, Hoffman M. Molecular biology and biochemistry of the coagulation factors and pathways of hemostasis. In: Williams Hematology. 7th ed. New York: McGraw-Hill; 2006. p. 1666 [chapter 106].

Other proteins that participate tangentially in coagulation, such as pre-kallikrein and high-molecular-weight kininogen, as well as anticoagulants and fibrinolytic factors, have not been assigned Roman numerals.

Progress in advancing the field of blood coagulation was dependent on the development of broadly accessible methods to assay the putative clotting factors. An early clotting test that was simple and reproducible was the prothrombin time, described by Armand Quick in 1935 [10]. This test was based on the observation that plasma anticoagulated with a salt such as sodium citrate rapidly formed a clot upon the addition of a suspension of acetone-dried brain tissue and calcium. The time required to form a clot, the prothrombin time, was prolonged in people with serious liver disease and those who were vitamin K deficient or taking drugs such as warfarin that antagonize vitamin K. These individuals have a decrease in FVII; the prothrombin time test is very sensitive to FVII deficiency. Although the test was useful in these patients, it was normal in individuals with hemophilia who are not deficient in FVII; a different method would be needed to study their disorder. Investigators soon discovered that extracting brain tissue with organic solvents provided phospholipids that modestly accelerated the clotting time of normal but not hemophilic plasma [11]. These phospholipids were considered "partial" thromboplastins, to distinguish them from the unextracted lipids used for the prothrombin time, and the test based on their use was called the partial thromboplastin time (PTT). Plasma from individuals with moderate to severe hemophilia had a prolonged PTT, but the test was not very sensitive; normal values were often observed in patients with mild disease and clotting factor concentrations above 30% of normal. To provide a more sensitive measurement of clotting factor levels, the test was modified by adding dilutions of patient plasma to plasma known to be deficient in the factor to be assayed. The clotting times of these mixtures were compared with those of mixtures in which normal plasma replaced the patient plasma. While this method was not technically demanding, it was not very accurate unless the normal plasma had a full complement of clotting factors, the plasma dilutions were tested immediately after preparation, and the deficient plasma had less than 1% of the clotting factor in question [12]. Because of these limitations, this type of assay is not sufficiently sensitive and specific for the scientific examination of FVIII; to develop a better method was the challenge undertaken by a brilliant English postgraduate medical student, Rosemary Biggs.

Doctor Biggs (1912–2001) was born in London and in her teens expressed a desire to become a physician [13]. Her parents discouraged this choice of occupation, and she was sent to study botany at London University. After receiving her undergraduate degree, she enrolled at the University of Toronto and wrote a dissertation on mycology for which she was awarded a PhD. She returned to England at the outbreak of World War II and entered the London School of Medicine for Women, receiving the MBBS in 1943. She then became a graduate assistant in the Department of Pathology at Radcliffe Infirmary, working under the tutelage of Professor R.G. Macfarlane (1907–87), who had already made several major contributions to the management of bleeding disorders. Doctor Biggs became fascinated by blood coagulation; she is quoted as saying "it's such a wonderful trick, isn't it? You tilt a tube and the contents fall out and then you do it again and it doesn't fall out. It's so sudden, and it's so easy to see" [14]. She recognized that there was no satisfactory test for quantifying the concentration of prothrombin in the plasma, and decided to make this the subject of her thesis for the MD degree.

Most of the assays extant at the time measured thrombin formation rather than prothrombin, and were performed by adding tissue extract (thromboplastin) and calcium to the plasma (modified from the quick prothrombin time). These clotted the plasma in 10–15 s; when calcium alone was added to plasma in the absence of tissue factor, there was a delay of about 5 min and then a strong coagulant appeared. Doctor Biggs suspected that this coagulant was a thromboplastin intrinsic to the blood and distinct from the extrinsic factor found in tissues. To study intrinsic thromboplastin, she prepared a mixture containing all the known clotting factors with the exception of prothrombin, fibrinogen, and calcium [15,16]. The mixture included serum, which contained most of the clotting factors except FV and FVIII; aluminum-hydroxide-absorbed plasma, which was a source of FV and FVIII but had none of the serum factors; and a partial phospholipid. Calcium was then added to initiate thromboplastin formation, and the amount formed over time was assessed by repeatedly adding small amounts of the mixture to normal plasma. As increasing amounts of thromboplastin were generated, the clotting times of the plasma became progressively shorter. To detect deficiencies of individual clotting factors, normal or patient absorbed plasma or serum was examined. Hemophilia due to FVIII deficiency could be clearly distinguished from the bleeding disorder due to FIX deficiency because thromboplastin generation was impaired with absorbed plasma, but not serum, from patients with FVIII deficiency, and was decreased with serum, but not absorbed plasma, from patients with FIX deficiency [17]. Doctor Biggs used a modification of this assay to measure the amount of FVIII in plasma and clotting factor concentrates; dilutions of the aluminum-hydroxide-absorbed plasma of the sample to be tested and a source of FV were included in the modified version [18]. With this assay, it was possible to accurately measure FVIII, and when the concentrates were given to patients with hemophilia, to correlate the plasma levels of FVIII with the hemostatic efficiency of the concentrate [19].

FVIII had been identified as a protein, but was it an enzyme or a substrate, and how did it interact with the other clotting factors? Several coagulation proteins had been discovered prior to and during the time Doctor Biggs was conducting her studies; FV was identified in 1947, and factors VII, IX, X, and XI in quick succession thereafter. Thrombin was known to be an enzyme; its action on fibrinogen to form a blood clot was comparable to the way that rennet acted on casein to form milk curds. Thrombin could be cleaved from prothrombin by enzymes such as trypsin [20], suggesting that similar prothrombin-converting enzymes might be generated during blood clotting. This speculation was given credence by Seegers et al. [8], who reported that prothrombin and factors VII, IX, and X all had similar structures; in addition, they could be activated to form serine proteases, enzymes that split the peptide bonds of proteins. This led to the concept that most clotting factors were enzyme precursors (proenzymes) that were converted to enzymes and then activated other clotting factors in a sequential fashion. But what was the initial trigger?

In 1955, Ratnoff and Colopy [21] described a man with a very prolonged clotting time that was shortened by the addition of normal plasma deficient in all the then-recognized clotting factors. This new factor, when activated by glass or other silicates, initiated coagulation by

FIG. 1.1   A sequence of small waterfalls cascading into a large cataract. *Photograph courtesy of the author.*

activating FXI. This suggested to Davie and Ratnoff [22] that coagulation occurred in a stepwise fashion, like a series of waterfalls (Fig. 1.1): activation of one factor led to the activation of the next factor in the sequence. Glass (or some physiological counterpart) activated FXII, which then activated FXI, which in turn activated FIX, etc. Additional mediators were needed for some reactions; for example, calcium and phospholipid for the activation of FVIII by activated FIX. The waterfall sequence provided a useful framework for understanding the interactions of the several clotting factors. In the same year (1964) that Davie and Ratnoff published their waterfall paper, R.G. Macfarlane [23] published a similar scheme that described coagulation as a sequence of proenzyme-enzyme transformations. He theorized that the clotting mechanism functioned like a photomultiplier tube; an initial stimulus that might generate only a small amount of activated FXII, would activate a larger amount of FXI. Subsequently, the amount of each precursor that was activated would progressively increase from one stage of coagulation to the next, culminating in a burst of thrombin sufficient to convert fibrinogen to fibrin and form an effective hemostatic clot. He calculated that if each enzyme activated 10 times its own weight of proenzyme, the overall gain in response would be 1×10^6 in the cascade. Once each enzyme made its contribution, it was rapidly extinguished by inhibitors, much as a spark that runs down a fuse leaves a dead wire.

The concept of a coagulation cascade was a major contribution to knowledge about blood coagulation, but it had important limitations. Coagulation was clearly activated when blood vessels were cut, suggesting that a tissue factor, rather than a glass-like substance, was the physiologic activator of clotting. Tissue factor had been shown to activate FVII, and the tissue factor-FVII complex was known to activate FX. However, this pathway short-circuits the cascade, bypassing factors XI, IX, and VIII. Not including FVIII and FIX was untenable, because patients with deficiencies of these factors have definite bleeding disorders. This conundrum was eventually resolved by the observation that the procoagulants generated by the tissue factor pathway are unable to sustain coagulation because they are rapidly neutralized by inhibitory proteins; effective hemostasis is achieved only if FVIII and FIX are activated. This put FVIII back into the mainstream of coagulation, but its exact function was still unclear.

Studies of FVIII were hampered by the difficulty of separating it from other plasma proteins. The plasma concentration of FVIII is less than 1 µg/mL (Table 1.2), and its biologic activity is quickly lost during purification procedures because it readily undergoes proteolytic degradation. In addition, the isolated protein is unstable at modestly altered temperatures and pHs, and even loses activity on storage. Nevertheless, many investigators attempted to purify and characterize FVIII, often with aberrant results. For example, in 1972, investigators reported that FVIII could be produced by treatment of albumin with succinic anhydride [24]; its detection in hemophilic and VWD plasma suggests that it was not authentic FVIII. A few years later, another group described the synthesis of FVIII by leukocytes cultured from healthy as well as hemophilic individuals [25]. This report was met with considerable skepticism; researchers thought the procoagulant was not FVIII but rather tissue factor derived from monocytes in the leukocyte cultures [26]. Other workers examined material presumed to be pure FVIII and described it as being composed of rod-shaped particles; they reported morphologic differences between the particles in normal and hemophilic isolates [27]. These particles might have been Weibel-Palade bodies, although the appearance of these organelles is similar in normal and hemophilic plasma.

The work of Vehar and Davie [28], published in 1977, considerably advanced knowledge about FVIII. They purified bovine FVIII approximately 10,000-fold and showed that its activity increased 25-fold after incubation with thrombin. They observed that activated FVIII, along with activated FIX, calcium, and phospholipid, converted FX to its active form. Activated **bovine** FVIII in the presence of $0.25\,M$ $CaCl_2$ was stable and could be inactivated by inhibitors of serine proteases, suggesting that it was a serine enzyme. However, thrombin-activated **human** FVIII proved to be very unstable and the primary structure of FVIII was not deduced until investigators applied the powerful techniques of immunoaffinity chromatography and gene cloning [29].

FVIII has held considerable fascination for investigators because of its critical role in hemostasis. In 1961, Rizza [30] reported that plasma FVIII concentrations rose considerably after exercise, and Cohen et al. [31] showed that the increase was due to beta-adrenergic stimulation. The exercise-induced increase in FVIII is mediated by nitric oxide; partial blockade of nitric oxide synthase blunts the exercise-induced increase in FVIII [32]. Beta-adrenergic stimulation also accounts for the rise in FVIII following the intravenous infusion of epinephrine; it is prevented by beta-adrenergic blockers [33]. FVIII levels are also increased by thyroid hormone, but this is due to enhanced synthesis of the protein rather than stimulation of the beta-adrenergic receptors [34].

VON WILLEBRAND FACTOR AND VON WILLEBRAND DISEASE

This eponymous disorder was first described by E.A. von Willebrand, a physician/scientist born in 1870 in Vaasa, a seaport in Finland located on the Gulf of Bothnia. Von Willebrand studied at the University of Helsinki, receiving a PhD degree in 1899 for a thesis entitled "Blood Changes After Venesection," and subsequently a medical degree [35]. He was appointed to the position of lecturer at the University and studied the hematologic changes that occur in muscles during exercise and the use of insulin for diabetes. Von Willebrand became a hospital director in 1922 and an honorary professor in 1930. He retired in 1935 and died in Finland in 1949.

In April 1924, a 5-year-old girl was admitted to von Willebrand's Internal Medicine Department because of a "malignant hemorrhagic diathesis" [36]. The patient had recurrent, lifelong epistaxis, bleeding into the skin and mucous membranes, and almost exsanguinated after a tooth extraction. Von Willebrand learned that the patient came from Föglö in the Åland Archipelago in the Gulf of Bothnia (Fig. 1.2), and that her siblings also had a bleeding tendency.

FIG. 1.2   The Åland Islands, located between Sweden and Finland. *Modified from World Atlas and reproduced with permission.*

The Åland Islands had remained relatively isolated until the end of the 19th century, and endogamy (marriage within a clan) exceeded 80% [37]. Von Willebrand went to the Islands to study members of the patient's family as well as two other families and reported that 23 of the 66 family members had hemorrhagic symptoms and several had bled to death. In 1926, he published his observations in Finska Läkaresällskapets Handlingar (68:87–112). He called the disorder hereditary pseudohemophilia, because it differed in a number of respects from hemophilia: women were affected more often than men, mucocutaneous bleeding was common, and joint hemorrhages were rare. Von Willebrand returned to the Islands frequently, and in 1931 reported detailed medical histories of the bleeder families and discussions of the likely pathogenesis of the bleeding disorder, concluding that there was a disturbance of platelet function and altered capillaries [38].

One of the earliest hemostatic defects identified in people with Von Willebrand Disease (VWD) was a decrease in FVIII [39]. The relationship between FVIII and VWF remained obscure for many decades. As early as 1957, immunologic studies using rabbit antisera raised against crude preparations of human FVIII disclosed that the anticoagulant activity of this antisera could be blocked by hemophilic plasma, leading to the conclusion that something was present in hemophilic plasma resembling FVIII but lacking its clotting activity [40]. These observations were confirmed and extended by Zimmerman et al. [41] and Stites et al. [42] using highly purified FVIII: their antisera detected an antigen in plasma samples from hemophiliacs but in decreased amounts from patients with VWD. This experiment was repeated several years later, with the same outcome;

in fact, a FVIII-like protein was identified in the plasma of each of more than 100 individuals with classical hemophilia [43].

There was additional evidence that FVIII and VWF might be distinct entities: although most individuals with VWD had decreased levels of FVIII, their response to transfusion differed from that observed in people with hemophilia [44]. Transfusions resulted in FVIII increases that were greater in patients with VWD than in those with hemophilia, and the levels remained elevated for as long as 2–3 days in contrast to only a few hours in hemophiliacs [45]. Furthermore, even the infusion of hemophilic plasma, completely lacking FVIII, produced a rise in FVIII activity and corrected the bleeding time [46].

Discrepancies between the levels of FVIII and what was designed FVIII-related antigen (Von Willebrand antigen) suggested that they might represent different gene products [47,48], and thermostability and dissociation studies favored the idea of two closely related but separate proteins constituting a FVIII/VWF complex [49,50]. This raised the question of whether the hemostatic activities of the complex were attributes of one protein that was synthesized under the direction of two genes, or two independent proteins linked together. The latter construct was eventually proven to be correct.

Studies of patients with VWD disclosed several abnormalities related to platelets. These individuals lacked a factor that promoted the adhesion of platelets to glass beads [51,52], and a factor needed for the aggregation of platelets by the antibiotic, ristocetin [53]. Lastly, it was recognized that most but not all patients with VWD had prolonged skin bleeding times, and this test was considered the sine qua non for the diagnosis. However, the bleeding time did not always correlate with tests of platelet adhesion, ristocetin-induced platelet aggregation, and levels of FVIII [54–56]. A 1977 review in the *New England Journal of Medicine* noted that the hemostatic defects in VWD were qualitative as well as quantitative [57].

Investigators suspected that endothelial cells were the source of the missing factor, a supposition confirmed by Booyse et al. [58] in 1981; they reported the presence of activities associated with VWF in the culture media of normal endothelial cells but not the media of cells from a patient with VWD. Shortly thereafter, the biosynthesis of VWF by endothelial cells and megakaryocytes was reported by Wagner et al. [59,60]. Further exploration of the structure and function of VWF awaited the development of molecular biological methods for identifying the gene and characterizing its products (Chapter 9).

A major contribution to the study of FVIII came in 1974, when Mannucci et al. [55] reported that desmopressin (1-deamino-8-D-arginine vasopressin, DDAVP), a synthetic analogue of 8-arginine vasopressin, increased levels of both FVIII and VWF. They gave slow intravenous infusions of desmopressin to healthy volunteers and patients with bleeding disorders, and observed a rise in the levels of FVIII/VWF within 30–90 min of infusion. The infusions increased factor levels in patients with moderately severe VWD, but in those with severe hemophilia only VWF levels were raised. This research demonstrated important differences between the two hemostatic factors, and introduced desmopressin as a valuable tool for the investigation of hemostasis and as an adjunctive therapy for various bleeding disorders.

SUMMARY

The protein missing from individuals with hemophilia A (classical hemophilia) was first identified in 1937 and called antihemophilic globulin (AHG). It became known as factor VIII (FVIII) in the 1950s, when a Committee of the International Society of Thrombosis and Hemostasis assigned Roman numerals to all the known clotting factors. One-stage and two-stage assays for clotting factors were described in 1953, and these methods enabled clinicians to distinguish hemophilia A (FVIII deficiency) from hemophilia B (FIX deficiency, Christmas disease), as well as monitor therapeutic clotting factor infusions.

In 1964, investigators in the United States and United Kingdom described the clotting of blood as a sequence of enzyme-substrate reactions in which each clotting factor was a proenzyme converted to an enzyme by the preceding factor. The initiating stimulus might have a small effect on the first clotting factor in the sequence, but as each proenzyme was converted to an enzyme, the effect was amplified so that eventually sufficient thrombin was formed to convert fibrinogen to the final clotting product, fibrin. The concept of a coagulation cascade held sway for many years until it was modified by contemporary molecular studies that provided more precise information about clotting protein interactions.

A unique hemorrhagic disorder affecting some inhabitants of the Åland Islands was described by Erik A von Willebrand in 1926. The condition was inherited in an autosomal dominant manner, and mucocutaneous, and menstrual bleeding were common. Laboratory studies showed decreased levels of FVIII and impaired platelet adhesion and aggregation, attributed to deficiency of a protein initially designated FVIII-related antigen and then renamed Von Willebrand Factor.

By the end of the 1970s, scientists had purified FVIII 10,000-fold, activated it with thrombin, and demonstrated that it played a key role in the activation of FX. The high-molecular-weight complex circulating in plasma could be dissociated into two components, FVIII and Von Willebrand Factor (VWF). Exercise and hormones such as epinephrine and thyroxine were shown to raise the plasma concentrations of the proteins, and a vasopressin analogue, desmopressin, was found to promote the release of FVIII/VWF from endothelial cells. This observation led to the use of desmopressin for the diagnosis and treatment of mild-to-moderately severe hemophilia A and VWD patients, and had a significant impact on the study of these and other bleeding disorders.

References

[1] Ratnoff OD. Why do people bleed? In: Wintrobe MM, editor. Blood, pure and eloquent. New York: McGraw-Hill; 1980. p. 602.

[2] Babington BG. Some considerations with respect to the blood founded on one or two very simple experiments on that fluid. Med Chir Trans 1830;16:293–319.

[3] Virchow R. Gesammette Abdhandlungen zur Wissenschaftlischen Medicin. Frankfurt, M: Meidinger Sohn; 1856. 1024 p.

[4] Schmidt A. Zur Blutlehre. Leipzig: Vogel; 1892. 270 p.

[5] Morawitz P. The chemistry of blood coagulation. Ergeb Physiol 1905;4:307–411. Reprinted in Hartmann RC, Guenther PF (trans). Springfield, IL: Charles C Thomas; 1958. 194 p.

[6] Patek Jr AJ, Stetson RH. Hemophilia. I. The abnormal coagulation of the blood and its relation to the blood platelets. J Clin Invest 1936;15:531–42.

[7] Patek Jr AJ, Taylor FHL. Hemophilia II. Some properties of a substance obtained from normal plasma effective in accelerating the clotting of hemophilic blood. J Clin Invest 1937;16:113–24.

[8] Seegers WH, Alkjaersig N, Johnson SA. On the nature of the blood coagulation mechanisms in certain clinical states. Am J Clin Pathol 1955;25:983–7.

[9] Wright IS. Nomenclature of blood clotting factors. Can Med Assoc J 1959;80:659–61.

[10] Quick AJ, Stanley-Brown M, Bancroft FW. A study of the coagulation defect in hemophilia and in jaundice. Am J Med Sci 1935;190:501–11.

[11] Langdell RD, Wagner RH, Brinkhous KM. Effect of antihemophilic factor on one-stage clotting tests: a presumptive test for hemophilia and a simple one-stage antihemophilic factor assay procedure. J Lab Clin Med 1953;41:637–47.

[12] Austen DEG, Rhymes IL. A laboratory manual of blood coagulation. Oxford: Blackwell Scientific Publications; 1975. p. 15.

[13] Hawgood BJ. Rosemary Biggs MD FRCP (1912–2001) and Katharine Dormandy MD FRCP (1926–78): from laboratory to treatment and care of people with haemophilia. J Med Biography 2013;21:41–8.

[14] Tansey T. Witnessing medical history: an interview with Dr Rosemary Biggs. Haemophilia 1998;4:769–77.

[15] Biggs R. Plasma thromboplastin. Nature (London) 1952;170:280.

[16] Biggs R, Douglas AS. The thromboplastin generation test. J Clin Pathol 1953;6:23–9.

[17] Biggs R, Douglas AS, Macfarlane RG, Dacie JV, Pitney WR, Merskey C, O'Brien JR. Christmas disease: a condition previously mistaken for haemophilia. Br Med J 1952;2:1378–82.

[18] Biggs R. Human blood coagulation, haemostasis and thrombosis. 2nd ed. Oxford: Blackwell Scientific Publications; 1976. p. 684–8.

[19] Biggs R, Macfarlane RG, editors. The treatment of haemophilia and other coagulation disorders. Oxford: Blackwell Scientific Publications; 1966.

[20] Eagle H, Harris TN. Studies in blood coagulation. V. The coagulation of blood by proteolytic enzymes (trypsin, papain). J Gen Physiol 1937;20:543–60.
[21] Ratnoff OD, Colopy JEA. Familial hemorrhagic trait associated with a deficiency of a clot-promoting fraction of plasma. J Clin Invest 1955;34:602–13.
[22] Davie EW, Ratnoff OD. Waterfall sequence for intrinsic blood clotting. Science 1964;145:1310–2.
[23] Macfarlane RG. An enzyme cascade in the blood clotting mechanism, and its function as a biochemical amplifier. Nature (London) 1964;202:498–9.
[24] Barrow EM, Lester RH, Johnson AM, Graham JB. Production of antihemophilic (factor VIII) activity from albumin. Am J Phys 1972;222:920–7.
[25] Blecher TE, Westby JC, Thompson MJ. Synthesis of procoagulant antihaemophilic factor in vitro. Lancet 1978;i:1333–6.
[26] Mohanty D, Hilgard P. Synthesis of procoagulant antihaemophilic factor in vitro. Lancet 1979;i:101.
[27] Tan HK, Andersen JC. Human factor VIII: morphometric analysis of purified material in solution. Science 1977;198:932–4.
[28] Vehar GA, Davie EW. Formation of a serine enzyme in the presence of bovine factor VIII (antihemophilic factor) and thrombin. Science 1977;197:374–6.
[29] Kaufman RJ, Antonarakis SE, Fay PJ. Factor VIII and hemophilia A. In: Colman RW, Marder VJ, Clowes AW, George JN, Goldhaber SZ, editors. Hemostasis and thrombosis: basic principles and clinical practice. 5th ed. Philadelphia: Lippincott Williams & Wilkins; 2006. p. 152 [chapter 8].
[30] Rizza CR. Effect of exercise on the level of antihaemophilic globulin in human blood. J Physiol 1961;156:128–35.
[31] Cohen RJ, Epstein SE, Cohen LS, Dennis LH. Alterations of fibrinolysis and blood coagulation induced by exercise, and the role of beta-adrenergic-receptor stimulation. Lancet 1968;2:1264–6.
[32] Jilma B, Dirnberger E, Eichler HG, Matulla B, Schmetterer L, Kapiotis S, Speiser W, Wagner OF. Partial blockade of nitric oxide synthase blunts the exercise-induced increase of von Willebrand factor antigen and of factor VIII in man. Thromb Haemost 1997;78:1268–71.
[33] Ingram GI, Jones RV. The rise in clotting factor VIII induced in man by adrenaline: effect of alpha- and beta-blockers. J Physiol 1966;187:447 54.
[34] Hoak JC, Wilson WR, Warner ED, Theilen EO, Fry GL, Benoit FL. Effects of triiodothyronine-induced hypermetabolism on factor 8 and fibrinogen in man. J Clin Invest 1969;48:768–74.
[35] Shampo MA, Kyle RA. Erik von Willebrand-von Willebrand's disease. Mayo Clin Proc 1996;71:1088.
[36] Lehmann W, Forsius HR, Eriksson AW. Von Willebrand-Jurgens Syndrome on Åland. In: Erikkson AW, Forsius HR, Nevanlinna HR, Workman PL, Norio RK, editors. Population structure and genetic disorders. Academic Press; 1980. p. 509–63.
[37] Eriksson AW, Fellman JO, Forsius HR. Some genetic and clinical aspects of the Åland Islanders. In: Erikkson AW, Forsius HR, Nevanlinna HR, Workman PL, Norio RK, editors. Population structure and genetic disorders. Academic Press; 1980. p. 509–63.
[38] Von Willebrand EA. Uber hereditare Pseudohämophilie. Acta Med Scand 1931;76:521–50.
[39] Alexander B, Goldstein R. Dual hemostatic defect in pseudohemophilia. J Clin Invest 1953;551(548–615):32.
[40] Shanberge JN, Gore I. Studies on the immunologic and physiologic activity of antihemophilic factor (AHF) (abstract). J Lab Clin Med 1954;50:954.
[41] Zimmerman TS, Ratnoff OD, Powell AE. Immunologic differentiation of classic hemophilia (factor VIII deficiency) and von Willebrand's disease. J Clin Invest 1971;50:244–54.

[42] Stites DP, Hershgold EJ, Perlman JD, Fudenberg HH. Factor VIII detection by hemagglutination inhibition: hemophilia A and von Willebrand's disease. Science 1971;171:196–7.

[43] Ratnoff OD. Antihemophilic factor (factor VIII). Ann Intern Med 1978;88:403–9.

[44] Rizza CR. Von Willebrand's disease. In: Biggs R, editor. The treatment of hemophilia A and B and Von Willebrand disease. Oxford: Blackwell Scientific Publications; 1978. p. 242 [chapter 8].

[45] Kernoff PBA, Rizza CR, Kaelin AC. Transfusion and gel filtration studies in von Willebrand's disease. Br J Haematol 1974;28:357–70.

[46] Cornu P, Larrieu MJ, Caen JP, Bernard J. Transfusion studies in von Willebrand's disease: effect on bleeding time and factor VIII. Br J Haematol 1963;9:189–202.

[47] Veltkamp JJ, van Tilburg NH. Detection of heterozygotes for recessive von Willebrand's disease by the assay of antihemophilic-factor-like antigen. N Engl J Med 1973;289:882–5.

[48] Meyer D, Jenkins CSP, Dreyfus M, Larrieu M-J. Willebrand-factor activity and antigen in von Willebrand's disease. Lancet 1974;i:512–3.

[49] Meyer D, Jenkins CSP, Dreyfus M, Fressinaud E, Larrieu M-J. Willebrand factor and ristocetin. II. Relationship between Willebrand factor, Willebrand antigen and factor-VIII activity. Br J Haematol 1974;28:579–99.

[50] Van Mourik JA, Bouma BN, LaBruyere WT, de Graaf S, Mochtar IA. Factor VIII, a series of homologous oligomers and a complex of two proteins. Thromb Res 1974;4:155–64.

[51] Papayannis AG, Wood JK, Israels MCG. Factor-VIII levels, bleeding-times, and platelet adhesiveness in patients with von Willebrand's disease and in their relatives. Lancet 1971;i:418–21.

[52] Rossi EC, Green D. A study of platelet retention by glass bead columns ('platelet adhesiveness' in normal subjects). Br J Haematol 1972;23:47–57.

[53] Weiss HJ, Hoyer LW, Rickles FR, Varma A, Rogers J. Quantitative assay of a plasma factor deficient in von Willebrand's disease that is necessary for platelet aggregation. Relationship to factor VIII procoagulant activity and antigen content. J Clin Invest 1973;52:2708–16.

[54] Ratnoff OD, Bennett B. Clues to the pathogenesis of bleeding in von Willebrand's disease. N Engl J Med 1973;289:1182–3.

[55] Mannucci PM, Pareti FI, Ruggieri ZM. Enhanced factor VIII activity in von Willebrand's disease. N Engl J Med 1974;290:1259.

[56] Weiss HJ. Relation of von Willebrand factor to bleeding time. N Engl J Med 1974;291:420.

[57] Gralnick HR, Sultan Y, Coller BS. Von Willebrand's disease. Combined qualitative and quantitative abnormalities. N Engl J Med 1977;296:1024–30.

[58] Booyse FM, Quarfoot AJ, Chediak J, Stemerman MB, Maciag T. Characterization and properties of cultured human von Willebrand umbilical vein endothelial cells. Blood 1981;58:788–96.

[59] Wagner DD, Marder VJ. Biosynthesis of von Willebrand protein by human endothelial cells. J Biol Chem 1982;258:2065–7.

[60] Sporn LA, Chavin SI, Marder VJ, Wagner DD. Biosynthesis of von Willebrand protein by human megakaryocytes. J Clin Invest 1985;76:1102–6.

Recommended Reading

[1] Ratnoff OD. Why do people bleed? In: Wintrobe MM, editor. Blood, pure and eloquent. New York: McGraw-Hill; 1980. p. 602.

[2] Patek Jr AJ, Taylor FHL. Hemophilia II. Some properties of a substance obtained from normal plasma effective in accelerating the clotting of hemophilic blood. J Clin Invest 1937;16:113–24.

[3] Langdell RD, Wagner RH, Brinkhous KM. Effect of antihemophilic factor on one-stage clotting tests: a presumptive test for hemophilia and a simple one-stage antihemophilic factor assay procedure. J Lab Clin Med 1953;41:637–47.

[4] Hawgood BJ. Rosemary Biggs MD FRCP (1912–2001) and Katharine Dormandy MD FRCP (1926–78): from laboratory to treatment and care of people with haemophilia. J Med Biography 2013;21:41–8.

[5] Biggs R, Douglas AS. The thromboplastin generation test. J Clin Pathol 1953;6:23–9.

[6] Davie EW, Ratnoff OD. Waterfall sequence for intrinsic blood clotting. Science 1964;145:1310–2.

[7] Macfarlane RG. An enzyme cascade in the blood clotting mechanism, and its function as a biochemical amplifier. Nature (London) 1964;202:498–9.

[8] Ratnoff OD. Antihemophilic factor (factor VIII). Ann Intern Med 1978;88:403–9.

[9] Vehar GA, Davie EW. Formation of a serine enzyme in the presence of bovine factor VIII (antihemophilic factor) and thrombin. Science 1977;197:374–6.

[10] Mannucci PM, Pareti FI, Ruggeri ZM. Enhanced factor VIII activity in von Willebrand's disease. N Engl J Med 1974;290:1259.

Hemostasis and FVIII

Having a tooth extracted is an unpleasant experience. Aside from the pain, there is swelling and bleeding from the socket. Packing with some gauzy material usually controls the bleeding and analgesic medications dull the pain. This enables the individual to pass a fairly uneventful night and greet the morning with nothing more than some mouth soreness. People with hemophilia have a different sequence of events. A few hours after going to bed, a large gelatinous mass of blood clot fills the socket and extends into the oral cavity. Expectorating the mass results in fresh bleeding from the extraction site and soon another big clot fills the mouth. A repeat visit to the dentist or oral surgeon, often in the middle of the night, results in new packing carefully pressed into the socket, with cessation of bleeding. But again, after a few hours, the packing material is extruded by an emerging clot. This cycle of bleeding and clotting, repacking of the socket and recurrence of bleeding, persists for days to weeks before healing finally occurs. This was the usual outcome of dental extractions before it was recognized that hemophilia is due to the lack of a clotting factor, and that provision of this clotting factor is the only way to ensure permanent control of bleeding. In this chapter, the stages of blood coagulation will be described and the consequences of FVIII deficiency delineated.

NORMAL HEMOSTASIS

Hemostasis is a complex process with a large number of actors. Many of the events occur simultaneously, are interdependent, and reactants are rapidly consumed or neutralized. Fifty years ago, investigators recognized that most clotting factors were zymogens that were converted to active enzymes during the course of thrombin generation; the sequence of reactions was called the coagulation cascade and described in Chapter 1. It was an oversimplification but aided in interpreting the diagnostic tests then available. More recently, Monroe and Hoffman [1] described three phases of coagulation. The *initiation phase* is represented by the de-encryption of TF, formation of the TF-VIIa complex, activation of FIX and FX, and formation of the prothrombinase complex (FXa, FVa, prothrombin). This complex generates small amounts of thrombin on TF-bearing cells. The *amplification phase* is characterized by thrombin-induced platelet activation, followed by the binding of FV, FVIII, and FXI to the platelet surface and their activation by thrombin. The last stage is the *propagation phase*, during which FXIa activates FIX, FIXa binds to FVIIIa on the platelet surface, and the tenase complex (FIXa-VIIIa-FX) generates FXa. FXa rapidly associates with FVa and prothrombin (prothrombinase complex) to produce a burst of thrombin sufficient to convert fibrinogen to fibrin. A more detailed view of coagulation, built upon the concepts of Monroe and Hoffman and supplemented with many additional features of hemostasis, is described as follows and summarized in Table 2.1.

Vasoconstriction Phase

The initial response to a breach in vascular integrity is vasoconstriction, mediated by the sympathetic adrenergic fibers of the autonomic nervous system. Injury stimulates vascular endothelial cells to release the contents of their specific storage granules, called Weibel-Palade bodies [2]. The components of these bodies that engage in hemostasis are endothelin-1, a vasoconstrictor; Von Willebrand Factor (VWF), required for platelet adhesion and aggregation; and P-selectin, a lectin that binds to leukocytes and platelets. The vasoconstriction is opposed by the endothelial cell vasodilators, nitric oxide and prostacyclin. In addition, as the large multimers of VWF are expressed, they are cleaved by ADAMTS13 (a disintegrin-like and metalloproteinase with thrombospondin type 1 motifs) released from the endothelial cells. The vasoconstriction phase is of brief duration and only temporarily stanches blood loss. Weibel-Palade bodies, VWF, and ADAMTS13 are discussed in more detail in Chapter 9.

Initiation Phase

Leukocytes participate in the next phase and play a major, previously unappreciated role in hemostasis. P-selectin released from the endothelial cell Weibel-Palade bodies binds to the leukocyte receptor, P-selectin glycoprotein ligand-1 (PSGL-1), and induces leukocytes to roll on the endothelium toward the injury; upon arrival at the site, these white cells migrate into the subendothelial connective tissue, subdue microorganisms, and remove debris. Activated neutrophils release elastase, cathepsin G, and platelet-activating factor that promote platelet activation and the formation of leukocyte-platelet aggregates [3]. Activation of

TABLE 2.1 An Outline of Normal Hemostasis

I. *Vasoconstriction phase*

A. Immediate response is vasoconstriction, mediated by sympathetic adrenergic fibers of the autonomic nervous system.

B. Weibel-Palade bodies of stimulated endothelial cells release endothelin, a vasoconstrictor, Von Willebrand factor, and P-selectin.

C. Vasoconstriction is modulated by the vasodilators, nitric oxide and prostacyclin

II. *Initiation phase (leukocytes)*

A. Selectins released by endothelial cells bind to leukocyte P-selectin glycoprotein ligand-1 (PSGL-1), initiating leukocyte rolling on the endothelium and diapedesis at the site of injury. Leukocyte enzymes contribute to platelet activation and formation of platelet-leukocyte aggregates.

B. When monocytes are activated, Ca^{2+} influx induces translocation of phosphatidyl-serine (PS) to the outer membrane and de-encryption of tissue factor (TF).

C. TF forms a complex with FVIIa-activating FIX and FX bound to PS. FXa activates FV forming the prothrombinase complex that converts prothrombin to thrombin.

D. Microvesicles form in regions of the membrane rich in lipid rafts and TF, are shed into the blood, and bind to platelets; they fuse with the platelet membrane.

III. *Amplification phase (platelets)*

A. Thrombin binds to protease-activating receptor-1 (PAR-1), activating platelets.

B. Activated platelets form tethers used to translocate to the injury site and attach to subendothelial collagen via their membrane GPVI receptors.

C. Membrane GP1b/IX/V receptor binds high-molecular-weight Von Willebrand factor released by endothelial cell Weibel-Palade bodies.

D. Adenosine diphosphate (ADP) released by platelet dense bodies binds to two receptors: $P2Y_1$ inducing shape change and $P2Y_{12}$ producing conformational change in $GP\alpha_{IIb}/\beta_3$. $GP\alpha_{IIb}/\beta_3$ binds fibrinogen, inducing platelet aggregation.

IV. *Propagation phase (clotting factors)*

A. FXI binds to GP1b and is activated by thrombin; FXIa activates FIX bound to PS on platelet membrane.

B. FIXa binds to FVIIIa that thrombin has released from the Von Willebrand factor and activated.

C. FIXa, FVIIIa, and FX attached to the platelet membrane form the tenase complex that produces FXa (FXa formed on leukocyte surface unable to reach platelets because it is bound by tissue factor pathway inhibitor).

D. FXa forms prothrombinase complex with FVa released from platelet granules, and this complex converts prothrombin to thrombin.

E. In addition to activating platelets and factors XI, VIII, and V, thrombin converts fibrinogen to fibrin & activates FXIII.

F. FXIIIa stabilizes fibrin by crosslinking aggregated fibrin polymers.

Continued

TABLE 2.1 An Outline of Normal Hemostasis—cont'd

Alternative pathway

A. Contact with negatively charged surfaces activates FXII; reciprocal activation of prekallikrein by FXIIa generates more FXIIa.

B. FXIIa activates FXI membrane-bound by high-molecular-weight kininogen; FXIa activates FIX, continuing the propagation phase.Polyphosphate, released from injured tissues and platelet dense granules, greatly accelerates the activation of FXI.

V. *Termination phase*

A. VWF is cleaved by ADAMTS13.

B. Tissue factor pathway inhibitor (TFPI) binds FX and inhibits TF-FVIIa complex.

C. Antithrombin inhibits thrombin and factors IXa, Xa, XIa, XIIa. Inhibition potentiated by heparin and vessel wall proteoglycans.

D. Thrombomodulin binds thrombin; thrombin-thrombomodulin complex activates protein C. Protein C, with cofactor free protein S, cleaves FVa and FVIIIa.

E. Fibrinolysis: Thrombin-thrombomodulin complex activates thrombin activatable fibrinolysis inhibitor (TAFI). TAFI cleaves lysines from fibrinogen, preventing binding of plasminogen and formation of plasmin.

F. Tissue plasminogen activator (t-PA) converts plasminogen to plasmin on the surface of fibrin; plasmin lyses fibrin. Plasminogen activator inhibitor-1 binds t-PA, preventing cleavage of plasminogen to plasmin; plasmin escaping from the clot is neutralized by circulating antiplasmin.

monocytes is accompanied by an influx of extracellular Ca^{2+}, triggering the translocation of phosphotidylserine (PS) from the inner to the outer leaflet of the cell membrane [4] and the de-encryption of tissue factor (TF), a transmembrane protein [5,6]. The decrypted TF forms a complex with activated clotting factor VII (FVIIa), and this TF-FVIIa complex activates FIX and FX [7]. The binding of FVIIa and FX to PS via their γ-carboxyglutamic acid domains greatly enhances the activation of FX by the TF-FVIIa complex [8]. FXa activates FV [9] and forms a 1:1 Ca^{2+}-dependent, stoichiometric complex with FVa and prothrombin that generates sufficient thrombin to activate platelets and clotting factors V, VIII, and XI. In addition, activated monocytes form microvesicles in regions of the cell membrane rich in lipid rafts and TF. These microvesicles are shed in the vicinity of the tissue injury and display PSGL-1 on their surface, enabling them to bind to P-selectin on activated platelets [10]. The microvesicles fuse with the platelet membrane, providing decrypted TF for subsequent reactions with clotting factors and formation of additional thrombin.

Amplification Phase

Platelets are the second most populous bodies, after erythrocytes, in the circulation. Thrombin-activated platelets produce membrane tethers that are used to translocate toward the site of injury [11], and platelets attach to subendothelial collagen via their membrane glycoproteins (GP) Ia/IIa and VI. In addition, platelet membrane GPIb/IX/V binds to surface immobilized VWF. Once becoming adherent to the subendothelial connective tissue, these platelets remain tightly attached to the site of injury despite high shear blood flow [12]. The activated platelets release adenosine diphosphate (ADP), which binds to the platelet $P2Y_1$ receptor, inducing shape change and greatly increasing the surface area of the platelet membrane. ADP also binds to the platelet $P2Y_{12}$ receptor, producing a conformational change in membrane $GP\alpha_{IIb}/\beta_3$. In its active conformation, $GP\alpha_{IIb}/\beta_3$ promotes platelet adhesion and aggregation by serving as a receptor for three adhesive proteins: VWF, fibronectin, and fibrinogen. Platelets aggregate when fibrinogen binds to activated $GP\alpha_{IIb}/\beta_3$ on adjacent platelets, forming bridges between them. In addition, $GP\alpha_{IIb}/\beta_3$ binds platelets to fibrinogen on the surface of developing thrombi. The binding of fibrinogen to $GP\alpha_{IIb}/\beta_3$ is stabilized by thrombin and thromboxane A_2 released from the platelets.

GPIb/IX/V of activated platelets is a binding site for FXI as well as VWF, and the GPIbα subunit promotes the activation of FXI by thrombin [13]. FXIa activates FIX that is bound by its γ-carboxyglutamic acid domain to the PS exposed on the platelet membrane. This FIXa, along with the

FIXa generated by the TF-FVIIa complex, binds to FVIIIa that has been released from the VWF and activated by thrombin. All these activated clotting factors are clustered together on a small region of the membrane of procoagulant platelets, and this close physical association facilitates their ability to interact [14]. FIXa and FVIIIa, together with FX bound to the platelet membrane, form the tenase complex whose main function is to generate FXa. This platelet-bound FXa is assembled with FVa and prothrombin to form the prothrombinase complex, converting prothrombin to thrombin. Note that the FXa that is formed by the TF-FVIIa complex on leukocytes during the initiation phase is rapidly inactivated by binding to tissue factor pathway inhibitor (TFPI), and this complex not only removes FXa but also inhibits the TF-FVIIa complex [15]. Because of the actions of TFPI, the tenase complex on platelets becomes the principal source of FXa for the prothrombinase complex. Fig. 2.1 displays the two major clotting complexes, tenase and prothrombinase.

Propagation Phase

Once thrombin is generated, it affects many aspects of hemostasis. It cleaves fibrinopeptides A and B from fibrinogen; the resultant fibrin monomer undergoes aggregation and polymerization. Thrombin cleaves the A-subunit from the B-(carrier) subunit of FXIII; the A-subunit then

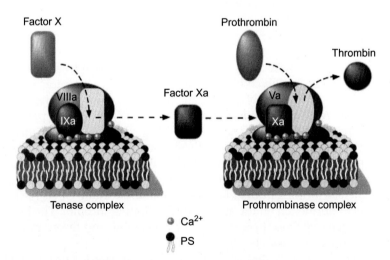

FIG. 2.1 Tenase and prothrombinase assembly on phosphatidyl serine-bearing cell membranes. *Reproduced with permission from Zwaal RFA, Comfurius P, Bevers EM. Scott syndrome, a bleeding disorder caused by defective scrambling of membrane phospholipids. Biochim Biophys Acta 2004;1636:119–28.*

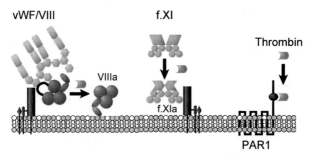

vWF/VIII   f.XI

Thrombin

VIIIa

f.XIa

PAR1

FIG. 2.2 Some actions of thrombin on the platelet surface: releasing FVIII from von Willebrand factor (vWF), activating FXI, cleaving the protease-activated receptor (PAR1). *Reproduced with permission from Monroe DM, Hoffman M. What does it take to make the perfect clot? Arterioscler Thromb Vasc Biol 2006;26:41–8.*

crosslinks the fibrin monomers by forming γ-glutamyl-lysine bridges, thereby stabilizing the clot.

Thrombin activates platelets by cleaving protease-activated receptor-1 (PAR-1) [16] and GP1bα, contributing to platelet aggregation and shape change (Fig. 2.2). Also shown in Fig. 2.2 are thrombin-activating FXI bound to GP1bα and releasing FVIII from VWF; the FVIII and FV originating from platelet alpha granules become activated when thrombin converts them from single-chain to two-chain molecules. FVIIIa binds to FIXa and FX that are anchored by Ca^{2+} bridges to the negatively charged PS exposed on the platelet membrane; there they form the tenase complex that generates FXa. FVa, FXa, and prothrombin form the prothrombinase complex that produces the copious amounts of thrombin needed to propagate coagulation.

ALTERNATIVE PATHWAY

An alternative pathway of coagulation is triggered when blood comes into contact with the negatively charged surfaces of prosthetic heart valves, stents, and other intravascular devices. Interaction with such surfaces activates FXII, and reciprocal activation of prekallikrein by FXIIa results in the formation of additional FXIIa [17]. FXIIa activates FXI membrane-bound by high-molecular-weight kininogen, and FXIa continues the series of reactions described in the preceding paragraph that culminate in the formation of thrombin. **Polyphosphate**, an anionic polymer secreted from the dense granules of activated platelets, weakly activates FXII but greatly accelerates the activation of FXI [18]. Other procoagulant activities of polyphosphate are the enhancement of FV conversion to FVa, blocking the anticoagulant activity of TFPI, and enhancing fibrin clot structure [19].

Termination Phase

ADAMTS13 on the surface of endothelial cells cleaves the high-molecular-weight multimers of VWF. Thrombin is inhibited by antithrombin, a serine protease inhibitor (serpin) that also neutralizes activated factors IX, X, XI, and XII. The inhibitory activity of antithrombin is greatly enhanced when heparin or vessel wall proteoglycans form trimolecular complexes with the clotting factor protease and antithrombin. Free factor XIa that has not been inhibited by antithrombin is inactivated by protease nexin. TFPI binds FXa and inhibits the TF-VIIa complex, as described previously. Feedback inhibition of thrombin formation occurs when thrombin binds to thrombomodulin; the bound thrombin activates protein C (APC), and APC with its cofactor, free protein S, cleaves FVa and FVIIIa, limiting further thrombin generation. Finally, rapid blood flow disperses platelets and accumulated activated clotting factors.

Fibrinolysis

The fibrinolytic system becomes activated when fibrin thrombi are formed. A circulating protein, plasminogen, binds to lysine receptors present on fibrin; the bound plasminogen is cleaved by plasminogen activators (tissue plasminogen activator, urokinase) to form the active enzyme, plasmin. Plasmin dissolves the fibrin thrombus, generating fibrin degradation products and crosslinked dimers from the D-domain of fibrin (D-dimers). Plasmin that escapes from the confines of the thrombus is promptly bound by circulating α_2-antiplasmin and α_2-macroglobulin. A robust fibrinolytic mechanism is essential for thrombus resolution and wound healing.

Fibrinolysis is regulated by proteins that inhibit tissue plasminogen activator and urokinase; these include plasminogen activator inhibitors 1 & 2 (PAI-1 & PAI-2), activated protein C-inhibitor, and thrombin activatable fibrinolysis inhibitor (TAFI). The thrombin-thrombomodulin complex described previously activates TAFI, and TAFI cleaves lysine residues from fibrin, preventing the binding of plasminogen to fibrin. Elevated levels of PAI-1 are found in obesity and the metabolic syndrome.

HEMOSTASIS IN THE ABSENCE OF FVIII

In bleeding disorders, the location of the hemorrhage often suggests the nature of the hemostatic defect. Prolonged bleeding after a minor injury to the skin and mucous membranes occurs in people with platelet disorders or Von Willebrand Disease (VWD) because an adherent mass of platelets (a platelet plug) fails to form at the site of the lesion. On the other hand, deep bleeding within muscles and joints is characteristic of clotting factor deficiencies. In individuals with hemophilia, persistent oozing of blood from small cuts on the tongue (a muscle) is typical in infancy, and repeated hemorrhages into the joints of boys and men are common and result in pain, swelling, deformities, and ultimately, crippling. However, these differences in the location of hemorrhages in platelet and clotting factor disorders are not iron-clad, and some people with severe clotting factor deficiencies experience large bruises, frequent nosebleeds during child-hood, and have occasional episodes of blood in the urine.

These protean manifestations of bleeding in people with hemophilia A (FVIII deficiency) attest to the critical role that FVIII plays in coagulation. In healthy individuals, there appears to be continuous activation of small amounts of FVII, keeping the clotting system primed should vascular integrity be compromised. However, in hemophiliacs with severe deficiencies of FVIII ($<2\%$ of normal), the basal levels of FVIIa and probably FIXa are decreased [20]. When injury occurs, the formation of FXa and the generation of small amounts of thrombin on TF-bearing cells are delayed. FXa attempting to diffuse from TF-bearing cells to platelets becomes irreversibly bound to TFPI or inactivated by antithrombin. The importance of TFPI and antithrombin in hemophilic bleeding is demonstrated by the observation that blocking or depleting these inhibitors improves hemostasis in people with hemophilia [21], and drugs based on this concept are currently being evaluated in clinical trials (Chapter 7).

When FVIII is deficient, the formation of the tenase complex and FXa on the platelet surface is abrogated, and the small amounts of thrombin formed by TF-bearing cells become bound to thrombomodulin and antithrombin, Although some thrombin might escape to convert fibrinogen to fibrin, this fibrin will be located mainly at the periphery of the lesion, in proximity to cells bearing the TF-FVIIa complex [1]. The clinical example of a dental extraction in a hemophiliac, described earlier, is consistent with this formulation: the individual's mouth fills with clotted blood but the tooth socket, the location of the lesion, continues to bleed because the vascular defect is not occluded by a fibrin-platelet meshwork. Another phenomenon observed in people with hemophilia is a prolonged secondary bleeding time [22]. If a small, shallow incision is made on the forearm, the wound will ooze for a mean of 7 min, as compared to 5 min in nonbleeders [23]. However, if the clot is wiped away from the incision, rebleeding from the wound occurs and persists for vastly longer in hemophiliacs than in people without a coagulopathy. An examination of the thrombus shows a dearth of platelet-bound fibrin strands in the wound, accounting for the prolonged bleeding. Studies of hemophilic mice report that they have repeated rebleeding from skin wounds and ultimately delayed healing [24].

Augmented fibrinolysis is another factor that affects bleeding and wound healing in individuals with FVIII deficiency. Fibrin clots become resistant to lysis when thrombin activates TAFI, an inhibitor of fibrinolysis, as previously described. Because thrombin formation is reduced in FVIII deficiency, formation of the thrombin-thrombomodulin complex is defective and less TAFI is formed. This results in unregulated plasmin production, the premature dissolution of thrombi, and accounts, in part, for the recurrent bleeding and impaired wound healing characteristic of hemophilia [25].

LABORATORY ASSESSMENT OF HEMOSTASIS

The evaluation of individuals with suspected bleeding disorders includes a thorough medical history; the questionnaire and scoring system (bleeding assessment tool) developed by the International Society of Thrombosis and Haemostasis provides a quantitative assessment of bleeding severity [26]. The higher the bleeding score, the more likely a specific diagnosis will be established by laboratory investigations. These include a few simple tests to examine the integrity of each of the phases of hemostasis, with the exception of the neural phase. Blood is collected in tubes containing sodium citrate to reversibly bind calcium and prevent clotting. The formed elements (erythrocytes, leukocytes, and platelets) are removed by centrifugation and the plasma stored frozen until testing is required. The prothrombin time is sensitive to defects in the initiation phase, and is performed by adding recombinant tissue factor and calcium to the plasma. Within 10–15 s, fibrin strands are observed, and this is the endpoint of the test. Prolongation of the time for fibrin formation occurs with deficiencies of factors VII, V, X, prothrombin, and fibrinogen, as well as with anticoagulants that inhibit these factors. The prothrombin time is very useful in assessing the effects of vitamin K antagonists, since these drugs alter the ability of factors VII, X, and prothrombin to bind to negatively charged cell membranes and form thrombi.

The activated partial thromboplastin time (aPTT) is performed by adding a silicate or other contact-activating substance, along with calcium in a phospholipid suspension, to the plasma. These reagents activate FXII and provide a negatively charged lipid surface to assess the amplification and propagation phases of coagulation. The normal clotting time of 35–45 s is prolonged with deficiencies of all the clotting factors except FVII. If the prothrombin time is normal, a prolonged aPTT indicates deficient factors XII, XI, IX, VIII, prekallikrein, and high-molecular-weight kininogen. The aPTT is also prolonged if the plasma contains heparin or other thrombin inhibitors, as well as inhibitors of any of these clotting factors.

Platelet numbers are measured by automated counters and platelet characteristics and function are evaluated by the platelet function analyzer (PFA), platelet aggregometry, and flow cytometry. A number of tests are available for assessing the VWF and are discussed in detail in Chapter 10. The most commonly used assays are the enzyme-linked immunosorbent assay for the VWF antigen, ristocetin-induced platelet aggregation, binding of VWF to modified glycoprotein Ib fragments, and PFA closure time [27].

Thrombin generation can be measured by calibrated automated thrombinography, and inhibitors of thrombin are assessed with the dilute thrombin time test; fibrinogen is assayed directly by quantitative methods. In addition, there are specific assays for the various clotting factors, including FXIII, and tests for the fibrinolytic factors, t-PA, plasminogen, and PAI-1. Thromboelastography is a global test of hemostasis that assesses clotting time (reaction time), platelet function, and clot dissolution. Selection of the appropriate clotting assays enables the clinician to identify the specific deficit and provide the most suitable remedy, which could be infusion of platelets or coagulation factor concentrates.

Laboratory studies show that the clotting of blood lacking FVIII is prolonged and thrombin generation is only 1/29th the rate in nonhemophilic blood [28]; the impairment in thrombin generation is closely associated with the frequency and extent of bleeding [29]. Bleeding is also related to the level of FVIII, but surprisingly hemophiliacs exhibit only a mild bleeding tendency if they have little more than 5% of the FVIII concentration typically present in nonhemophiliacs. This very efficient use of the small amounts of available FVIII might be possible because the clotting factors are concentrated on the surface of platelets and other cells, enabling them to interact to form the requisite clotting complexes and generate thrombin. Recognition of the relationship between bleeding and FVIII levels is relevant when using exogenous clotting factor concentrates to control and prevent hemophilic bleeding, as will be discussed in Chapter 5.

The use of the prothrombin time and activated partial thromboplastin time in the diagnosis of hemophilia can be illustrated by the following clinical example. An 18-year-old boy fell from his bicycle earlier in the day and presents at the emergency room with a large bruise on his thigh (Fig. 2.3).

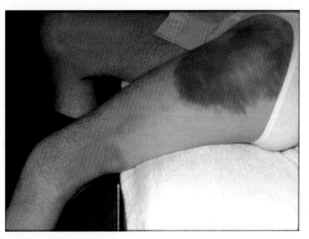

FIG. 2.3   Large bruise sustained after minor trauma (falling from a bicycle). *Reproduced from Green D. Linked by blood: hemophilia and AIDS, Fig. 2.1.*

He had a past history of recurrent nose bleeds, and had previously sustained hematomas after trauma. Laboratory studies revealed an aPTT of 78 s (normally, <34 s) and a prothrombin time of 12 s (normally, <13 s). A prolonged aPTT occurs with all factor deficiencies except FVII; the normal prothrombin time excludes deficiencies of factors V, X, prothrombin, and fibrinogen. Of the remaining clotting factors, FXII, prekallikrein, and high-molecular-weight kininogen deficiencies are possible but unlikely since decreased levels of these factors are not associated with a hemorrhagic disorder. This leaves factors VIII, IX, and XI as possible candidates; other possibilities are anticoagulant drugs such as heparin or a circulating anticoagulant. Therefore, the next step in the evaluation would be a mixing study to determine whether adding normal plasma to the patient's plasma corrects the aPTT; complete correction, especially after incubation, would exclude the presence of an anticoagulant. To distinguish between deficiencies of factors VIII, IX, and XI, the patient's plasma is mixed with plasmas known to be deficient in these factors; failure of mutual correction indicates the specific deficiency. Alternatively, the patient's plasma could be assayed for each of these factors, and the severity of the deficiency determined. That would be useful for estimating the amount of clotting factor replacement required to control bleeding, as will be discussed in Chapter 6.

SUMMARY

Hemostasis can be conceptualized as occurring in several coordinated phases, each with principal players: vasoconstriction (endothelial cells), initiation (leukocytes and platelets), amplification and propagation (clotting factors), and termination (anticoagulant and fibrinolytic factors). In the absence of VWF, platelet adherence and aggregation are impaired and mucosal bleeding is characteristic. If FVIII is deficient, thrombin generation is decreased, fibrin formation is delayed, and fibrinolysis is excessive. The consequence is unremitting bleeding from various sites and accumulation of blood in muscles and joints. The evaluation of hemostasis includes a thorough medical history and completion of a questionnaire that provides data for the calculation of a bleeding score. The medical history is supplemented by laboratory tests that evaluate specific phases of hemostasis as suggested by the patient examination. For suspected hemophilia A, the FVIII assay provides an assessment of the severity of the disorder and genotyping identifies the relevant mutations. For VWD, the evaluation usually includes the platelet function analyzer (PFA) closing time, the FVIII assay, and assays of VWF using immunologic and functional tests to enable categorization of the disease as types 1, 2, or 3. Inhibitors of coagulation are recognized by prolonged clotting times that fail to correct with normal plasma.

References

[1] Monroe DM, Hoffman M. What does it take to make the perfect clot? Arterioscler Thromb Vasc Biol 2006;26:41–8.
[2] Rondaij MG, Bierings R, Kragt A, van Mourik JA, Voorberg J. Dynamics and plasticity of Weibel-Palade bodies in endothelial cells. Arterioscler Thromb Vasc Biol 2006;26:1002–7.
[3] Swystun L, Liaw PC. The role of leukocytes in thrombosis. Blood 2016;128:753–62.
[4] Zwaal RFA, Comfurius P, Bevers EM. Scott syndrome, a bleeding disorder caused by defective scrambling of membrane phospholipids. Biochim Biophys Acta 2004;1636:119–28.
[5] Bach RR. Tissue factor encryption. Arterioscler Thromb Vasc Biol 2006;26:456–61.
[6] Rao LVM, Pendurthi UR. Regulation of tissue factor coagulant activity on cell surfaces. J Thromb Haemost 2012;10:2242–53.
[7] Bouchard BA, Tracy PB. The participation of leukocytes in coagulant reactions. J Thromb Haemost 2003;I:464–9.
[8] Ansari SA, Pendurthi UR, Sen P, Rao LVM. The role of putative phosphatidylserine-interactive residues of tissue factor on its coagulant activity at the cell surface. PLoS ONE 2016;11(6):e0158377. https://doi.org/10.1371/journal.pone.0158377.
[9] Schuijt TJ, Bakhtiari K, Daffre S, DePonte K, Wielders SHG, Marquart JA, Hovius JW, van de Poll T, Fikrig E, Bunce MW, Camire RM, Nicolaes GAF, Meijers JCM, van't Veer C. Factor Xa activation of factor V is of paramount importance in initiating the coagulation system. Circulation 2013;128:254–66.
[10] Del Conde I, Shrimpton CN, Thiagarajan P, Lopez JA. Tissue-factor–bearing microvesicles arise from lipid rafts and fuse with activated platelets to initiate coagulation. Blood 2005;106:1604–11.
[11] Jackson SP. The growing complexity of platelet aggregation. Blood 2007;109:5087–95.
[12] Kuwahara M, Sugimoto M, Tsuji S, Matsui H, Mizuno T, Miyata S, Yoskioka A. Platelet shape changes and adhesion under high shear flow. Arterioscler Thromb Vasc Biol 2002;22:329–34.

[13] Baglia FA, Badelino KO, Li CQ, Lopez JA, Walsh PN. Factor XI binding to the platelet glycoprotein Ib-IX-V complex promotes factor XI activation by thrombin. J Biol Chem 2002;277:1662–8.

[14] Podoplelova NA, Sveshnikova AN, Kotova YN, Eckly A, Receveur N, Nechipurenko DY, Obydennyi SI, Kireev II, Gachet C, Ataullakhanov FI, Mangin PH, Panteleev MA. Coagulation factors bound to procoagulant platelets concentrate in cap structures to promote clotting. Blood 2016;128:1745–55.

[15] Broze Jr GJ, Girard TJ. Tissue factor pathway inhibitor: structure-function. Front Biosci (Landmark Ed) 2012;17:262–80.

[16] Grimsey N, Lin H, Trejo J. Endosomal signaling by protease-activated receptors. Methods Enzymol 2014;535:389–401.

[17] Renne T, Schmaier AH, Nickel KF, Blomback M, Mass C. In vivo roles of factor XII. Blood 2012;120:4296–303.

[18] Morrissey JH, Smith SA. Polyphosphate as modulator of hemostasis, thrombosis, and inflammation. J Thromb Haemost 2015;13(Suppl. 1):S92–7.

[19] Morrissey JH, Choi SH, Smith SA. Polyphosphate: an ancient molecule that links platelets, coagulation, and inflammation. Blood 2012;119:5972–9.

[20] Wildgoose P, Nemerson Y, Hansen LL, Nielsen FE, Glazer S, Hedner U. Measurement of basal levels of factor VIIa in hemophilia A and B patients. Blood 1992;80:25–8.

[21] Nordfang O, Valentin S, Beck TC, Hedner U. Inhibition of extrinsic pathway inhibitor shortens the coagulation time of normal plasma and of hemophilia plasma. Thromb Haemost 1991;66:464–7.

[22] Borchgrevink CF, Waaler BA. The secondary bleeding time; a new method for the differentiation of hemorrhagic diseases. Acta Med Scand 1958;162:361–74.

[23] Eyster ME, Gordon RA, Ballard JO. The bleeding time is longer than normal in hemophilia. Blood 1981;58:71923.

[24] Hoffman M, Monroe DM. Wound healing in haemophilia-breaking the vicious cycle. Haemophilia 2010;16(Suppl. 3):13–8.

[25] Broze Jr GJ, Higuchi DA. Coagulation-dependent inhibition of fibrinolysis: role of carboxypeptidase-U and the premature lysis of clots from hemophilic plasma. Blood 1996;88:3815–23.

[26] Rodeghiero F, Tosetto A, Abshire T, Arnold DM, Coller B, James P, Neunert C, Lillicrap D. ISTH/SSC bleeding assessment tool: a standardized questionnaire and a proposal for a new bleeding score for inherited bleeding disorders. J Thromb Haemost 2010;8:2063-5.

[27] Castaman G, Montgomery RR, Meschengieser SS, Haberichter SL, Woods AI, Lazzari MA. Von Willebrand's disease diagnosis and laboratory issues. Haemophilia 2010;16:67–73.

[28] Cawthern KM, van't Veer C, Lock JB, ME DL, Branda RF, Mann KG. Blood coagulation in hemophilia A and hemophilia C. Blood 1998;91:4581–92.

[29] Brummel-Ziedins KE, Whelihan MF, Gissel M, Mann KG, Rivard GE. Thrombin generation and bleeding in haemophilia A. Haemophilia 2009;15:1118–25.

Recommended Reading

[1] Monroe DM, Hoffman M. What does it take to make the perfect clot? Arterioscler Thromb Vasc Biol 2006;26:41–8.

[2] Swystun L, Liaw PC. The role of leukocytes in thrombosis. Blood 2016;128:753–62.

[3] Bach RR. Tissue factor encryption. Arterioscler Thromb Vasc Biol 2006;26:456–61.

[4] Del Conde I, Shrimpton CN, Thiagarajan P, Lopez JA. Tissue-factor–bearing microvesicles arise from lipid rafts and fuse with activated platelets to initiate coagulation. Blood 2005;106:1604–11.

[5] Broze Jr GJ, Girard TJ. Tissue factor pathway inhibitor: structure–function. Front Biosci (Landmark Ed) 2012;17:262–80.

[6] Morrissey JH, Smith SA. Polyphosphate as modulator of hemostasis, thrombosis, and inflammation. J Thromb Haemost 2015;13(Suppl. 1):S92–7.

[7] Brummel-Ziedins KE, Whelihan MF, Gissel M, Mann KG, Rivard GE. Thrombin generation and bleeding in haemophilia A. Hemophilia 2009;15:1118–25.

[8] Rodeghiero F, Tosetto A, Abshire T, Arnold DM, Coller B, James P, Neunert C, Lillicrap D. ISTH/SSC bleeding assessment tool: a standardized questionnaire and a proposal for a new bleeding score for inherited bleeding disorders. J Thromb Haemost 2010;8:2063-5.

3

FVIII Anatomy and Physiology

FVIII has been discovered among the coagulation proteins of all jawed vertebrates; the amino acid sequence of the protein found in puffer fish and chickens has a 50%–60% homology with that of humans [1]. Davidson and colleagues [2] present biochemical, molecular cloning, and sequence data consistent with FVIII appearing at least 430 million years ago, when the first round of whole-genome duplication generated the FVIII signature domain structure of A1-A2-A3. These observations suggest that FVIII has played an essential role in hemostasis during the evolution of creatures as diverse as fish, birds, and man.

ISOLATION AND STRUCTURE

In Chapter 1, it was noted that in 1937 Patek and coworkers determined that FVIII was a globular protein [3]. Other workers attempted to confirm this report, but it was not until 1963 that Barkhan et al. [4], using the technique of starch-block electrophoresis, were able to demonstrate that factor VIII migrated with the α-2 globulins. To isolate FVIII from other proteins, Ratnoff et al. [5] used gel chromatography and reported that their partially purified FVIII had a molecular weight of at least two million. By the early 1970s, several investigators were using relatively simple methods to achieve low yields of human factor VIII of fairly high purity [6,7]. To isolate sufficient quantities of the protein for chemical analyses, Schmer et al. [8] purified the FVIII protein in bovine plasma approximately 10,000-fold, and reported that their preparation was physically heterogeneous; the smallest species had a molecular weight of 1.1 million. They concluded that the heterogeneity was due to aggregation of a chemically pure species. However, evidence from immunologic studies suggested otherwise.

If antibodies are raised to "purified" FVIII, it would be anticipated that immunodiffusion studies would show a single precipitin line against normal plasma, but no line if tested against hemophilic plasma known to be devoid of FVIII. However, Meyer et al. [9] reported that the plasma of

33

FIG. 3.1   Immunodiffusion of normal plasma (PN), hemophilic plasma with/without material reacting with a human FVIII antibody (HA$^+$/HA$^-$), hemophilic plasma with an FVIII antibody (HA.AC), and Von Willebrand Disease (VWD) plasma, against an antibody prepared by immunization of rabbits with FVIII purified by gel chromatography (in the center well). A single precipitin line is seen with all except VWD plasma. *Reproduced from Meyer D, Lavergne J-M, Larrieu M-J, Josso F. Cross-reacting material in congenital factor VIII deficiencies (haemophilia A and von Willebrand's disease). Thromb Res 1972;1:183–96, Pergamon Press.*

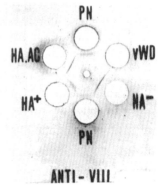

ANTI - VIII

every hemophilic patient examined (83 in total) contained a protein that was precipitated by these antibodies (Fig. 3.1).

On the other hand, these and other investigators found that this material in hemophilic plasma was decreased or absent in the plasma of patients with Von Willebrand Disease (VWD) [10,11]. This bleeding disorder had been described by Eric von Willebrand in 1926 and differs from hemophilia in a number of respects. While classical hemophilia is a disorder of boys and men and is characterized by bleeding into the muscles and joints, VWD affects women as well as men, and bleeding usually occurs from the gastrointestinal and genitourinary tracts. Persons with VWD lack a protein, the Von Willebrand Factor (VWF), which normally controls bleeding from mucous membranes. However, in both disorders, FVIII levels are decreased. Furthermore, when patients with VWD are transfused with normal plasma, the levels of both FVIII and VWF increase, but the increase in FVIII exceeds that of VWF and persists longer [12]; it appears to be free of VWF and has a much lower molecular weight [13]. This phenomenon was interpreted as indicating that the transfused VWF stimulated the release of stored FVIII.

The scientific characterization of FVIII required that it be free of VWF. By developing a monoclonal antibody specifically directed against FVIII coagulant activity and using it for affinity chromatography, investigators were able to separate FVIII from VWF [14]. This FVIII was then digested with trypsin, and the amino acid sequence of the resulting peptides used to prepare DNA probes for identifying FVIII genomic clones, as will be described in Chapter 4. The protein deduced from these clones had 2351 amino acids of which 19 were in the leader sequence, giving a mature protein of 2332 amino acids with a calculated molecular weight of 264,763. When FVIII is synthesized in vivo, the large single-chain precursor undergoes proteolysis at the B-A3 junction plus additional cleavages within the B domain [15], producing a light chain with a molecular weight of 80,000 Da and a heavy chain whose size ranges from 90,000 to 210,000 Da (Fig. 3.2).

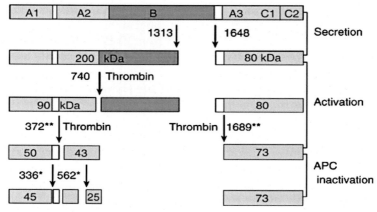

FIG. 3.2 Domain structure, activation, and inactivation of factor VIII. *APC*, activated protein C; *kDa*, kilodaltons. *Modified from Hoffman R, Benz Jr. EJ, Shattil SJ, Furie B, Cohen HJ, Silberstein LE, McGlave P, editors. Hematology: basic principles and practice. 4th ed. Elsevier; 2005, Fig. 114-01.*

The amino acid sequence of FVIII shows homology with clotting FV, but unexpectedly also with ceruloplasmin, a copper-binding protein [16,17]. In fact, FVIII has 1 molecule of Cu^{2+} per molecule of active protein; elimination of the copper-binding site results in failure of the light and heavy chains to associate and results in an inactive factor [18].

RELATIONSHIP TO THE VON WILLEBRAND FACTOR

To clarify the relationship of FVIII to the VWF, investigators had to physically separate the two proteins. This separation became possible when it was discovered that high concentrations of salts, either 0.25 M $CaCl_2$ or 1 M NaCl [19,20], dissociated the higher-molecular-weight VWF from the lower-molecular-weight FVIII [21]. The two molecules could also be separated using antibodies raised in rabbits to the FVIII/VWF complex, and an antibody present in the serum of a multiply-transfused patient with hemophilia that was specific for only FVIII. Each type of antibody was coupled to agarose beads and used to bind and remove the proteins from a partially purified mixture of FVIII and VWF. The rabbit antibody removed a greater proportion of VWF than FVIII, and the human antibody bound and removed only FVIII. This differential binding physically segregated FVIII from VWF [22].

Once the two proteins were separated, each could be analyzed and their features compared. Copurified FVIII and VWF factor have similar physicochemical properties when they are bound together; both are inactivated by heating to 56°C, have similar storage stability, and virtually identical pH stability from pH 6 to 10. However, when FVIII is dissociated from VWF, its activity is lost after 30 min of incubation at 37°C and at pH's above 8.4. By contrast, VWF dissociated from FVIII retains activity after heating at 56°C for up to 90 min and has the same pH profile as when it is associated with FVIII [23].

Other characteristics of FVIII and VWF are shown in Table 3.1.

Perhaps most striking is the difference in the plasma concentrations of FVIII and VWF; the level of FVIII is about 1/100th that of VWF. The ratio of 1 molecule of FVIII to 50–100 VWF subunits is consistently observed in patients with varying degrees of VWF deficiency, suggesting that the VWF controls the plasma level of FVIII [24]. FVIII and VWF are held together by noncovalent bonds [25], and their association is mediated primarily by the FVIII C1 domain (Fig. 3.2) [26], with secondary binding sites in the acidic a3 peptide located at the A3 terminus and a site in the C2 domain [27]. The major binding site on VWF is located within the first 272 amino acids of the mature VWF protein, in the D'D3 region [28]. As will be discussed in subsequent chapters, the binding to VWF affects FVIII assembly, secretion, activation, and survival in the circulation.

TABLE 3.1   Characteristics of Factor VIII and Von Willebrand Factor[a]

Feature	Factor VIII	Von Willebrand Factor
Chromosomal location	Xq28	12p13.3
Size of gene (kb; exons)	186; 26	178; 52
mRNA (bp)	9030	9000
Primary product		
No. of amino acids	2351	2813
Molecular weight	330,000	310,000
Mature subunit		
No. of amino acids	2332	2050
Molecular weight	264,000	170,000
Domains	3A, 1B, 2C	3A, 3B, 2C, 4D
Plasma concentration	0.1–0.2 µg/mL	10 µg/mL

[a]References: Kaufman RJ, Antonarakis SE, Fay PJ. Factor VIII and hemophilia A. In: Colman RW, Clowes AW, Goldhaber SZ, Marder VJ, George JN, editors. Hemostasis and thrombosis. 5th ed. Philadelphia: Lippincott Williams & Wilkins; 2006. p. 151–75 [chapter 8]; Johnsen J, Ginsburg D. von Willebrand disease. In: Lichtman MA, Kipps TJ, Kaushansky K, Beutler E, Seligsohn U, Prchal JT. Williams hematology. 7th ed. New York: McGraw-Hill; 2006. p. 1929–45 [chapter 118]; Kaufman RJ, Antonarakis SE. Structure, biology, and genetics of factor VIII. In: Hoffman R, Benz Jr. EJ, Shattil SJ, Furie B, Cohen HJ, Silberstein LE, McGlave P. Hematology: basic principles and practice. 4th ed. Elsevier; 2005. p. 2011–30 [chapter 114].

CELL OF ORIGIN, SECRETION, AND CLEARANCE

In the past, considerable controversy surrounded the origin of FVIII. Initially, it was assumed that the protein was made in the liver, since nearly all the clotting factors owe their synthesis to this organ. Doubts arose because clinical experience showed that patients with severe liver disease and very low levels of prothrombin and fibrinogen often had normal or high levels of FVIII [29]. Enlightenment came when it was reported that the hepatocytes, which are vulnerable to injury by liver pathogens, do not synthesize FVIII; rather, it is the more resistant endothelial cells of the hepatic sinusoids that are a locus of FVIII production [30].

That the liver is not the only site of FVIII synthesis was shown by experiments replacing the livers of normal dogs with those of hemophilic animals; low normal levels of FVIII activity persisted in the transplanted dogs [31,32]. Some of the alternative candidate organs for FVIII production were the kidneys [33], lungs [34], and spleen [35–37]. In fact, investigators became so convinced that the spleen was an important source of the clotting factor that they transplanted a normal spleen into a patient with hemophilia, with only transient benefit [38]. Some studies used antibodies to identify FVIII in tissue samples, but it seems likely that these antibodies recognized VWF rather than FVIII [39]. However, in 1986, researchers used a very specific panel of monoclonal antibodies to show that FVIII was present in sporadic mononuclear phagocytes in lymph nodes, lung, and spleen as well as in hepatic sinusoidal endothelial cells [40]. Subsequent studies confirmed the presence of FVIII protein as well as mRNA expression in hepatic fenestrated sinusoidal endothelial cells and lymphatic postcapilllary high endothelial venules, as well as in the glomerular endothelium of the kidney [41,42]. Most recently, Fahs et al. [43] knocked-out the FVIII gene in hepatocytes, hematopoietic cells, and endothelial cells of mice. Their data were consistent with endothelial cells being the principal and possibly only source of plasma FVIII.

The major sites for protein synthesis within cells are the ribosomes that stud sheets of the rough endoplasmic reticulum (ER) [44]. Within the ER, the newly minted FVIII takes at least three temporary dance partners: the immunoglobulin-binding protein BiP, calnexin, and calreticulin. Binding of the nascent FVIII to these proteins is thought to facilitate copper ion incorporation and proper disulfide bond formation in the final product [24]. Correctly folded and processed FVIII is then released from BiP by adenosine triphosphate and from calnexin and calreticulin by glucosidase II. It then binds to the ER cargo receptor complex LMAN1-MCFD2 [45]; these proteins were discovered when the genetic defect in individuals with combined FV and FVIII deficiency was investigated; they transport both FV and FVIII [46]. FVIII bound to LMAN1 (lectin mannose binding1) and MCFD2 (multiple coagulation factor deficiency gene 2) enters the fused vesicles that comprise the ER-Golgi intermediate compartment

(ERGIC), where it is decorated with carbohydrate and sulfate moieties [24,47]. Most of the carbohydrate moieties are added to the B-domain and function to prevent aggregation of folded intermediates, enabling the interaction of the newly synthesized chains with ER chaperones and enzymes [48]. The B-domain facilitates the transport of the single-chain molecule to the Golgi compartment where it undergoes extensive proteolysis. Most of the B-domain is removed, leaving a 2-chain structure consisting of a heavy chain and a light chain, but a remnant of the B-domain remains attached to the heavy chain.

In glomerular and umbilical vein endothelial cells, mature FVIII is tightly packed into Weibel-Palade bodies along with the VWF, and the two proteins are secreted together [49]. FVIII transduced into umbilical vein endothelial cells also enters the Weibel-Palade bodies and can be released along with the VWF [50]. But lymphatic endothelial cells secrete FVIII without VWF [42], and platelet α-granules store and release FVIII independent of VWF [51]. These observations are consistent with secretion of FVIII usually, but not always, in conjunction with VWF.

Endothelial cells are generally oriented with their base resting on the subendothelial connective tissue and their apex facing the vessel lumen. In polarized human umbilical vein endothelial cells, VWF is constitutively secreted basolaterally; in contrast, stimulated and continuous basal release occurs apically from the Weibel-Palade bodies [52]. Agonists, such as desmopressin (1-desamino-8-D-arginine vasopressin, DDAVP), induce the corelease of both FVIII and VWF into the blood stream and the two proteins circulate as a stable complex.

The plasma concentration of FVIII is 0.1–0.2 µg/mL [53]. As early as 1960, it was observed that FVIII levels rise with age, especially after age 40 [54]. These increases in FVIII occur in parallel with increases in VWF and are affected by ABO blood group; however, adjusting for VWF reduces the influence of blood group on the FVIII level from 10.7% to 0.6% [55]. The age-related increases in both proteins are significantly greater in type non-O individuals [56]. In those over 65, blacks have higher levels than whites, and women have higher values than men at all ages [57]. Interestingly, plasma FVIII continues to increase with age so that it is higher in centenarians than in those less than age 100 (1.65 U/mL vs 1.34 U/mL, $P = .005$), suggesting that modestly elevated FVIII is not incompatible with a long life [58].

FVIII in the complex with VWF has a plasma half-life of 12 h, but without VWF, FVIII is rapidly captured by macrophages [59] and its half-life is only 2 h [60]. FVIII binds to the D'D3 domains of VWF; binding to this isolated fragment is sufficient to slow the loss of FVIII from the circulation [61]. Recently, the mechanism underlying the removal of FVIII bound to VWF has been elucidated. Studies show that the complex undergoes phagocytosis by macrophages in vascular beds exposed to high shear

TABLE 3.2 Clearance of FVIII

	Circulating FVIII	Thrombin-activated FVIII[a]
In complex with:	Von Willebrand Factor	Factors IXa and X
Binding requires:	High shear	Exposed cell surface phosphatidylserine
Receptors:	LRP-1 Heparan sulfate proteoglycans	LRP-1 LDLR
Cells/organs:	Macrophages in liver and spleen; kupffer cells of liver	Endothelial cells, platelets, macrophages

[a]*Activated factor VIII is cleaved by FXa and other serine proteases, as well as by activated protein C and protein S.*
Abbreviations: LRP-1, lipoprotein-related receptor protein-1; LDLR, low-density lipoprotein receptor.

stress [59,62]. The receptor on the macrophage for the FVIII/VWF complex is low-density lipoprotein receptor-related protein-1 (LRP-1) [63,64], and binding is facilitated by cell surface heparan sulfate proteoglycans [65]. The binding sites on FVIII are the C1 domain for LRP-1 [66] and the A2 domain for the heparan sulfate proteoglycans [63]. A second ligand for FVIII is the low-density lipoprotein receptor (LDLR) [67]. FVIII, bound to these receptors and free of VWF, is endocytosed and degraded by lysosomes. The various fates of FVIII are shown in Table 3.2.

ACTIVATION AND INACTIVATION

A conceptual advance in demystifying blood coagulation was the recognition that most clotting factors function as members of enzyme-cofactor complexes bound to cellular membranes [68]. This conclusion emanated from the studies of Kenneth G. Mann, Professor and Chair of Biochemistry at the University of Vermont. Doctor Mann and his group showed that the catalytic efficiency of enzyme complexes bound to membranes is increased by a factor of 10^4–10^5 as compared to enzyme components in solution. They suggested that three enzyme complexes participate in thrombin generation: one containing tissue factor and FVIIa; a second that includes factors VIII, IX, and X; and a third comprising factors V, X, and prothrombin. In order to become membrane bound and participate in its complex, FVIII must be activated by thrombin.

Small amounts of thrombin are generated when tissue injury releases tissue factor, followed by activation of factors VII and X; the activated FX converts some prothrombin to thrombin before being inactivated by tissue factor pathway inhibitor (TFPI). Thrombin cleaves the heavy chain of FVIII beyond residues 372 and 740, and the light chain after residue

1689, as shown in Fig. 3.2 [69]. The cleavage at residue 1689 removes the site of VWF binding, the acidic a3 peptide of the A3 domain, resulting in the release of activated FVIII (FVIIIa) [70]. This permits the C2 domain on the FVIIIa light chain to bind to soluble fibrin bound to the platelet $\alpha_{IIb}\beta_3$ receptor; the free FVIIIa can also bind to phosphotidylserine exposed on superactivated platelets [71]. The adjacent C1 domain of FVIIIa also contributes to platelet binding [72,73]. In addition, FVIIIa can bind to the surface of endothelial cells; the binding site on the FVIIIa is the region between Ala2318 and Tyr2332 of the C2 domain [74]. Once it is membrane bound, FVIIIa is primed to form a complex with factors IXa and X. Fig. 3.3 shows the formation of the FVIIIa, FIXa, FX complex on a phospholipid surface.

The affinity of activated FIX (FIXa) for FVIIIa is increased 2000-fold when FVIIIa is membrane bound; the site on FVIIIa that binds FIXa is in the A3 domain, between residues 1803 and 1818 [75,76], and there is evidence for a second site in the A2 domain [77]. When bound to FVIIIa, the catalytic activity of FIXa toward FX is increased approximately 10^6-fold.

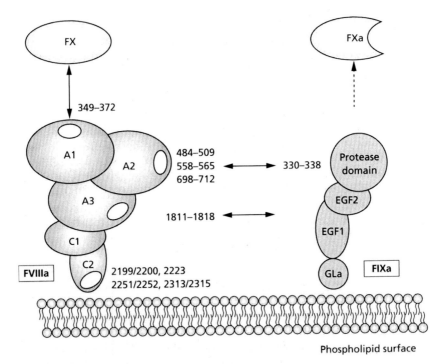

FIG. 3.3   Components of the Xase complex showing residues involved in FVIIIa binding to FX, FIXa, and the phospholipid bilayer. *Reproduced with permission from Lee CA, Berntorp EE, Hoots WK, editors. Textbook of hemophilia. Malden, MA: Blackwell Publishing Ltd; 2005, p. 30 (Fig. 5.2).*

FX binds to the FVIIIa light chain (A3C1C2 domains) [78], and undergoes hydrolytic cleavage by the hydroxyl side chain of Ser365 of FIXa [79]. Activation of FX depends on the binding of FVIIIa to the membranes of either platelets or endothelial cells, as well as the binding of FIXa to FVIIIa, and the binding of FX to the factor VIIIa/IXa complex; mutations of key residues at any of these binding sites result in a greatly prolonged clotting time, characteristic of hemophilia. Recently, it was shown that a bispecific antibody (emicizumab) that is capable of binding to factors IXa and X can substitute for FVIIIa, and shorten the clotting time in patients with hemophilia (Chapter 8) [80].

Following activation on the membrane surface, activated FVIII (FVIIIa) undergoes rapid decay. This is due to the reversible dissociation of the A2 subunit from the A1 subunit of the A1/A3-C1-C2 dimer [81], but stabilization of the A2 domain can prolong FVIII cofactor activity [82]. FIXa is capable of stabilizing and modulating FVIIIa, but prolonged reaction of FVIIIa with FIXa results in proteolysis at Arg336 of the A1 subunit and inactivation of the cofactor [83]. FXa also cleaves FVIIIa at Arg336 and Lys36 in the A1 subunit, suggesting the regulation of FVIIIa activity by feedback inhibition [84]. Thrombin-activated FVIII, attached to phosphatidylserine on cell surfaces, binds to LRP-1 via its A2 domain and is internalized [85].

FVIIIa is also inactivated by activated protein C (APC), which cleaves FVIIIa at Arg 336 and Arg562 [86]; both sites must be cleaved to achieve FVIIIa inactivation [87] (Fig. 3.2). Cofactors for this cleavage are protein S and FV [88,89]. The cleavage at Arg562 is inhibited by VWF and FIXa [90], but this inhibition is prevented by protein S [91]. However, FXa can inactivate protein S [92], thereby preserving the procoagulant activity of the FVIIIa, FIXa complex.

SUMMARY

FVIII has played an essential role in hemostasis during the evolution of creatures as diverse as fish, birds, and man. The FVIII in plasma is bound to VWF; when dissociated from VWF, it consists of a single chain of 2332 amino acids with a calculated molecular weight of 264,763. FVIII is synthesized in the ribosomes of endothelial cells, binds to the LMAN1-MCFD2 receptor complex, and enters the ERGIC compartment where it is decorated with carbohydrate moieties. The mature FVIII is packed into Weibel-Palade bodies where it forms noncovalent bonds with VWF, and the two proteins are usually secreted together. The concentrations of FVIII and VWF in plasma are 0.1–0.2 and 10 µg/mL, respectively. FVIII has a half-life of 12 h and is cleared from the circulation by macrophages in the liver and spleen.

Thrombin activates FVIII by cleavages at amino acids 372 and 1689 to form a two-chain molecule that binds to phospholipids on cell membranes. The activated FVIII forms a bridge between FIXa and FX to constitute the tenase complex, which generates hemostatic concentrations of FXa. Activated FVIII undergoes rapid decay due to proteolysis by FXa and is also cleaved by activated protein C. Researchers have developed a bispecific antibody (emicizumab) to replace activated FVIII in the tenase complex; this agent was shown to shorten the clotting time of patients with hemophilia.

References

[1] Davidson CJ, Hirt RP, Lal K, Snell P, Elgar G, Tuddenham EG, McVey JH. Molecular evolution of the vertebrate blood coagulation network. Thromb Haemost 2003;89:420–8.
[2] Davidson CJ, Tuddenham EG, McVey JH. 450 million years of hemostasis. J Thromb Haemost 2003;1:1487–94.
[3] Patek Jr AJ, Taylor FHL. Hemophilia II. Some properties of a substance obtained from normal plasma effective in accelerating the clotting of hemophilic blood. J Clin Invest 1937;16:113–24.
[4] Barkhan P, Lai M, Stevenson M. Antihaemophilic factor (factor VIII): an α-2 globulin. Br J Haematol 1963;9:499–505.
[5] Ratnoff OD, Kass L, Lang PD. Studies on the purification of antihemophilic factor (factor VIII). II. Separation of partially purified antihemophilic factor by gel filtration. J Clin Invest 1969;48:957–62.
[6] Van Mourik JA, Mochtar IA. Purification of human antihemophilic factor (factor VIII) by gel chromatography. Biochim Biophys Acta 1970;221:677–9.
[7] Green D. A simple method for the purification of factor VIII (antihemophilic factor) employing snake venom. J Lab Clin Med 1971;77:153–8.
[8] Schmer G, Kirby EP, Teller DC, Davie EW. The isolation and characterization of bovine factor VIII (antihemophilic factor). J Biol Chem 1972;2512–21.
[9] Meyer D, Lavergne J-M, Larrieu M-J, Josso F. Cross-reacting material in congenital factor VIII deficiencies (haemophilia A and von Willebrand's disease). Thromb Res 1972;1:183–96.
[10] Zimmerman TS, Ratnoff OD, Powell AE. Immunologic differentiation of classic hemophilia (factor 8 deficiency) and von Willebrand's disease, with observations on combined deficiencies of antihemophilic factor and proaccelerin (factor V) and on an acquired circulating anticoagulant against antihemophilic factor. J Clin Invest 1971;50:244–54.
[11] Stites DP, Hershgold EJ, Perlman JD, Fudenberg MM. Factor 8 detection by hemagglutination inhibition: hemophilia A and von Willebrand's disease. Science 1971;171:196–7.
[12] Nilsson IM, Blomback M, Blomback B. v. Willebrand's disease in Sweden; its pathogenesis and treatment. Acta Med Scand 1959;164:263–78.
[13] Bloom AL, Giddings JC, Peake IR. Low molecular weight factor VIII. Lancet 1973;i:661–2.
[14] Vehar GA, Keyt B, Eaton D, Rodriguez H, O'Brien DP, Rotblat F, Oppermann H, Keck R, Wood WI, Harkins RN, Tuddenham EGD, Lawn RM, Capon DJ. Structure of human factor VIII. Nature 1984;312:337–42.

[15] Fay PJ, Anderson MT, Chavin SI, Marder VJ. The size of human factor VIII heterodimers and the effects produced by thrombin. Biochim Biophys Acta 1986;871:268–78.
[16] Church WR, Jernigan RL, Toole J, Hewick RM, Knopf J, Knutson GJ, Nesheim ME, Mann KG, Fass DN. Coagulation factors V and VIII and ceruloplasmin constitute a family of structurally related proteins. Proc Natl Acad Sci 1984;81:6934–7.
[17] Kane WH, Davie EW. Blood coagulation factors V and VIII: structural and functional similarities and their relationship to hemorrhagic and thrombotic disorders. Blood 1988;71:539–55.
[18] Tagliavacca L, Moon N, Dunham WR, Kaufman RJ. Identification and functional requirement of Cu(I) and its ligands within coagulation factor VIII. J Biol Chem 1997;272:27428–34.
[19] Owen WG, Wagner RH. Separation of an active antihemophilic factor fragment following dissociation by salts or detergents. Thromb Diath Haemorrh 1972;27:502–15.
[20] Weiss HJ, Kochwa S. Molecular forms of antihaemophilic globulin in plasma, cryoprecipitate and after thrombin activation. Br J Haematol 1970;18:89–100.
[21] Zimmerman TS, Edgington TS. Factor VIII coagulant activity and factor VIII-like antigen: independent molecular entities. J Exp Med 1973;138:1015–20.
[22] Weiss HJ, Hoyer LW. Dissociation of antihemophilic factor procoagulant activity from the von Willebrand factor. Science 1973;182:1149–51.
[23] Green D. The relationship of factor VIII to von Willebrand factor [Doctoral dissertation]. Chicago: Northwestern University; 1974.
[24] Kaufman RJ, Antonarakis SE, Fay PJ, Factor VIII, Hemophilia A. In: Colman RW, Clowes AW, Goldhaber SZ, Marder VJ, George JN, editors. Hemostasis and thrombosis. 5th ed. Philadelphia: Lippincott Williams & Wilkins; 2006. p. 151–75 [chapter 8].
[25] Poon M-C, Ratnoff OD. Evidence that functional subunits of antihemophilic factor (factor VIII) are linked by noncovalent bonds. Blood 1976;48:87–94.
[26] Yee A, Oleskie AN, Dosey AM, Kretz CA, Gildersleeve RD, Dutta S, Su M, Ginsburg D, Skiniotis G. Visualization of an N-terminal fragment of von Willebrand factor in complex with factor VIII. Blood 2015;126:939–42.
[27] Chiu P-L, Bou-Assaf GM, Chhabra ES, Chambers MG, Peters RT, Kulman JD, Walz T. Mapping the interaction between factor VIII and von Willebrand factor by electron microscopy and mass spectrometry. Blood 2015;126:935–8.
[28] Foster PA, Fulcher CA, Mari T, Titani K, Zimmerman TS. A major factor VIII binding domain resides within the amino-terminal 272 amino acid residues of von Willebrand factor. J Biol Chem 1987;262:8443–6.
[29] Van Outryve M, Baele G, de Weerdt GA, Barbier F. Antihaemophilic factor A (FVIII) and serum fibrin-fibrinogen degradation products in hepatic cirrhosis. Scand J Haemat 1973;11:148–52.
[30] Shahani T, Covens K, Lavend'Homme R, Jazouli N, Sokal E, Peerlinck K, Jacquemin M. Human liver sinusoidal endothelial cells but not hepatocytes contain factor VIII. J Thromb Haemost 2014;12:36–42.
[31] Penick GD, Webster WP, Peacock EE, Hutchin P, Zukoski CF. Organ transplantation in animal hemophilia. In: Brinkhous KM, editor. Hemophilia and new hemorrhagic states. Chapel Hill: U of North Carolina Press; 1971. p. 97–105.
[32] Veltkamp JJ, van de Torren K, Schalm SW, van der Does J, Epstein RB, Terpstra JL. Organ transplantation in canine hemophilia. Haemophilia 1971;19–25.
[33] Rall LB, Bell GI, Caput D, Truett MA, Masiarz FR, Najarian RC, Valenzuela P, Anderson HD, Din N, Hansen B. Factor VIII:C synthesis in the kidney. Lancet 1985;325:44.
[34] Veltkamp JJ, Asfaou E, van de Torren K, van der Does JA, van Tilburg NH, EKJ P. Extra-hepatic factor VIII synthesis. Transplantation 1974;18:56–62.

[35] Norman JC, Covelli VH, Sise HS. Transplantation of the spleen: experimental cure of hemophilia. Surgery 1968;64:1–14.

[36] Dodds WH. Hepatic influence on splenic synthesis and release of coagulation activities. Science 1969;166:882–3.

[37] Aronovich A, Tchorsh D, Katchman H, Eventov-Friedman S, Shezen E, Martinowitz U, Blazar BR, Cohen S, Tal O, Reisner Y. Correction of hemophilia as a proof of concept for treatment of monogenic diseases by fetal spleen transplantation. Proc Natl Acad Sci U S A 2006;103:19075–80.

[38] Hathaway WE, Mull MM, Githens JH, Groth CG, Marchioro TL, Starzl TE. Attempted spleen transplant in classical hemophilia. Transplantation 1969;7:73–5.

[39] Hoyer LW, de los Santos RP, Hoyer J. Antihemophilic factor antigen. Localization in endothelial cells by immunofluorescent microscopy. J Clin Invest 1973;52:2737–44.

[40] Van der Kwast TH, Stel HV, Cristen E, Bertina RM, Veerman ECI. Localization of factor VIII-procoagulant antigen: an immunohistological survey of the human body using monoclonal antibodies. Blood 1986;67:222–7.

[41] Hollestelle MJ, Thinnes T, Crain K, Stiko A, Kruijt JK, van Berkel TJC, Loskutoff DJ, van Mourik JA. Tissue distribution of factor VIII gene expression *in vivo*—a closer look. Thromb Haemost 2001;86:855–61.

[42] Pan J, Dinh TT, Rajaraman A, Lee M, Scholz A, Czupalla CJ, Kiefel H, Zhu L, Xia L, Morser J, Jiang H, Santambrogio L, Butcher EC. Patterns of expression of factor VIII and von Willebrand factor by endothelial cell subsets in vivo. Blood 2016;128:104–9.

[43] Fahs SA, Hille MT, Shi Q, Weiler H, Montgomery RR. A conditional knockout mouse model reveals endothelial cells as the principal and possibly exclusive source of plasma factor VIII. Blood 2014;123:3706–13.

[44] Schwarz DS, Blower MD. The endoplasmic reticulum: structure, function and response to cellular signaling. Cell Mol Life Sci 2016;73:79–94.

[45] Khoriaty R, Vasievich MP, Ginsburg D. The COPII pathway and hematologic disease. Blood 2012;120:31–8.

[46] Ginsburg D, Nichols WC, Zivelin A, Kaufman RJ, Seligsohn U. Combined factor V and VIII deficiency-the solution. Haemophilia 1998;4:677–82.

[47] Zheng C, Zhang B. Combined deficiency of coagulation factors V and VIII: an update. Semin Thromb Hemost 2013;39:613–20.

[48] Pipe SW. Functional roles of the factor VIII B domain. Haemophilia 2009;15:1187–96.

[49] Turner NA, Moake JL. Factor VIII is synthesized in human endothelial cells, packaged in Weibel-Palade bodies and secreted bound to ULVWF strings. PLoS ONE 2015;10.

[50] Rosenberg JB, Greengard JS, Montgomery RR. Genetic induction of a releasable pool of factor VIII in human endothelial cells. Arterioscler Thromb Vasc Biol 2000;20:2689–95.

[51] Yarovoi H, Nurden AT, Montgomery RR, Nurden P, Poncz M. Intracellular interaction of von Willebrand factor and factor VIII depends on cellular context: lessons from platelet-expressed factor VIII. Blood 2005;105:4674–6.

[52] Lopes da Silva M, Cutler DF. Von Willerbrand factor multimerization and the polarity of secretory pathways in endothelial cells. Blood 2016;128:277–85.

[53] Furie B, Furie BC, Benz Jr EJ. Molecular basis of blood coagulation. In: Hoffman R, Shattil SJ, Furie B, Cohen HJ, Silberstein LE, McGlave P, editors. Hematology basic principles and practice. Philadelphia: Elsevier Churchill Livingstone; 2005. p. 1937 [chapter 109].

[54] Cooperberg AA, Teitelbaum J. The concentration of antihaemophilic globulin (AHG) related to age. Br J Haematol 1960;6:281–5.

[55] Song J, Chen F, Campos M, Bolgiano D, Houck K, Chambless LE, Wu KK, Folsom AR, Couper D, Boerwinkle E, Dong JF. Quantitative influence of ABO blood groups on factor VIII and its ratio to von Willebrand factor, novel observations from an ARIC study of 11,673 subjects. PLoS ONE 2015;10.

[56] Albanez S, Ogiwara K, Michels A, Hopman W, Grabell J, James P, Lillicrap D. Aging and ABO blood type influence von Willebrand factor and factor VIII levels through interrelated mechanisms. J Thromb Haemost 2016;14:953–63.

[57] Tracy RP, Bovill EG, Fried LP, Heiss G, Lee MH, Polak JF, Psaty BM, Savage PJ. The distribution of coagulation factors VII and VIII and fibrinogen in adults over 65 years. Results from the Cardiovascular Health Study. Ann Epidemiol 1992;2:509–19.

[58] Mari D, Mannucci PM, Coppola R, Bottasso B, Bauer KA, Rosenberg RD. Hypercoagulability in centenarians: the paradox of successful aging. Blood 1995;85:3144–9.

[59] Castro-Nunez L, Dienava-Verdoold I, Herczenik E, Mertens K, Meijer AB. Shear stress is required for the endocytic uptake of the factor VIII-von Willebrand factor complex by macrophages. J Thromb Haemost 2012;10:1929–37.

[60] Tuddenham EG, Lane RS, Rotblat F, Johnson AJ, Snape TJ, Middleton S, Kernoff PB. Response to infusions of polyelectrolyte fractionated human factor VIII concentrate in human haemophilia A and von Willebrand's disease. Br J Haematol 1982;52:259–67.

[61] Yee A, Gildersleeve RD, Gu S, Kretz CA, McGee BM, Carr KM, Pipe SW, Ginsburg D. A von Willebrand factor fragment containing the D'D3 domains is sufficient to stabilize coagulation factor VIII in mice. Blood 2014;124:445–52.

[62] Rastegarlari G, Pegon JN, Casari C, Odouard S, Navarrete AM, Saint-Lu N, van Vlijmen BJ, Legendre P, Christophe OD, Denis CV, Lenting PJ. Macrophage LRP1 contributes to the clearance of von Willebrand factor. Blood 2012;119:2126–34.

[63] Saenko EL, Yakhyaev AV, Mikhailenko I, Strickland DK, Sarafanov AG. Role of the low density lipoprotein-related protein receptor in mediation of factor VIII catabolism. J Biol Chem 1999;274:37685–92.

[64] Lenting PJ, Neels JG, van den Berg BM, Clijsters sPP, Meijerman DW, Pannekoek H, van Mourik JA, Mertens K, van Zonneveld AJ. The light chain of factor VIII comprises a binding site for low density lipoprotein receptor-related protein. J Biol Chem 1999;274:23734–9.

[65] Sarafanov AG, Ananyeva NM, Shima M, Saenko EL. Cell surface heparan sulfate proteoglycans participate in factor VIII catabolism mediated by low density lipoprotein receptor-related protein. J Biol Chem 2001;276:11970–9.

[66] Bloem E, van den Biggelaar M, Wroblewska A, Voorberg J, Faber JH, Kjalke M, Stennicke HR, Mertens K, Meijer AB. Factor VIII C1 domain spikes 2092-2093 and 2158-2159 comprise regions that modulate cofactor function and cellular uptake. J Biol Chem 2013;288:29670–9.

[67] Bovenschen N, Mertens K, Hu L, Havekes LM, van Vijmen BJM. LDL receptor cooperates with LDL receptor-related protein in regulating plasma levels of coagulation factor VIII in vivo. Blood 2005;106:906–12.

[68] Mann KG, Nesheim ME, Church WR, Haley P, Krishnaswamy S. Surface-dependent reactions of the vitamin K-dependent enzyme complexes. Blood 1990;76:1–16.

[69] Pittman DD, Millenson M, Marquette K, Bauer K, Kaufman RJ. A2 domain of human recombinant-derived factor VIII is required for procoagulant activity but not for thrombin cleavage. Blood 1992;79:389–97.

[70] Lollar P, Hill-Eubanks DC, Parker CG. Association of the factor VIII light chain with von Willebrand factor. J Biol Chem 1988;263:10451–5.

[71] Gilbert GE, Novakovic VA, Shi J, Rasmussen J, Pipe SW. Platelet binding sites for factor VIII in relation to fibrin and phosphatidylserine. Blood 2015;126:1237–44.

[72] Hsu T-C, Pratt KP, Thompson AR. The factor VIII C1 domain contributes to platelet binding. Blood 2008;111:200–8.

[73] Meems H, Meijer AB, Cullinan DB, Mertens K, Gilbert GE. Factor VIII C1 domain residues Lys 2092 and Phe 2093 contribute to membrane binding and cofactor activity. Blood 2009;114:3938–46.

[74] Brinkman H-JM, Mertens K, van Mourik JA. Phospholipid-binding domain of factor VIII is involved in endothelial cell-mediated activation of factor X by factor IXa. Arterioscler Thromb Vasc Biol 2002;22:511–6.

[75] Lenting PJ, van de Loo JW, Donath MJ, et al. The sequence Glu1811-Lys1818 of human blood coagulation factor VIII comprises a binding site for activated factor IX. J Biol Chem 1996;271:1935–40.

[76] Bovenschen N, Boertjes RC, van Stempvoort G, et al. Low density lipoprotein receptor-related protein and factor IXa share structural requirements for binding to the A3 domain of coagulation factor VIII. J Biol Chem 2003;278:9370–7.

[77] Bajaj SP, Schmidt AE, Mathur A, et al. Factor IXa:factor VIIIa interaction. Helix 330-338 of factor IXa interacts with residues 558-565 and spatially adjacent regions of the a2 subunit of factor VIIIa. J Biol Chem 2001;276:16302–9.

[78] Takeyama M, Wakabayashi H, Fay PJ. Factor VIII light chain contains a binding site for factor X that contributes to the catalytic efficiency of factor Xase. Biochemistry 2012;51:820–8.

[79] Bajaj SP, Thompson AR. Molecular and structural biology of factor IX. In: Colman RW, Clowes AW, Goldhaber SZ, Marder VJ, George JN, editors. Hemostasis and thrombosis. 5th ed. Philadelphia: Lippincott Williams & Wilkins; 2006. p. 140 [chapter 7].

[80] Shima M, et al. Factor VIII-mimetic function of humanized bispecific antibody in hemophilia A. N Engl J Med 2016;374:2044–53.

[81] Fay PJ, Smudzin TM. Characterization of the interaction between the A2 subunit and A1/A3-C1-C2 dimer in human factor VIII. J Biol Chem 1992;267:13246–50.

[82] Leong L, Sim D, Patel C, Tran K, Liu P, Ho E, Thompson T, Kretchmer PJ, Wakabayashi H, Pay PJ, Murphy JE. Noncovalent stabilization of the factor VIII A2 domain enhances efficacy in hemophilia A mouse vascular injury models. Blood 2015;125:392–8.

[83] Lamphear BJ, Fay PJ. Proteolytic interactions of actor IXa with human factor VIII and factor VIIIa. Blood 1992;80:3120–6.

[84] Nogami K, Wakabayashi H, Fay PJ. Mechanisms of factor Xa-catalyzed cleavage of the factor VIIIa A1 subunit resulting in cofactor inactivation. J Biol Chem 2003;278:16502–9.

[85] Bovenschen N, van Stempvoort G, Voorberg J, Mertens K, Meijer AB. Proteolytic cleavage of factor VIII heavy chain is required to expose the binding-site for low-density lipoprotein receptor-related protein within the A2 domain. J Thromb Haemost 2006;4:1487–93.

[86] Fay PJ, Smudzin TM, Walker FJ. Activated protein C-catalyzed inactivation of human factor VIII and factor VIIIa. Identification of cleavage sites and correlation of proteolysis with cofactor activity. J Biol Chem 1991;266:20139–45.

[87] Amano K, Michnick DA, Moussalli M, Kaufman RJ. Mutation at either Arg 336 or Arg562 in factor VIII is insufficient for complete resistance to activated protein C (APC)-mediated inactivation: implications for the APC resistance test. Thromb Haemost 1998;79:557–63.

[88] Lu D, Kalafatis M, Mann KG, Long GL. Comparison of activated protein C/protein S-mediated inactivation of human factor VIII and factor V. Blood 1996;87:4708–17.

[89] Gale AJ, Cramer TJ, Rozenshteyn D, Cruz JR. Detailed mechanisms of the inactivation of factor VIIIa by activated protein C in the presence of its cofactors, protein S and factor V. J Biol Chem 2008;283:16355–62.

[90] Rick ME, Esmon NL, Krizek DM. Factor IXa and von Willebrand factor modify the inactivation of factor VIII by activated protein C. J Lab Clin Med 1990;115:415–21.

[91] Regan LM, Lamphear BJ, Huggins CF, Walker FJ, Fay PJ. Factor IXa protects factor VIIIa from activated protein C. Factor IXa inhibits activated protein-C catalyzed cleavage of factor VIIIa at Arg 562. J Biol Chem 1994;269:9445–52.

[92] Long GL, Lu D, Xie RL, Kalafatis M. Human protein S cleavage and inactivation by coagulation factor Xa. J Biol Chem 1998;273:11521–6.

Recommended Reading

[1] Zimmerman TS, Ratnoff OD, Powell AE. Immunologic differentiation of classic hemophilia (factor 8 deficiency) and von Willebrand's disease, with observations on combined deficiencies of antihemophilic factor and proaccelerin (factor V) and on an acquired circulating anticoagulant against antihemophilic factor. J Clin Invest 1971;50:244–54.

[2] Schmer G, Kirby EP, Teller DC, Davie EW. The isolation and characterization of bovine factor VIII (antihemophilic factor). J Biol Chem 1972;2512–21.

[3] Weiss HJ, Hoyer LW. Dissociation of antihemophilic factor procoagulant activity from the von Willebrand factor. Science 1973;182:1149–51.

[4] Vehar GA, Keyt B, Eaton D, Rodriguez H, O'Brien DP, Rotblat F, Oppermann H, Keck R, Wood WI, Harkins RN, Tuddenham EGD, Lawn RM, Capon DJ. Structure of human factor VIII. Nature 1984;312:337–42.

[5] Kane WH, Davie EW. Blood coagulation factors V and VIII: structural and functional similarities and their relationship to hemorrhagic and thrombotic disorders. Blood 1988;71:539–55.

[6] Mann KG, Nesheim ME, Church WR, Haley P, Krishnaswamy S. Surface-dependent reactions of the vitamin K-dependent enzyme complexes. Blood 1990;76:1–16.

[7] Davidson CJ, Tuddenham EG, McVey JH. 450 million years of hemostasis. J Thromb Haemost 2003;1:1487–94.

[8] Bovenschen N, Mertens K, Hu L, Havekes LM, van Vijmen BJM. LDL receptor cooperates with LDL receptor-related protein in regulating plasma levels of coagulation factor VIII in vivo. Blood 2005;106:906–12.

[9] Fahs SA, Hille MT, Shi Q, Weiler H, Montgomery RR. A conditional knockout mouse model reveals endothelial cells as the principal and possibly exclusive source of plasma factor VIII. Blood 2014;123:3706–13.

4

FVIII Genetics

Prior to the modern era, many bleeding disorders, especially in children, were called hemophilia. A more stringent definition was a bleeding condition that was X-linked, affecting males but transmitted by females. However, FIX deficiency is also an X-linked bleeding disorder that clinically mimics FVIII deficiency, and was not recognized as being distinct from classical or hemophilia A until 1952 [1]. Therefore, much of what was called hemophilia in the older literature might not have been FVIII deficiency. For example, it was assumed that the bleeding disorder that originated with Queen Victoria and eventually affected several of the Royal Families of Europe was hemophilia A. But an analysis of the DNA from skeletal remains of her granddaughter, Alexandra, and great-grandson, Alexei, showed that the defect was in the *F9*, not the *F8* gene [2]. This chapter examines the genetics of FVIII and hemophilia A.

THE FVIII GENE

More than 50 years ago, McCusick listed 55 disorders very likely, and 17 possibly, linked to the X-chromosome [3]; by 1980, it was known that the genes for color blindness and glucose-6-phosphate dehydrogenase were clustered with the gene for classical hemophilia (FVIII deficiency) at the distal end of the X-chromosome [4]. In 1984, Gitschier and colleagues [5] and Wood and colleagues [6] isolated and characterized the *F8* gene. They recovered FVIII clones from a bacteriophage library enriched for the X-chromosome, using DNA from a lymphoblast cell line derived from an individual with four X-chromosomes (49,XXXXY). The DNA was screened with a 36-base oligonucleotide probe representing one codon of a peptide recovered from purified FVIII. The clones were expanded to contain 200 kb of the human genome encompassing the entire *F8* gene. The gene they described consists of 186,000 base-pairs containing 26 exons and 25 introns; two nested genes, *F8A* and *F8B*, are located in intron 22 (they feature in a hemophilia-inducing mutation, intron

22 inversion). The order of the domains in the parent gene is A1-A2-B-A3-C1-C2, and this is the same domain order found in the protein product. The messenger RNA (mRNA) for FVIII is 9 kb in length. The location and the structure of the gene are shown in Figs. 4.1 and 4.2.

Recombinant FVIII was prepared by inserting the reconstructed gene into a plasmid that transcribed the heterologous sequences upon transfection into a hamster kidney cell line [6]. The FVIII appearing in the

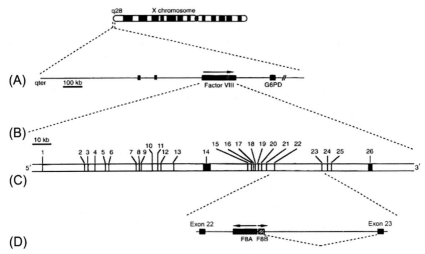

FIG. 4.1 Chromosomal localization and structure of the *F8* gene. (A) The gene is on the X chromosome. (B) It is approximately 1 Mb from the Xq telomere (qter). (C) It contains 26 exons. (D) Two nested genes, F8A and F8B, are located between exons 22 and 23. *Reprinted from Kaufman RJ, Antonarakis SE. Structure, biology, and genetics of factor VIII. In: Hoffman R, Benz Jr. EJ, Shattil SJ, Furie B, Cohen HJ, Silberstein LE, McGlave P, editors. Hematology: basic principles and practice. 4th ed. Elsevier; 2005, Fig. 114.4 [chapter 114], with permission.*

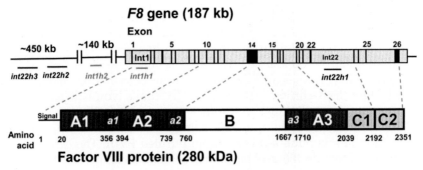

FIG. 4.2 *F8* gene and FVIII protein structure. *Int1h* and *Int22h* are recombination sites. *Reprinted with permission from Swystun LL, James PD. Genetic diagnosis in hemophilia and von Willebrand disease. Blood Rev 2017;31:47–56.*

supernatant of the cell cultures was confirmed to be authentic FVIII by demonstrating that it corrected the clotting defect of hemophilic plasma, accelerated the rate of FX activation by FIXa, and neutralized FVIII antibodies. Commercially produced recombinant FVIII is currently the treatment of choice for replacing the missing FVIII in patients with hemophilia A.

FVIII MUTATIONS

Once the gene for FVIII had been identified, it became possible to search for mutations that might result in hemophilia A. As of August 2016, 2573 mutations were listed in the Human Gene Mutation Database at the Institute of Medical Genetics in Cardiff [7]. A listing of the individual mutations with references can be found at the Online Mendelian Inheritance in Man site (https://www.OMIM.org), updated daily, and edited at the McKusick-Nathans Institute of Genetic Medicine at Johns Hopkins University School of Medicine. The *F8* genomic coordinates are 154,835,787 to 155,022,722, and it is 1.1 Mb from the telomeric end of the X-chromosome. Hemophilia A is listed as OMIM 306700. As shown in Fig. 4.3, most mutations are of the missense/nonsense type, 17% are small deletions, and most of the rest are gross deletions, splicing, and small insertions. A category not displayed is regulatory, constituting just 0.2%.

Missense/Nonsense Mutations

The majority of mutations are of the missense and nonsense variety (Fig. 4.3). Missense mutations occasionally produce nonfunctional FVIII, termed crossreacting material (CRM), because the nonfunctional protein

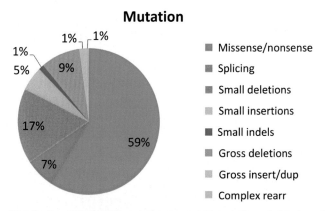

Mutation

- Missense/nonsense
- Splicing
- Small deletions
- Small insertions
- Small indels
- Gross deletions
- Gross insert/dup
- Complex rearr

FIG. 4.3 Distribution (%) of FVIII mutation types (abbreviations: *indels*, insertions/deletions; *insert/dup*, insertions/duplications; *rearr*, rearrangements).

in the patient's plasma is still able to react with most FVIII antibodies [8,9]. CRM is generally found in patients with mild-to-moderate hemophilia [10], and the mutations often occur at critical cleavage sites [11]. For example, single base substitutions (C → T) that replace arginine with cysteine at residues 372 or 1689, two thrombin cleavage sites, result in a CRM$^+$, dysfunctional FVIII [12,13]. Another mutation, at the cleavage site for activated protein C (residue 336), truncates the mature protein to a 335-amino-acid fragment [14].

CRM is also detected when missense mutations affect the role of activated FVIII (FVIIIa) in coagulation. They might hinder the interaction of FVIIIa with FIXa ([1]R527W, I566T, V634A, V634M) [15], or severely decrease the catalytic activity of the Xase complex (S558F, V559A, D560A, Q565R, R562A) [16]. Other mutations affect the binding of FVIII to Von Willebrand Factor (VWF). These mutations are located within the C1 domain and include I2098S, S2119Y, N2129S, R2150H, and P2153Q [17]. In some C1 mutations (Q2087E, R2150C), FVIII coagulant activity and antigen are reduced to the same extent, suggesting that the protein is more rapidly cleared from the circulation [18].

Mutations such as R531H, A284E/P, S289L, N694I, and R698W/L lie at the interface of the three FVIII A-domains and increase the rate of dissociation of the A2 domain from thrombin-activated FVIII, resulting in a loss of coagulant activity [19,20]. Patients with these mutations generally have mild-to-moderately severe hemophilia, but their FVIII levels are higher when measured by 1-stage assays than by 2-stage and chromogenic assays. These latter methods include an incubation step that facilitates the dissociation of the A2 domain, destabilizing the molecule and lowering FVIII levels [21]. The range of FVIII levels is broad with the 1-stage assay, and some patients might have values within the normal range. Nevertheless, all patients with these mutations, even those with normal 1-stage FVIII levels, have aberrant thrombin generation [22]. Trossaert et al. [23] report that automated chromogenic assays correlate better with thrombin generation and bleeding risk, so that if mild hemophilia is suspected but a normal FVIII level is reported, the patient should be re-examined with the chromogenic assay. Less frequently encountered is the reverse situation, a lower 1-stage than 2-stage FVIII level. This has been observed with the following mutations: I388T, E739K, E1989G, and P2146S [24]. Overall, discrepancies between 1-stage and 2-stage assays are reported in 10% of families with members having mild-to-moderate hemophilia [25].

---

[1]Amino acid letter abbreviations: A, alanine; C, cysteine; D, aspartic acid; E, glutamic acid; F, phenylalanine; G, glycine; H, histidine; I, isoleucine; K, lysine; L, leucine; M, methionine; N, asparagine; P, proline; Q, glutamine; R, arginine; S, serine; T, threonine; V, valine; W, tryptophan; Y, tyrosine.

Nonsense mutations can create a signal for messenger RNA (mRNA) to skip exons in which they are located [24]. For example, an E1987X mutation in exon 19 led to all detectable mRNA lacking this exon, and an R2116X mutation in exon 22 resulted in 50% of mRNA without this exon [26]. In the Canadian Database, 45 of 380 (12%) mutations were of the nonsense variety and were usually associated with a severe bleeding phenotype [27].

Complex Rearrangements (Intron-22 Inversions)

Among the complex rearrangements, the most important is the intron-22 inversion (OMIM IVS22 = 300841.0067), observed in 42% of patients with severe hemophilia [28] and 23% overall [29].

First described in 1993 [30,31], this anomaly occurs when one or more upstream copies of the *F8* gene undergo homologous recombination with an intronless copy of the gene nested within intron-22 (Fig. 4.4). Because the orientation of upstream copy is usually opposite to that of the intron 22-copy, there is an inversion of the intervening DNA sequences and exons 1–22 become separated from and are in the opposite orientation to exons 23–26. The most common types of inversions involve either the

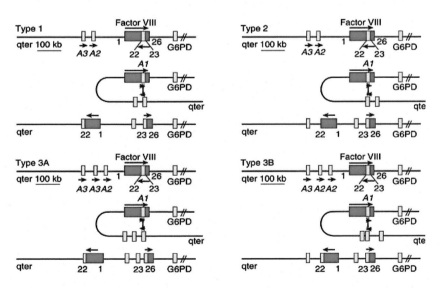

FIG. 4.4 FVIII inversions in classical hemophilia. Exons 1–22 become separated from exons 23–26 because of intrachromosomal crossing-over between the homologous regions A1, A2, A3. The three types of inversions are shown. *Reproduced with permission from Kaufman RJ, Antonarakis SE. Structure, biology, and genetics of factor VIII. In: Hoffman R, Benz Jr. EJ, Shattil SJ, Furie B, Cohen HJ, Silberstein LE, McGlave P, editors. Hematology: basic principles and practice. 4th ed. Philadelphia: Elsevier; 2005, Fig. 114.6 [chapter 114].*

distal (type 1) or the proximal (type 2) upstream copies, but rarely patients have three or more extragenic copies that recombine (type 3). Although the gene is inverted, two truncated protein products are synthesized: one contains the first 2124 amino acids of the FVIII protein corresponding to exons 1–22, and the second has 216 amino acids corresponding to exons 23–26 [32]. These protein fragments have been identified within peripheral blood mononuclear cells and liver sections from patients with the intron-22 inversion. However, the affected patients have no circulating fully functional FVIII molecules and therefore a severe bleeding phenotype. Another common inversion breaks the FVIII gene at intron 1 and affects about 5% of patients with severe hemophilia [33].

Miscellaneous Mutations

The A3-domain N1922S substitution is associated with moderate-to-severe hemophilia. Summers et al. [34] report that the intracellular levels of the mutant protein are similar to wild-type FVIII, but the protein displays abnormal folding in the A3 and C1 domains. As a consequence, N1922S protein accumulates in the endoplasmic reticulum, is unable to transit to the Golgi apparatus, and is not secreted.

Sequencing of the FVIII gene has revealed four single-nucleotide polymorphisms whose haplotypes encode six distinct proteins, designated H1 to H6. In 24% of black individuals, valine replaces methionine (M2238V) in the C2 domain, or histidine replaces arginine (R484H) in the A2 domain, giving rise to H3 or H4 rather than H1 and H2 FVIII proteins. Viel et al. [35] raised the possibility that these differences might explain the higher incidence of factor VIII antibodies (inhibitors) in blacks, because therapeutic recombinant FVIII derived from whites is always of the H1 or H2 haplotype. However, an extensive study of African-Americans with hemophilia that used three different approaches showed no correlation of inhibitor status with patient haplotype [36].

Retrotransposons

Retrotransposons are long interspersed DNA elements (LINE-1) that have been transcribed into RNA and then reverse-transcribed into complementary DNA (cDNA). The cDNA is then reinserted into the genome at a new location, where it might truncate the protein product of the gene [37]. Kazazian et al. [38] observed that a full-length LINE-1 was present on chromosome 22 of the mother of a boy with hemophilia. A portion of the LINE-1 had been inserted into exon 14 of his F8 gene, disrupting the synthesis of FVIII. This genetic aberration was identified in two of 240 individuals with hemophilia.

Functional Classification of Mutations

Mutations can be classified as null or nonnull, depending on whether FVIII activity can be detected on clotting assays. Null mutations include inversions, large deletions, nonsense mutations, small deletions/insertions outside poly-A runs, or splice-site mutations involving conserved nucleotides, and are associated with a severe bleeding phenotype. Nonnull mutations comprise missense mutations, small deletions/insertions within poly-A runs, or splice-site mutations involving nonconserved nucleotides, and are usually associated with a mild clinical phenotype. Studies have shown that the mutation type, null or nonnull, is a better predictor of bleeding than the level of FVIII [39]. Hemophiliacs with nonnull mutations and FVIII levels below the current limits of detection have their first hemorrhages at a later age than those with null mutations, suggesting that very low levels of FVIII, not detectable with current clotting assays, provide some protection against bleeding [40]. Features of various gene mutations are displayed in Table 4.1.

FVIII Mutations in Women

Hemophilia A is rare in women because mutations are usually limited to the FVIII gene on one of the two X-chromosomes, and the unmutated gene on the other X-chromosome provides adequate amounts of FVIII.

TABLE 4.1 Summarizes the Prevalence, Bleeding Phenotype, and Selected Characteristics of Some FVIII Gene Mutations

Mutation	Prevalence	Bleeding phenotype	Comment
Intron-22 inversion[a]	25% of all mutations but 42% of severe hemophilia	Severe	Truncated protein within mononuclear cells
Missense mutations[b]	57% of all mutations	Mild to moderate	Presence of crossreacting material
Missense mutations at the interface of the A-domains[b]	30% of mild-moderate hemophilia	Mild to moderate	Discrepant 1-stage and 2-stage assays
N1922S[a]	Rare	Severe	Misfolded protein unable to be secreted
R484H, M2238V[b]	24% of black hemophiliacs	Does not affect bleeding severity	No correlation with inhibitor development

[a] Null mutation.
[b] Nonnull mutation.

In each cell, the unmutated X-chromosome, as well as the X-chromosome bearing the hemophilic mutation, undergoes random inactivation [41]. Chromosomal inactivation occurs when Xist, a long noncoding RNA, recruits one of the X-chromosomes to the nuclear lamina, where Xist spreads over the actively transcribed genes, excluding RNA polymerase II [42]. To accomplish this goal, Xist recruits polycomb repressive complexes (PRC), and PRC2 leads to deposition of methylated histones and gene inactivation [43]. Because sufficient numbers of X-chromosomes with unmutated genes have avoided inactivation in the female carriers of hemophilia, levels of FVIII are usually high enough to prevent bleeding. However, true hemophilia occurs if the woman inherits the mutant alleles from both her father (a hemophiliac) and her mother (a carrier of the hemophilia mutation). Female hemophilia is also possible in a female carrier with the monosomy X syndrome (Turner Syndrome, XO). On the other hand, FVIII levels are increased in women with an extra X-chromosome (trisomy X syndrome) [44].

There are several reasons for reduced levels of FVIII in women. By far, the most common explanation is Von Willebrand Disease and this should be the focus of diagnostic evaluations; other causes should be sought if levels of VWF are within the normal range. Occasionally, female carriers of hemophilia have decreased FVIII levels because of nonrandom or skewed inactivation of the X-chromosome bearing the unmutated gene, resulting in dominance of the chromosome with the mutated gene [45–47]. Carriers with lower levels of FVIII often have prolonged bleeding from small wounds and surgical procedures [48]. Recent publications have called attention to "symptomatic carriers," and urge that they receive a diagnosis of hemophilia and appropriate management [49,50]. Acquired causes of decreased FVIII should also be considered; these include FVIII autoantibodies (Chapter 8), hypothyroidism and the acquired Von Willebrand Syndrome [51] (Chapter 12), and disseminated intravascular coagulation.

APPLICATION OF GENETIC ANALYSES

Genotyping patients with hemophilia provides information for carrier detection and prenatal diagnosis, risk of inhibitor formation, and bleeding severity. Several years ago, the American Thrombosis and Hemophilia Network (ATHN) established a registry for collecting clinical information and blood samples from people with bleeding disorders. *MyLifeOurFuture* (MLOF) is a collaboration between ATHN, the National Hemophilia Foundation, BloodworksNW, and Biogen to provide hemophilia genotype analysis and create a research repository for future studies [52]. DNA variants were detected in 2357 of 2401 (98.1%) of consented patients with hemophilia A, and more than one variant was reported in 40 patients, including 10 females. The collaboration plans to genotype more than 6000 US hemophilia patients.

Carrier Detection and Prenatal Diagnosis

Sisters of hemophiliacs and other women with a family history of hemophilia often wish to learn whether they are carriers of this disorder. Daughters of men with hemophilia are obligate carriers, but laboratory testing is required for other women with a family history of hemophilia. Although the level of FVIII is statistically lower in carriers than in controls, this evaluation alone is inadequate because the levels vary considerably in carriers, making FVIII measurements too imprecise for carrier detection. For example, a large study observed that the median clotting factor level of carriers was 0.60 IU/mL compared with 1.02 IU/mL for noncarriers [53]. A more accurate method of identifying carriers is the measurement of VWF as well as FVIII, since the two factors are generally in a 1:1 (0.84–1.49) ratio in healthy individuals and 0.5 (0.13–1.04) ratio in carriers [54,55]. With the use of high precision factor assays and discriminant analysis, up to 97% of obligate carriers can be identified [56,57]. This method greatly improved carrier detection but eventually became superseded by direct genetic analyses. If an informative mutation was not previously identified in the family, the analysis should start with a search for intron-22 inversions [58–60].

Prenatal diagnosis usually begins with a determination of whether the fetus is male using amniotic fluid or chorionic villus samples obtained from the mother. These techniques are well established but invasive and associated with a small, but definite risk of miscarriage. Alternatively, by using sensitive techniques, fetal chromosome Y DNA sequences can be detected in maternal plasma as early as the 7th week of gestation [61]. The subsequent effort to identify FVIII mutations in the male fetus is greatly aided by knowledge of the specific mutation present in other family members with hemophilia or female carriers of the disorder. Fetal cells are obtained from confluent amniocytic cells or chorionic villus tissue, and examined for known mutations, intron inversions, deletions, or duplications.[2] To avoid placing the fetus at risk, noninvasive approaches have been described based on the observation that the small amount of fetal DNA present in maternal blood is either of the wild or mutant type [62]. The maternal plasma contains balanced amounts of wild-type and mutant DNA; during the pregnancy the addition of the fetal DNA will

---

[2]Such analyses are performed by the Clinical and Molecular Diagnostic Laboratory at City of Hope (cmdl@coh.org).

increase the amount of wild-type or mutant DNA, depending on which type of DNA the fetus has inherited. Detecting more mutant than wild-type DNA in the maternal blood indicates that the fetus probably has hemophilia. Recently, direct detection of *F8* sequence variants in the maternal plasma of hemophilia carriers has become possible, using droplet digital polymerase chain reaction and targeted massively parallel sequencing [63]. Maternally inherited *F8 int22h-related* inversions were correctly identified in 15 at-risk pregnancies in samples obtained from 8 to 42 weeks of gestation.

Knowledge of whether the fetus has hemophilia can inform decisions about early termination, or if the pregnancy is carried to term, the management of labor and delivery. A nationwide survey among female carriers in the Netherlands found that 54% elected to undergo prenatal diagnosis; hemophilia was diagnosed in 26 pregnancies and 18 were terminated early [64]. In the event that the pregnancy is carried to term, the diagnosis of hemophilia in the infant can be confirmed by testing uncontaminated cord blood and assay results interpreted using age (gestation)-adjusted normal ranges [65].

SUMMARY

The *F8* gene is located at position Xq28 on the X-chromosome, and contains 186,000 base-pairs, 26 exons, 25 introns, and two nested genes, *F8A* and *F8B*, located between exons 22 and 23. Gene mutations are present in 98% of individuals with hemophilia, and more than 2500 distinct variants have been described; most are of the missense/nonsense variety and affect the binding of FVIIIa to FIXa, the activation of FX, or the dissociation of thrombin-activated FVIII. These mutations invariably result in decreased thrombin generation. While the bleeding phenotype might be mild to moderate with missense mutations, it is usually severe with nonsense mutations and complex rearrangements. The most common genetic alteration is the intron-22 inversion, in which a copy of the *F8* gene undergoes homologous recombination with a segment of intron 22, separating exon 22 from exons 23–26 and producing a truncated, nonfunctional protein. The female relatives of hemophiliacs should have genetic analyses to determine whether they have inherited mutant *F8* genes; such women often experience excessive bleeding and might require FVIII prophylaxis.

The cloning of the *F8* gene enabled the preparation of the recombinant protein and the elucidation of FVIII structure, synthesis, activation, and clearance from the circulation. It also led to advances in carrier detection and prenatal diagnosis, informed decisions about when to initiate prophylaxis therapy for hemophilia, clarified the risk of inhibitor formation, and made gene therapy possible. The use of recombinant FVIII for the treatment of hemophilia was a seminal event, and its success encouraged the application of other recombinant proteins for the management of several diseases; for example, recombinant insulin for the treatment of diabetes and recombinant erythropoietin (epoetin alfa) for the management of anemia.

References

[1] Biggs R, Douglas AS, Macfarlane RG, Dacie JV, Pitney WR, Merskey C, O'Brien JR. Christmas disease: a condition previously mistaken for haemophilia. Br Med J 1952;2:1378–82.

[2] Rogaev EI, Grigorenko AP, Fashutdinova G, Kittler EL, Moliaka YK. Genotype analysis identifies the cause of the "Royal Disease". Science 2009;326:817.

[3] McCusick VA. On the X chromosome of man. Ann Intern Med 1962;56:991–6.

[4] McCusick VA. The anatomy of the human genome. Am J Med 1980;69:267–76.

[5] Gitschier J, Wood WI, Goralka TM, Wion KL, Chen EY, Eaton DH, Vehar GA, Capon DJ, Lawn RM. Characterization of the human factor VIII gene. Nature 1984;312:326–30.

[6] Wood WI, Capon DJ, Simonsen CC, Eaton DH, Gitschier J, Keyt B, Seeburg PH, Smith DH, Hollingshead P, Wion KL, Delwart E, Tuddenham EGD, Vehar GA, Lawn RM. Expression of active human factor VIII from recombinant DNA clones. Nature 1984;312:330–7.

[7] http://www.hgmd.cf.ac.uk/ac/gene.php?gene=F8.

[8] Denson KWE, Biggs R, Haddon ME, Borrett R, Cobb K. Two types of haemophilia (A$^+$ and A$^-$): a study of 48 cases. Br J Haematol 1969;17:163–71.

[9] Feinstein D, Chong MNY, Kasper CK, Rapaport SI. Hemophilia A: polymorphism detectable by a factor VIII antibody. Science 1969;163:1071–2.

[10] Lechner K. Inactive factor VIII in hemophilia A and Willebrand's disease. A study of 117 cases. Acta Haematol 1972;48:257–68.

[11] Cutler JA, Mitchell MJ, Smith MP, Savidge GF. The identification and classification of 41 novel mutations in the factor VIII gene (F8C). Hum Mutat 2002;19:274–8.

[12] Shima M, Ware J, Yoshioka A, Fukui H, Fulcher CA. An arginine to cysteine amino acid substitution at a critical thrombin cleavage site in a dysfunctional factor VIII molecule. Blood 1989;74:1612–7.

[13] Arai M, Higuchi M, Antonarakis SE, Kazazian Jr HH, Phillips III JA, Janco RL, Hoyer LW. Characterization of a thrombin cleavage site mutation (Arg 1689 to Cys) in the factor VIII gene of two unrelated patients with cross-reacting material-positive hemophilia A. Blood 1990;75:384–9.

[14] Gitschier J, Kogan S, Levinson B, Tuddenham EGD. Mutations of factor VIII cleavage sites in hemophilia A. Blood 1988;72:1022–8.

[15] Amano K, Sarkar R, Pemberton S, Kemball-Cook G, Kazazian Jr HH, Kaufman RJ. The molecular basis for cross-reacting material-positive hemophilia A due to missense mutations within the A2-domain of factor VIII. Blood 1998;91:538–48.

[16] Jenkins PV, Freas J, Schmidt KM, Zhou Q, Fay PJ. Mutations associated with hemophilia A in the 558-565 loop of the factor VIIIa A2 subunit alter the catalytic activity of the factor Xase complex. Blood 2002;100:501–8.

[17] Jacquemin M, Lavend'homme R, Benhida A, Vanzieleghem B, d'Oiron R, Lavergne J-M, Brackmann HH, Schwaab R, Vanden Driessche T, Chuah MKL, Hoylaerts M, Giles JGG, Peerlinck K, Vermylen J, Saint-Remy J-MR. A novel cause of mild/moderate hemophilia A: mutations scattered in the factor VIII C1 domain reduce factor VIII binding to von Willebrand factor. Blood 2000;96:958–65.

[18] Liu M-L, Shen BW, Nakaya S, Pratt KP, Fujikawa K, Davie EW, Stoddard BL, Thompson AR. Hemophilic factor VIII C1- and C2-domain missense mutations and their modeling to the 1.5-angstrom human C2-domain crystal structure. Blood 2000;96:979–87.

[19] Pipe SW, Saenko EL, Eickhorst AN, Kemball-Cook G, Kaufman RJ. Hemophilia A mutations associated with 1-stage/2-stage activity discrepancy disrupt protein-protein interactions within the triplicated A domains of thrombin-activated factor VIIIa. Blood 2001;97:685–91.

[20] Hakeos WH, Miao H, Sirachainan N, Kemball-Cook G, Saenko EL, Kaufman RJ, Pipe SW. Hemophilia A mutations within the factor VIII A2-A3 subunit interface destabilize factor VIIIa and cause one-stage/two-stage activity discrepancy. Thromb Haemost 2002;88:781–7.

[21] Pipe SW, Eickhorst AN, McKinley SH, Saenko EL, Kaufman RJ. Mild hemophilia A caused by increased rate of factor VIII A2 subunit dissociation: evidence for nonproteolytic inactivation of factor VIIIa in vivo. Blood 1999;93:176–83.

[22] Gilmore R, Harmon S, Gannon C, Byrne M, O'Donnell JS, Jenkins PV. Thrombin generation in haemophilia A patients with mutations causing factor VIII assay discrepancy. Haemophilia 2010;16:671–4.

[23] Trossaert M, Lienhart A, Nougier C, Fretigny M, Sigaud M, Meunier S, Fouassier M, Ternisien C, Negrier C, Dargaud Y. Diagnosis and management challenges in patients with mild haemophilia A and discrepant FVIII measurements. Haemophilia 2014;20:550–8.

[24] Dietz HC, Valle D, Francomano CA, Kendzior Jr RJ, Pyeritz RE, Cutting GR. The skipping of constitutive exons in vivo induced by nonsense mutations. Science 1993;259:680–3.

[25] Trossaert M, Boisseau P, Quemener A, Sigaud M, Fouassier M, Ternisien C, Lefrancois-Bettembourg A, Tesson C, Thomas C, Bezieau S. Prevalence, biological phenotype and genotype in moderate/mild hemophilia A with discrepancy between one-stage and chromogenic factor VIII activity. J Thromb Haemost 2011;9:524–30.

[26] Kaufman RJ, Antonarakis SE, Fay PJ. Factor VIII and hemophilia A. In: Colman RW, Clowes AW, Goldhaber SZ, Marder VJ, George JN, editors. Hemostasis and thrombosis. 5th ed. Philadelphia: Lippincott Williams & Wilkins; 2006. p. 166 [chapter 8].

[27] Natalia R, Jayne L, Shawn T, James P, Lillicrap D. The Canadian "National Program for hemophilia mutation testing" database: a ten-year review. Am J Hematol 2013;88:1030–4.

[28] Antonarakis SE, Rossiter JP, Young M, Horst J, de Moerloose P, Sommer SS, Ketterling RP, Kazazian Jr HH, Négrier C, Vinciguerra C, Gitschier J, Goossens M, Girodon E, Ghanem N, Plassa F, Lavergne JM, Vidaud M, Costa JM, Laurian Y, Lin SW, Lin SR, Shen MC, Lillicrap D, Taylor SA, Windsor S, Valleix SV, Nafa K, Sultan Y, Delpech M, Vnencak-Jones CL, Phillips III JA, Ljung RC, Koumbarelis E, Gialeraki A, Mandalaki T, Jenkins PV, Collins PW, Pasi KJ, Goodeve A, Peake I, Preston FE, Schwartz M, Scheibel E, Ingerslev J, Cooper DN, Millar DS, Kakkar VV, Giannelli F, Naylor JA, Tizzano EF, Baiget M, Domenech M, Altisent C, Tusell J, Beneyto M, Lorenzo JI, Gaucher C, Mazurier C, Peerlinck K, Matthijs G, Cassiman JJ, Vermylen J, Mori PG, Acquila M, Caprino D, Inaba H, et al. Factor VIII gene inversions in severe hemophilia A: results of an international consortium study. Blood 1995;86:2206–12.

[29] Tizzano E, Domenech M, Altisent C, Tusell J, Baiget M. Inversions in the factor VIII gene in Spanish hemophilia A patients. Blood 1994;83:3826.

[30] Naylor JA, Green PM, Rizza CR, Giannelli F. Factor VIII gene explains all cases of haemophilia A patients. Lancet 1992;340:1066–7.

[31] Lakich D, Kazazian Jr HH, Antonarakis SE, Gitschier J. Inversions disrupting the factor VIII gene are a common cause of severe haemophilia A. Nat Genet 1993;5:236–41.

[32] Pandey GS, Yanover C, Miller-Jenkins LM, Garfield S, Cole SA, Curran JE, Moses EK, Rydz N, Simhadri V, Kimchi-Sarfaty C, Lillicrap D, Viel KR, Przytycka TM, Pierce GF, Howard TE, Sauna ZE, PATH (Personalized Alternative Therapies for Hemophilia) Study Investigators. Endogenous factor VIII synthesis from the intron-22-inverted F* locus may modulate the immunogenicity of replacement therapy for hemophilia A. Nat Med 2013;19:1318–24.

[33] Bagnall RD, Waseem N, Green PM, Giannelli F. Recurrent inversion breaking intron 1 of the factor VIII gene is a frequent cause of severe hemophilia A. Blood 2002;99:168–74.

[34] Summers RJ, Meeks SL, Healey JF, Brown HC, Parker ET, Kempton CL, Doering CB, Lollar P. Factor VIII A3 domain substitution N1922S results in hemophilia A due to domain-specific misfolding and hyposecretion of functional protein. Blood 2011;117:3190–8.

[35] Viel KR, Ameri A, Abshire TC, Iyer RV, Watts RG, Lutcher C, Channell C, Cole SA, Fernstrom KM, Najaya S, Kasper CK, Thompson AR, Almasy L, Howard TE. Inhibitors of factor VIII in black patients with hemophilia. N Engl J Med 2009;360:1618–27.

[36] Gunasekera D, Ettinger RA, Fletcher SN, James EA, Liu M, Barrett JC, Withycombe J, Matthews DC, Epstein MS, Hughes RJ, Pratt KP, on behalf of the Personalized Approaches to Therapies for Hemophilia (PATH) Study Investigators. Factor VIII gene variants and inhibitor risk in African American hemophilia A patients. Blood 2015;126:895–904.

[37] Kazazian Jr HH, Moran JV. Mobile DNA in health and disease. N Engl J Med 2017;377:361–70.

[38] Kazazian Jr HH, Wong C, Youssoufian H, Scott AF, Phillips DG, Antonarakis SE. Haemophilia A resulting from de novo insertion of L1 sequences represents a novel mechanism for mutation in man. Nature 1988;332:164–6.

[39] Santagostino E, Mancuso ME, Tripodi A, Chantarangkul V, Clerici M, Garagiola I, Mannucci PM. Severe hemophilia with mild bleeding phenotype: molecular characterization and global coagulation profile. J Thromb Haemost 2010;8:737–43.

[40] Carcao MD, van den Berg HM, Ljung R, Mancuso ME, for the PedNet and the Rodin Study Group. Correlation between phenotype and genotype in a large unselected cohort of children with severe hemophilia A. Blood 2013;121:3946–52.

[41] Lyon MF. X-chromosome inactivation and human genetic disease. Acta Paediatr Suppl 2002;91:107–12.

[42] Chen C-K, Blanco M, Jackson C, Aznauryan E, Ollikainen N, Surka C, Chow A, Cerase A, McDonel P, Guttman M. Xist recruits the X chromosome to the nuclear lamina to enable chromosome-wide silencing. Science 2016;354:468–71.

[43] Almeida M, Pintacuda G, Masui O, Koseki Y, Gdula M, Cerase A, Brown D, Mould A, Innocent C, Nakayama M, Schermelleh L, Nesterova TB, Koseki H, Brockdorff N. PCGF3/5-PRC1 initiates polycomb recruitment in X chromosome inactivation. Science 2017;356:1081–3.

[44] Mantle DJ, Pye C, Hardisty RM, Vessey MP. Plasma factor-VIII concentrations in XXX women. Lancet 1971;i:58–9.

[45] Valliex S, Vinciguerra C, Lavergne J-M, Leuer M, Delpech M, Negrier C. Skewed X-chromosome inactivation in monochorionic diamniotic twin sisters results in severe and mild hemophilia A. Blood 2002;100:3034–6.

[46] Favier R, Lavergne JM, Costa JM, Caron C, Mazurier C, Viemont M, Delpech M, Valleix S. Unbalanced X-chromosome inactivation with a novel FVIII gene mutation resulting in severe hemophilia A in a female. Blood 2000;96:4373–5.

[47] Zheng J, Ma W, Xie B, Zhu M, Zhang C, Li J, Wang Y, Wang M, Jin Y. Severe female hemophilia A patient caused by a nonsense mutation (G1686X) of *F8* gene combined with skewed X-chromosome inactivation. Blood Coagul Fibrinolysis 2015;26:977–80.

[48] Plug I, Mauser-Bunschoten EP, Brocker-Vriends AHJT, van Amstel HKP, van der Bom JG, van Diemen-Homan JEM, Willemse J, Rosendaal FR. Bleeding in carriers of hemophilia. Blood 2006;108:52–6.

[49] Wyman-Collins LE. Women's bleeding disorders. Dateline Fed 2017;17:8–9.

[50] Burns CB. Clearing a path: women need a new diagnosis. The FactorNet 2017;28:8–10.

[51] Stuijver DJ, Piantanida E, van Zaane B, Galli L, Romualdi E, Tanda ML, Meijers JC, Buller HR, Gerdes VE, Squizzato A. Acquired von Willebrand syndrome in patients with overt hypothyroidism: a prospective cohort study. Haemophilia 2014;20:326–32.

[52] Johnsen J, Fletcher SN, Huston H, Roberge S, Martin BK, Kircher M, Josephson NC, Shendure J, Ruuska S, Koerper MA, Meltzer L, Pierce GF, Aschman D, Konkle B. Novel approach to and results of genetic analysis of 3000 hemophilia patients enrolled in the *MyLifeOurFuture* initiative. Blood 2016;128:205 [abstract].

[53] Rapaport SI, Patch MJ, Moore FJ. Antihemophilic globulin levels in carriers of hemophilia A. J Clin Invest 1960;39:1619–25.

[54] Bennett E, Huehns ER. Immunological differentiation of three types of haemophilia and identification of some female carriers. Lancet 1970;2:956–8.

[55] Zimmerman TS, Ratnoff OD, Littell AS. Detection of carriers of classic hemophilia using an immunologic assay for antihemophilic factor (factor VIII). J Clin Invest 1975;50:255–8.

[56] Klein HG, Aledort LM, Bouma BN, Hoyer LW, Zimmerman TS, DeMets DL. A co-operative study for the detection of the carrier state of classic hemophilia. N Engl J Med 1977;296:959–62.

[57] Ratnoff OD, Jones PK. Diagnosis of the carrier state in hemophilia. Am J Hematol 1988;28:132.

[58] Jenkins PV, Collins PW, Goldman E, McCraw A, Riddell A, Lee CA, Pasi KJ. Analysis of intron 22 inversions of the factor VIII gene in severe hemophilia A: implications for genetic counseling. Blood 1994;84:2197–201.

[59] Windsor S, Taylor SAM, Lillicrap D. Direct detection of a common inversion mutation in the genetic diagnosis of severe hemophilia A. Blood 1994;84:2202–5.

[60] Poon M-C, Low S, Sinclair GD. Factor VIII gene rearrangement analysis and carrier determination in hemophilia A. J Lab Clin Med 1995;125:402–6.

[61] Lo YMD, Tein MS, Lau TK, Haines CJ, Leung TN, Poon PM, Wainscoat JS, Johnson PJ, Chang AM, Hjelm NM. Quantitative analysis of fetal DNA in maternal plasma and serum: implications for noninvasive prenatal diagnosis. Am J Hum Genet 1998;62:768–75.

[62] Tsui NBY, Kadir RA, Chan KCA, Chi C, Mellars G, Tuddenham EG, Leung TY, Lau TK, Chiu RWK, Lo YMD. Noninvasive prenatal diagnosis of hemophilia by microfluidics digital PCR analysis of maternal plasma DNA. Blood 2011;117:3684–91.

[63] Hudecova I, Jiang P, Daviews J, Lo YMD, Kadir RA, Chiu RWK. Noninvasive detection of F8 int22h-related inversions and sequence variants in maternal plasma of hemophilia carriers. Blood 2017. https://doi.org/10.1182/blood-2016-12-755017.

[64] Balak DM, Gouw SC, Plug I, Mauser-Bunschoten EP, Vriends AH, Van Diemen-Homan JE, Rosendaal FR, van der Bom JG. Prenatal diagnosis for haemophilia: a nation-wide survey among female carriers in the Netherlands. Haemophilia 2012;18:584–92.

[65] Chalmers E, Williams M, Brennand J, Liesner R, Collins P, Richards M, on behalf of the Paediatric Working Party of the United Kingdom Haemophilia Doctors' Organization. Guideline on the management of haemophlila in the fetus and neonate. Br J Haematol 2011;154:208–15.

Recommended Reading

[1] Wood WI, Capon DJ, Simonsen CC, Eaton DH, Gitschier J, Keyt B, Seeburg PH, Smith DH, Hollingshead P, Wion KL, Delwart E, Tuddenham EGD, Vehar GA, Lawn RM. Expression of active human factor VIII from recombinant DNA clones. Nature 1984;312:330–7.

[2] Antonarakis SE, Rossiter JP, Young M, et al. Factor VIII gene inversions in severe hemophilia A: results of an international consortium study. Blood 1995;86:2206–12.

[3] Plug I, Mauser-Bunschoten EP, Brocker-Vriends AHJT, van Amstel HKP, van der Bom JG, van Diemen-Homan JEM, Willemse J, Rosendaal FR. Bleeding in carriers of hemophilia. Blood 2006;108:52–6.

[4] Chalmers E, Williams M, Brennand J, Liesner R, Collins P, Richards M, on behalf of the Paediatric Working Party of the United Kingdom Haemophilia Doctors' Organization. Guideline on the management of haemophlila in the fetus and neonate. Br J Haematol 2011;154:208–15.

[5] Carcao MD, van den Berg HM, Ljung R, Mancuso ME, for the PedNet and the Rodin Study Group. Correlation between phenotype and genotype in a large unselected cohort of children with severe hemophilia A. Blood 2013;121:3946–52.

[6] Swystun LL, James PD. Genetic diagnosis in hemophilia and von Willebrand disease. Blood Rev 2017;31:47–56.

[7] Kazazian Jr HH, Moran JV. Mobile DNA in health and disease. N Engl J Med 2017;377:361–70.

5

Congenital Deficiency of Factor VIII: Hemophilia A

HISTORICAL ASPECTS

The earliest reference to what might have been hemophilia is attributed to Rabbi Judah the Patriarch and recorded in the 5th century Talmud [1]. He taught that a woman should not have her third son circumcised if her first two sons had exsanguinated after this procedure. In subsequent discussion, the 12th century physician Moses Maimonides explained that only a child free of disease can be circumcised because danger to life overrides every other consideration. Early in the 18th century, Consbruch [2] provided the first description of a hemorrhagic disorder that only affected males, and in 1802, Otto [3] reported that "Although the females are exempt, they are still capable of transmitting it to their male children." The most famous example of this mode of inheritance was the family tree of Queen Victoria (1813–1901), who passed the condition to one of her sons and two of her daughters. The daughters, in turn, transmitted the disorder to progeny of the Prussian, Russian, and Spanish royal houses. It was assumed for many years that all these individuals were affected by hemophilia A, but DNA studies of tissues obtained from descendants of the Russian royal family showed that they had hemophilia B, also known as Christmas disease [4].

In 1872, Doctor J. Wickham Legg defined hemophilia as a hemorrhagic diathesis that lasted throughout an individual's lifetime, accompanied by a tendency to swelling of the joints, although it is not clear if he recognized that these were joint bleeds (hemarthroses) [5]. He described five patients, all male, with recurrent hemorrhages; male siblings of these patients also had a history of bleeding. One of these patients was a 23-year-old man who presented with chronic swelling and pain in his right knee. He was 4 years old when the knee first became symptomatic; recurrent episodes of swelling resulted in joint deformity and crippling. Nose bleeding (epistaxis) occurred at age 2 and persisted into adulthood. He bruised easily and had excessive bleeding from cuts; a small laceration of the palm bled for 3 weeks. The family history revealed a maternal uncle who bled to death after breaking his leg, and a brother with a history of recurrent hemorrhages and death at age 7. This case history appears compatible with moderately severe hemophilia, but proof of defective blood coagulation in patients such as the one described by Legg had to await the studies of AE Wright, who reported in 1893 that the clotting time of hemophilic blood was prolonged [6]. The reader is referred to the work of Ratnoff [7], who has written a superb history of the bleeding disorders.

EPIDEMIOLOGY

Hemophilia A affects every population on the globe, but the severity of the disorder varies according to the specific mutation in the FVIII gene (Chapter 4), and this mutation determines the level of FVIII. In healthy individuals, FVIII levels range from 0.5 to 1.5 IU/mL. Severe hemophilia is associated with levels that are ≤0.02 IU/mL; in moderate hemophilia, the levels are 0.02–0.05 IU/mL; and in mild hemophilia, >0.05–0.5 IU/mL. These levels do not always correspond with the severity of bleeding; other factors also influence the ability of patient plasma to generate thrombin. For example, the low incidence of bleeding in hemophilic neonates might be related to their physiologically low levels of antithrombin and tissue factor pathway inhibitor [8]. Decreased bleeding severity and enhanced thrombin generation are observed in hemophiliacs with coincident FV Leiden mutations associated with resistance to activated protein C [9,10]. Conversely, the co-inheritance of the 353Q allele of the F7 gene is associated with a severe bleeding phenotype [11]. When the bleeding phenotype does not correspond with the FVIII level, the thrombin generation assay might be a better indicator of disease severity [12].

The World Federation of Hemophilia (WFH) Global Survey of 2015 collected data on 151,000 individuals with hemophilia A from 111 countries. In high-income countries where the blood levels of nearly all hemophiliacs are routinely measured, the prevalence of severe hemophilia was 45%. The total number of hemophilia A individuals reported from the United States was 18,596, or roughly 1 in 5000 male births. Prospectively collected data from federally funded Hemophilia Treatment Centers report severe disease in 74.5% of 2078 individuals born during the interval from 1983 to 1992 [13]. Most were white (62.3%); African-Americans (16.4%), Hispanic (13.8%), Asian (3.2%), and other (4.3%) constituted the remainder. The first visit to a center occurred at ≤2 years of age for 69.1%, and 82.9% were frequent center visitors. Home infusion of clotting factor concentrates was initiated before age 6 in 45.7%, and eventually 73.9% were on a home infusion regimen. During the era when hemophilia was treated with unpasteurized blood products, blood-borne infectious agents were transmitted to many hemophiliacs [14]. Currently, the WFH Global Survey reports that in the United States, 1274 hemophiliacs are infected with the human immunodeficiency virus (HIV) and 4137 with hepatitis C virus (HCV); these are among the highest numbers worldwide because virally contaminated clotting factor concentrates were administered more frequently in the United States than in any other country.

CLINICAL MANIFESTATIONS OF HEMOPHILIA

As described in Chapter 2, the formation of FXa and generation of thrombin on tissue factor-bearing cells is delayed in the absence of FVIII. The small amounts of thrombin and fibrin formed are located mainly at the periphery of lesions where they are less effective in establishing hemostasis. In addition, the thrombin deficit impairs activation of the thrombin-activatable fibrinolysis inhibitor (TAFI), resulting in excessive fibrinolysis and premature dissolution of thrombi. Table 5.1 displays the frequency and location of hemorrhages in individuals with severe hemophilia.

The earliest indication of hemophilia is often prolonged oozing from small cuts on the tongue or frenulum; applying pressure to the site transiently controls the bleeding, but as soon as the pressure is relieved, bleeding resumes and can continue for hours. Small traumas that occur during crawling and walking are associated with large ecchymoses, and the child has frequent nosebleeds. As noted previously, the intensity of bleeding generally corresponds to the FVIII level; spontaneous hemorrhages occur when levels are $\leq 0.02\,IU/mL$, and less severe bleeding is associated with higher levels. However, even patients with mild hemophilia are prone to excessive bleeding after tooth extraction.

In general, having a tooth extracted is an unpleasant experience. Aside from the pain, there is swelling and bleeding from the socket. Packing with some gauzy material usually controls the bleeding and analgesic medications dull the pain. This enables the individual to pass a fairly uneventful night and greet the morning with nothing more than some mouth soreness. People with hemophilia have a different sequence of events. A few hours after going to bed, a large gelatinous mass of blood clot fills the socket and extends into the oral cavity. Expectorating the mass results in fresh bleeding from the extraction site and soon another big clot fills the mouth. A repeat visit to the dentist or oral surgeon, often in the middle of the night, results in new packing carefully pressed into the socket, with cessation of bleeding. But again, after a few hours, the packing material is extruded by an emerging clot. This cycle of bleeding and clotting, repacking of the socket and recurrence of bleeding, persists for days to

TABLE 5.1   Types of Hemorrhages in Severe Hemophilia (% Patients Affected)

Common (80%–100%)

    Hemarthroses

    Muscle hematoma

    Skin bruising

    Children: epistaxis, oral bleeding

Less common (20%–30%)

    Upper gastrointestinal bleeding, rectal bleeding

    Renal bleeding (hematuria)

Infrequent (5%–20%)

    Bleeding into airways

    Intracranial hemorrhage

    Compression of nerve root, compartment syndrome

Rare (<5%)

    Pseudotumors

    Intramural bowel hematoma

    Spinal hemorrhage

Data from Forbes CD. Clinical aspects of the genetic disorders of coagulation. In: Ratnoff OD, Forbes CD, editors. Disorders of hemostasis. 3rd ed. Philadelphia: WB Saunders Company; 1996. p. 138–51.

weeks before healing finally occurs. This was the usual outcome of dental extractions before it was recognized that hemophilia is due to the lack of a clotting factor, and that provision of this clotting factor is the only way to ensure permanent control of bleeding (Chapters 6 and 7).

Hemarthroses

Hemarthroses involve the knees, elbows, ankles, shoulders and hips, in that order; bleeding into the small joints of the hands and feet is unusual. The episodes of joint bleeding begin when the child starts to crawl or walk and are nearly unique to people with FVIII or FIX deficiency. They are most frequent in those with FVIII levels $\leq 0.02\,U/mL$, and very rare with levels $\geq 0.12\,U/mL$ [15]. As many as 35% of severe hemophiliacs report having ≥ 5 joint hemorrhages in a 6-month time period [13]. Joints are lined by a thin membrane, the synovium, whose functions are maintenance of an intact nonadherent tissue surface, lubrication of cartilage, control of synovial fluid volume and composition, and nutrition of chondrocytes within joints [16]. A dense net of small blood vessels are located just beneath the synovium and are vulnerable to hemorrhage. Ordinary activities such as bending and stretching appear to disrupt these vessels; leakage of blood can only be prevented if the breach is sealed with fibrin. Because fibrin formation is impaired in hemophilia, erythrocytes and other blood constituents gain access to the joint space. Initially, this cellular debri is removed by synovial macrophages, but soon the phagocytic capabilities of these cells are overwhelmed by fresh bleeding [17]. The hemoglobin in the shed red blood cells is degraded to hemosidern, an iron-containing compound that elicits a powerful inflammatory response, typically associated with joint swelling, erythema, and pain [18]. The release of a potent inflammatory and proliferative mediator, tumor necrosis factor alpha, promotes synovial membrane hyperplasia, neovascularization, and the production of enzymes that destroy cartilage and erode the underlying bone [19]. The development of chronic synovitis leads to muscle wasting, stiffness and limited motility, and to the complete destruction of the joint. Prior to the use of prophylactic replacement of FVIII, extensive joint destruction involving knees, ankles, and elbows was common, and crippling was a typical manifestation of hemophilia in boys and young men.

The unique susceptibility to bone and joint destruction in people with hemophilia might be due to dysregulation of cell proliferation. Studies in mice show that the decreased thrombin generation associated with FVIII deficiency reduces proteinase-activated receptor (PAR1) signaling, decreasing the ratio of bone to tissue volume and the number of bony trabeculae [20]. Furthermore, the iron deposited in the joint space triggers expression of mdm2, a p53-binding protein that increases synovial cell proliferation [21]. How these and other factors contribute to the pathobiology of hemophilic arthropathy are subjects for ongoing study.

Joint hemorrhages are associated with intense pain; other descriptors are burning, bubbling, and tingling if nerves are compressed. Fig. 5.1 shows an acute hemarthrosis in the left knee; the joint is swollen, tender, and painful.

Joint pain is reported by the majority of hemophiliacs, mostly in the knees, ankles, and elbows, and can persist for weeks despite the administration of clotting factor concentrate to control bleeding [22]. A recent survey found that pain was continually present in more than half of hemophiliacs aged 18–30 [23]. The pain was of moderate intensity in 68% and severe in 5%, and usually interfered with daily activities. In addition, one-third of those surveyed reported moderate anxiety or depression. Pain curtails employment opportunities, especially in older hemophiliacs with impaired mobility [24].

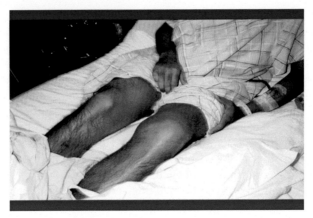

FIG. 5.1   Acute hemarthrosis of the left knee. *Photograph courtesy of the author.*

Muscle Hemorrhages

Muscles are vascular and a common site of bleeding, and muscle hematomas develop spontaneously or after trauma, and can even be provoked by intramuscular injections. The hematomas are painful and can become quite large, as shown in Fig. 5.2.

If bleeding occurs within the myofascial compartment, the pressure within the compartment becomes elevated and neurovascular structures are compressed. Failure to relieve elevated pressures within 6 h can result in paralysis and gangrene of the limb. Another expression of FVIII deficiency is hemorrhage into the iliopsoas muscle; this is associated with severe pain and numbness of the anterior thigh due to compression of the femoral nerve.

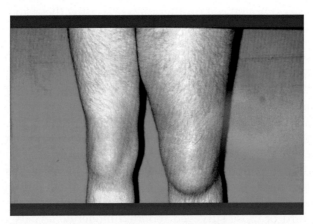

FIG. 5.2   Hematoma of the left quadriceps in a hemophiliac. *Reproduced with permission from Green D. What is hemophilia. Linked by blood: hemophilia and AIDS. London: Academic Press-Elsevier; 2016. p. 11–25 [chapter 2].*

Abdominal Hemorrhages and Pseudotumors

Hemophiliacs also experience bleeding within the abdomen that is often misconstrued as an acute surgical emergency. Occasionally, an intramural hematoma involving the bowel wall simulates an acute bowel obstruction. Although the clinical presentation suggests a "surgical" abdomen, imaging studies usually reveal the correct diagnosis and should abrogate consideration of invasive procedures. A pseudotumor is another hemorrhagic complication of hemophilia that can be mistaken for a surgical lesion. It develops when the membrane of a hemorrhagic cyst erodes bony structures, forming a large, multiloculated mass, mimicking an osteosarcoma. Pseudotumors are usually asymptomatic and often do not become apparent until there is fracture of the affected bone.

Central Nervous System Bleeding

The most dreaded forms of bleeding in hemophiliacs are hemorrhages within the central nervous system. In one survey, intracranial bleeding occurred in 106 of 1410 (7.5%) of patients coming to medical attention between the years 1960 and 1991 [25]. The mean age of these patients was 14.8 years, most had severe hemophilia, and a third had recurrent episodes. The majority of hemorrhages were spontaneous; only 39.7% had a history of recent trauma. The most frequent sequelae were seizure disorders and motor impairment, and the overall mortality rate was 29.2%. Intraspinal hemorrhage was rare; there were only two episodes.

Bleeding in Carriers of Hemophilia

As noted in Chapter 4, up to 40% of hemophilia carriers have prolonged bleeding from small wounds and surgical procedures [26]. Recent research suggests that a bleeding tendency is present in carriers even when FVIII levels are within the normal range [27]. A study of 168 carriers reported musculoskeletal bleeding as well as mucocutaneous hemorrhages and significantly higher scores on the bleeding assessment tool (BAT) than controls [28]. Further studies are needed to clarify why carriers with apparently adequate FVIII levels are at risk for bleeding.

Disorders Other Than Bleeding

The medical landscape in people with hemophilia is not confined to bleeding. Arthritis and arthropathy secondary to childhood hemarthroses can progress and severely compromise joint function in older hemophiliacs; these individuals are usually referred to orthopedists for joint repair or replacement [29]. However, hemophiliacs living in developing countries still must contend with the crippling effects of joint hemorrhages.

Another common disorder is depression. Iannone et al. [30] report that 15 of 47 (37%) of hemophilic adults had scores ≥ 5 on the Patient Health Questionnaire-9, and 8 had values >10, indicative of moderate-to-severe depressive symptoms. Higher scores were significantly associated with lack of social support and unemployment. Young adults with hemophilia might have difficulty obtaining an appropriate occupation because of physical limitations and psychosocial issues related to anxiety, trust, maturity, and parental attachment [31].

Chronic infection with hepatitis or human immunodeficiency viruses is a legacy of the era before the sterilization of clotting factor concentrates. The outcome of chronic hepatitis is occasionally cirrhosis or hepatocellular carcinoma. When these occur, patients are referred for liver transplantation, which has greatly improved the outlook for long-term survival [32,33]. Untreated HIV infections are associated with various malignancies as well as life-threatening infections. These have become much less common with the advent of effective antiretroviral therapy; although the drugs are not curative, they can halt disease progression and provide durable remissions.

Cardiovascular, renal disease, and cancer, diseases characteristic of older populations, are also observed in people with hemophilia [34,35]. Atherosclerosis occurs to the same degree as in the general population [36], probably because of similar risk factors such as smoking, diet, and obesity. Surveys reveal that 20% of children, 28% of teens, and 50% of hemophilic men are overweight or obese [37]; it is unfortunate that only one-third of Hemophilia Treatment Centers have a nutritionist [38]. In addition, those with joint disease are often unable to exercise regularly and have a sedentary lifestyle [39], further increasing their risk for atherosclerosis. However, hemophiliacs have decreased cardiovascular mortality, possibly related to a protective effect of the clotting factor deficiency on thrombus formation [40] (Chapter 13).

Hypertension is prevalent in the hemophilia population. At every age from 18 to ≥ 75, more individuals with hemophilia are hypertensive than in other cohorts in the United States [41]. Independent risk factors are age, body mass index, diabetes, and renal function. Elevated blood pressure tends to be more prevalent in individuals with severe and moderately severe hemophilia than with mild disease. A European study of hemophiliacs over age 40 reported that 239 of 532 (45%) were hypertensive [42]. The cause of the hypertension has been unclear; the usual cardiovascular risk factors and recurrent hematuria do not explain the association [43,44]. Other possibilities needing investigation are chronic viral infection (HCV, HIV) and the medications used to treat these infections, and the heavy protein load imposed by repeated infusions of FVIII concentrates [45]. It has been suggested that the observational studies cited here have exaggerated the prevalence of hypertension in the hemophilia population;

a recent cross-sectional analysis of hospital discharges found less hypertension in hemophiliacs than in controls (39% vs 56%; $P < .001$) [46]. However, as the authors of this study note, hospitalized people without hemophilia might not be the optimal control group, and prospective studies of community-based populations are needed.

DIAGNOSIS OF HEMOPHILIA

Individuals with hemophilia have a prolonged **whole blood clotting time** and **activated partial thromboplastin time (aPTT)**, but a normal prothrombin time. Their skin bleeding time is normal, but if the clot that forms at the wound edges is wiped away, the rebleeding time is prolonged. The platelet count and the thrombin time are normal, but the **thrombin generation assay** shows several abnormalities, including a prolonged lagtime, a decrease in peak thrombin generation, and a reduction in the endogenous thrombin potential [47]. The **thromboelastogram** (TEG), a point-of-care method for assessing coagulation, is also abnormal, displaying a prolonged reaction time and decreased maximum amplitude. The specific diagnosis of hemophilia A requires performance of an assay that reveals the extent of the FVIII deficiency. Most clinical laboratories use either clot-based or chromogenic FVIII assays [48]. One-stage **clot-based assays** are widely accessible and relatively simple to perform, but can be affected by problems with venepuncture, incomplete filling of tubes containing the citrate anticoagulant, storage conditions, and the presence of residual heparin in catheters used to obtain blood samples. Furthermore, the fidelity of the assay requires substrate plasma that is completely deficient in FVIII, but without excessive amounts of other clotting factors. Because of the variability in substrate plasmas and other factors, the interlaboratory variation in FVIII levels measured by these assays is considerable. Two-stage methods avoid many of these drawbacks, but are technically more difficult and require specialized reagents.

Chromogenic assays rely on the generation of FXa in mixtures containing FIXa, FX, and a sample of the patient's plasma that has been activated to form FVIIIa. The FXa that is generated cleaves a chromogenic substrate; the color intensity of the resulting product is directly proportional to the patient's plasma FVIII concentration [48]. Assay results are less variable than those using clot-based methods, but the cost of reagent kits is higher [49]. There are at least six different chromogenic assay kits available from commercial sources. **Enzyme-linked immunosorbent assays (ELISA)** measure the amount of FVIII bound to a FVIII-specific antibody precoated onto a microplate, using an avidin/biotin detection system [50]. Commercial assays report a detection range between 0.15 and 10 ng/mL, an intra-assay coefficient of variation (CV) <8%, and an interassay CV of <10%

[51]. In healthy individuals, FVIII measured by ELISA and clotting assays are closely correlated, but in patients with mutations producing nonfunctional FVIII, they will be disparate. Because treatment decisions are usually based on the patient's level of FVIII, it is recommended that multiple measurements be performed, assay results correlated with the clinical phenotype, and all patients have genotyping [52].

Lupus anticoagulants prolong the clotting times of one-stage assays, which could be interpreted that FVIII as well as other clotting factors are decreased and potentially implying that the patient has an FVIII autoantibody. The two-stage assay is very helpful in this situation, distinguishing lupus patients with normal or elevated FVIII levels from those with a true decrease in the factor. Alternatives to the two-stage clotting assay are chromogenic and ELISA methods, which are usually insensitive to the lupus anticoagulant. There are rare reports of patients harboring lupus anticoagulants as well as FVIII autoantibodies, and descriptions of procedures for differentiating them [53].

Differential Diagnosis

Usually the diagnosis of hemophilia is not difficult, but differentiating classical hemophilia from hemophilia B or Christmas disease requires performance of a specific assay for FVIII and FIX. FVIII levels remain constant throughout life in people with moderate or severe hemophilia, but an age-related increase (median, 8%) occurs in those with mild hemophilia [54]. In these patients, FVIII levels should always be checked prior to invasive procedures, and preoperative management modified accordingly.

Von Willebrand Disease (VWD) is occasionally misdiagnosed as hemophilia. Patients with severe type 1 or type 3 VWD not only have mucosal bleeding but also can have joint hemorrhages, and their levels of FVIII can be as low as 0.01 U/mL. They can be distinguished from hemophilia on the basis of the family history (affecting both males and females because it is not X-linked), by decreased levels of Von Willebrand Factor (VWF) antigen or activity, and by an increased ratio of FVIII to VWF [55]. Type 2N VWD is characterized by a failure of VWF to bind and stabilize FVIII; as a consequence, FVIII levels are decreased and the disorder easily mistaken for mild-to-moderate hemophilia [56]. However, the inheritance pattern is autosomal recessive, increases in FVIII persist after infusion of FVIII/VWF concentrates, and there is decreased binding of FVIII to the mutant VWF. The various types of VWD are discussed in Chapter 10.

Another congenital disorder that might be mistaken for hemophilia is combined FV and FVIII deficiency. Patients with this rare autosomal recessive bleeding disorder (1:1,000,000) have a lifelong history of excessive blood loss after surgery or trauma [57]. The aPTT is prolonged and

FVIII levels are usually in the 5%–20% range [58], but in contradistinction to hemophilia, the PT is also prolonged and FV is decreased. These patients have a genetic defect affecting proteins that chaperone factors V and VIII from the endoplasmic reticulum to the Golgi apparatus; the Lectin Mannose Binding Protein 1 (LMAN1) is mutated in 70% of patients, mostly those from the Middle East, and the MCFD2 mutation occurs in persons from India and Europe [59]. Treatment of mild bleeding episodes can be attempted with desmopressin, although this agent does not increase factor V levels; the management of more serious bleeds requires an FVIII concentrate as well as fresh frozen plasma to raise the levels of both FV and FVIII [60]. Sufficient FVIII should be infused to achieve concentrations of 50–70 IU/dL, and repeated every 12 h; FFP is given in an initial dose of 15–20 mL/kg followed by smaller doses of 5 mL/kg every 12 h to maintain levels of 15 IU/dL [61].

The differential diagnosis of excessive bleeding in women includes carriership for FVIII or FIX hemophilia, VWD, FXI deficiency, and platelet disorders. Most of these conditions are associated with heavy menstrual bleeding and bleeding complications during pregnancy and delivery. Typical symptoms include menstrual bleeding lasting longer than 7 days, passing large clots, necessity to change tampons/pads frequently, and bleeding after dental or other invasive procedures. A complete evaluation of hemostasis, including using the bleeding assessment tool (BAT) and taking a detailed family history, is required for an accurate diagnosis, and is needed to guide therapy. For assistance and advice, consult the Women and Girls Bleeding Disorders Directory at http://www.fwgbd.org; the Directory provides the location of the expert services available at the 140 federally funded Hemophilia Treatment Centers in the United States.

Some acquired diseases that have decreased FVIII levels are autoimmune hemophilia, acquired VWD, and disseminated intravascular coagulation (consumption coagulopathy). Autoimmune hemophilia is due to antibodies to FVIII that occasionally arise spontaneously in elderly individuals or appear in patients with cancer (especially those of lymphoid origin), lupus, and rheumatoid arthritis, or during treatment with some drugs such as penicillin, sulfa, or interferons. The diagnosis and management of this disorder is discussed in Chapter 8. Acquired VWD is another rare entity with multiple etiologies; it might be on an autoimmune basis or secondary in patients with certain cardiac disorders (Heyde's Syndrome), and is discussed in Chapter 12. Acute disseminated intravascular coagulation occurs in the setting of infection, obstetrical accidents, and major trauma, and is associated with hypotension, microvascular ischemia, and organ dysfunction. Laboratory studies reveal thrombocytopenia, hypofibrinogenemia, and decreases in anticoagulant proteins. FVIII levels are also impacted; however, the low levels of fibrinogen associated with these disorders affect the clotting times of most clotting assays, and must be taken into consideration when interpreting assay data. A decrease in FVIII is also observed when fibrinolysis is excessive, such as in acute pro-myelocytic leukemia.

SUMMARY

Hemophilia is an ancient malady; fatal bleeding after circumcision occurring in more than one family member was described 10 centuries ago. The incidence is 1 in 5000 male births, and 45% will have severe disease (FVIII ≤0.02 IU/mL). In the absence of FVIII, the formation of FXa is delayed and the fibrin formation that does occur is at the periphery of the wound. Bleeding persists or stops only briefly, and active clot lysis results in recurrent hemorrhage. The common sites of bleeding are the joints, muscles, nasal mucosa, and mouth; hemarthrosis is the principal cause of pain and disability, and intracranial bleeding is the main cause of death. Carriers can experience heavy menstrual bleeding and peripartum hemorrhage, as well as musculoskeletal bleeding. The legacy of repeated joint bleeding is arthropathy and arthritis. Individuals with hemophilia are not protected from the development of atherosclerosis, although they have decreased cardiovascular mortality. Hypertension appears to be more frequent than in the general population, and obesity is common, perhaps because of a sedentary lifestyle.

The diagnosis of hemophilia is usually not difficult; the disorder is suspected if patients have a prolonged activated partial thromboplastin time and a normal prothrombin time. Specific factor assays are required to differentiate FVIII from FIX deficiency. To assess the severity of the hemophilia, the level of FVIII should be measured using one of several available laboratory methods; these include one-stage clotting, chromogenic substrate, and enzyme-linked immunosorbent assays. In addition, genotyping is required for disease management, prognosis, and family counseling. The differential diagnosis includes VWD and combined FV and FVIII deficiency, as well as acquired conditions such as autoimmune hemophilia, acquired VWD, and disseminated intravascular coagulation.

Future research activities might include the following:

- Determine why individuals with FVIII or FIX deficiency are uniquely susceptible to joint bleeding. Factors VIII and IX are components of the tenase complex; is it the complex that maintains the integrity of the synovial membrane?
- Elucidate the pathogenesis of spontaneous brain hemorrhages in hemophiliacs.

- Clarify why carriers with apparently adequate FVIII levels are at risk for bleeding; are specific mutations associated with a disposition toward hemorrhage?
- Investigate the pathophysiology of hypertension in patients with moderate-to-severe hemophilia.

References

[1] Rosner F. Hemophilia in the Talmud and Rabbinic writings. Ann Intern Med 1969;70:833–7.

[2] Bulloch W, Fildes P. Hemophilia. In: Treasury of human inheritance, Parts V and VI, Sect. XIVa. London: Galton Laboratory, University of London London, Cambridge University Press; 1911. p. 169–347.

[3] Otto JC. An account of an hemorrhagic disposition existing in certain families. Med Repository 1803;6:1–4.

[4] Rogaev EI, Grigorenko AP, Faskhutdinova G, Kittler EL, Moliaka YK. Genotype analysis identifies the cause of the "Royal Disease" Science 2009;326:817.

[5] Legg JW. Treatise on haemophilia. London: HK Lewis; 1972. p.158.

[6] Wright AE. On the method of determining the condition of blood coagulability for clinical and experimental purposes, and on the effect of the administration of calcium salts in haemophilia and actual or threatened hemorrhage. Br Med J 1893;2:223–5.

[7] Ratnoff OD. Why do people bleed? In: Wintrobe MM, editor. Blood, pure and eloquent. New York: McGraw-Hill, Inc; 1980. p. 625–31 [chapter 18].

[8] Fritsch P, Cvirn G, Cimenti C, Baier K, Gallistl S, Koestenberger M, Roschitz B, Leschnik B, Muntean W. Thrombin generation in factor VIII-depleted neonatal plasma: nearly normal because of physiologically low antithrombin and tissue factor pathway inhibitor. J Thromb Haemost 2006;4:1071–7.

[9] Nichols WC, Amano K, Cacheris PM, Figueiredo MS, Michaelides K, Schwaab R, Hoyer L, Kaufman RJ, Ginsburg D. Moderation of hemophilia A phenotype by the factor V R506Q mutation. Blood 1996;88:1183–7.

[10] van't Veer C, Golden NJ, Kalafatis M, Simioni P, Bertina RM, Mann KG. An in vitro analysis of the combination of hemophilia A and factor V^{LEIDEN}. Blood 1997;90:3067–72.

[11] Jayandharan GR, Nair SC, Poonnoose PM, Thomas R, John J, Keshav SK, Cherian RS, Devadarishini M, Lakshmi KM, Shaji RV, Viswabandya A, George B, Mathews V, Chandy M, Srivastava A. Polymorphism in factor VII gene modifies phenotype of severe haemophilia. Haemophilia 2009;15:1228–36.

[12] Trossaert M, Regnault V, Sigaud M, Boisseau P, Fressinaud E, Lecompte T. Mild hemophilia A with factor VIII assay discrepancy: using thrombin generation assay to assess the bleeding phenotype. J Thromb Haemost 2008;6:486–93.

[13] Mazepa MA, Monahan PE, Baker JR, Riske BK, Soucie JM. Men with severe hemophilia in the United States: birth cohort analysis of a large national database. Blood 2016;127:3073–81.

[14] Green D. Blood: vital but potentially dangerous. In: Linked by blood: hemophilia and AIDS. London: Academic Press-Elsevier; 2016. p. 27–37. [chapter 3].

[15] den Uijl IE, Mauser Bunschoten EP, Roosendaal G, Schutgens RE, Biesma DH, Grobbee DE, Fischer K. Clinical severity of haemophilia A: does the classification of the 1950s still stand? Haemophilia 2011;17:849–53.

[16] Smith MD. The normal synovium. Open Rheumatol J 2011;5:100–6.

[17] Valentino LA, Hakobyan N, Enockson C, Simpson ML, Kakodkar NC, Cong L, Song X. Exploring the biological basis of haemophilic joint disease: experimental studies. Haemophilia 2012;18:310–8.

[18] Blobel CP, Haxaire C, Kalliolias GD, DiCarlo E, Salmon J, Srivastava A. Blood-induced arthropathy in hemophilia: mechanisms and heterogeneity. Semin Thromb Hemost 2015;41:832–7.

[19] Haxaire C, Blobel CP. With blood in the joint-what happens next? Could activation of a pro-inflammatory signaling axis leading to iRhom2/TNF-α-convertase-dependent release of TNFα contribute to haemophilic arthropathy. Haemophilia 2014;20 (Suppl. 4):11–4.

[20] Aronovich A, Nur Y, Shezen E, Rosen C, Klionsky YA, Milman I, Yarimi L, Hagin D, Rechavi G, Martinowitz U, Nagasawa T, Frenette PS, Tchorsh-Yutsis D, Reisner Y. A novel role for factor VIII and thrombin/PAR1 in regulating hematopoiesis and its interplay with the bone structure. Blood 2013;122:2562–71.

[21] Hakobyan N, Kazarian T, Jabbar AA, Jabbar KJ, Valentino LA. Pathobiology of hemophilic synovitis I: overexpression of mdm2 oncogene. Blood 2004;104:2060–4.

[22] Rambod M, Forsyth K, Sharif F, Khair K. Assessment and management of pain in children and adolescents with bleeding disorders: a cross-sectional study from three haemophilia centres. Haemophilia 2016;22:65–71.

[23] Witkop M, Guelcher C, Forsyth A, Hawk S, Curtis R, Kelley L, Frick N, Rice M, Rosu G, Cooper DL. Treatment outcomes, quality of life, and impact of hemophilia on young adults (aged 18-30 years) with hemophilia. Am J Hematol 2015;90:S3–S10.

[24] Forsyth AL, Witkop M, Lambing A, Garrido C, Dunn S, Cooper DL, Nugent DJ. Associations of quality of life, pain, and self-reported arthritis with age, employment, bleed rate, and utilization of hemophilia treatment center and health care provider services: results in adults with hemophilia in the HERO study. Patient Prefer Adherence 2015;9:1549–60.

[25] De Tezanos Pinto M, Fernandez J, Perez Bianco PR. Update of 156 episodes of central nervous system bleeding in hemophiliacs. Haemostasis 1992;22:259–67.

[26] Plug I, Mauser-Bunschoten EP, Brocker-Vriends AHJT, van Amstel HKP, van der Bom JG, van Diemen-Homan JEM, Willemse J, Rosendaal FR. Bleeding in carriers of hemophilia. Blood 2006;108:52–6.

[27] Paroskie A, Gailani D, DeBaun MR, Sidonio Jr RF. A cross-sectional study of bleeding phenotype in haemophilia A carriers. Br J Hematol 2015;170:223–8.

[28] James PD, Mahlangu J, Bidlingmaier C, Mingot-Castellano ME, Chitlur M, Fogarty PF, Cuker A, Mancuso ME, Holme PA, Grabell J, Satkunam N, Hopman WM, Mathew P, the Global Emerging Hemostasis Experts Panel (GEHEP). Evaluation of the utility of the ISTH-BAT in haemophilia carriers: a multinational study. Haemophilia 2016;22:912–8.

[29] Rana NA, Shapiro GR, Green D. Long-term follow-up of prosthetic joint replacement in hemophilia. Am J Hematol 1986;23:329–38.

[30] Iannone M, Pennick L, Tom A, Cui H, Gilbert M, Weihs K, Stopeck AT. Prevalence of depression in adults with haemophilia. Haemophilia 2012;18:868–74.

[31] Molter D, Frey M, Forsyth A, Tran D, Compton M, Iyer N, Holot N, Cooper D. Unmet needs in young adults with hemophilia: exploring career options and securing employment. Am J Hematol 2015;90(Suppl. 2):S20–22.

[32] Bontempo FA, Lewis JH, Gorenc TJ, Spero JA, Ragni MV, Scott JP, Starzl TE. Liver transplantation in hemophilia A. Blood 1987;69:1721–4.

[33] Ragni MV, Humar A, Stock PG, Blumberg EA, Eghtesad B, Fung JJ, Stosor V, Nissen N, Wong MT, Sherman KE, Stablein DM, Barin B. Hemophilia liver transplantation observational study (HOTS). Liver Transpl 2017;23:762–8.

[34] Konkle BA, Kessler C, Aledort L, Andersen J, Fogarty P, Kouides P, Quon D, Ragni M, Zakarija A, Ewenstein B. Emerging clinical concerns in the ageing haemophilia patient. Haemophilia 2009;15:1197–209.

[35] Mannucci PM, Mauser-Bunschoten EP. Cardiovascular disease in haemophilia patients: a contemporary issue. Haemophilia 2010;16(Suppl. 3):58–66.

[36] Kamphuisen PW, ten Cate H. Cardiovascular risk in patients with hemophilia. Blood 2014;123:1297–301.

[37] Ullman M, Zhang QC, Brown D, Grant A, Soucie JM. Association of overweight and obesity with the use of self and home-based infusion therapy among haemophilic men. Haemophilia 2014;20:340–8.

[38] Adams E, Deutsche J, Okoroh E, Owens-McAlister S, Majumdar S, Ullman M, Damiano ML, Recht M. An inventory of healthy weight practices in federally funded haemophilia treatment centres in the United States. Haemophilia 2014;20:639–43.

[39] Baumgardner J, Elon L, Antun A, Stein S, Ribeiro M, Slovensky L, Kempton CL. Physical activity and functional abilities in adult males with haemophilia: a cross-sectional survey from a single US haemophilia treatment centre. Haemophilia 2013;19:551–7.

[40] Makris M, van Veen JJ. Reduced cardiovascular mortality in hemophilia despite normal atherosclerotic load. J Thromb Haemost 2012;10:20–2.

[41] Von Drygalski A, Kolaitis NA, Bettencourt R, Bergstrom J, Kruse-Jarres R, Quon DV, Wassel C, Li MC, Waalen J, Elias DJ, Mosnier LO, Allison M. Prevalence and risk factors for hypertension in hemophilia. Hypertension 2013;62:209.

[42] Holme PA, Combescure C, Tait RC, Berntorp E, Rauchensteiner S, de Moerloose P. Hypertension, haematuria and renal functioning in haemophilia-a cross-sectional study in Europe. Haemophilia 2016;22:248–55.

[43] Barnes RF, Cramer TJ, Sait AS, Kruse-Jarres R, Quon DV, von Drygalski A. The hypertension of hemophilia is not explained by the usual cardiovascular risk factors: results of a cohort study. Int J Hypertens 2016;2016:2014201.

[44] Sun HL, Yang M, Sait AS, von Drygalski A, Jackson S. Haematuria is not a risk factor of hypertension or renal impairment in patients with haemophilia. Haemophilia 2016;22:549–55.

[45] Esposito P, Rampino T, Gregorini M, Fasoli G, Gamba G, dal Canton A. Renal diseases in haemophilic patients: pathogenesis and clinical management. Eur J Haematol 2013;91:287–94.

[46] Seaman CD, Apostolova M, Yabes J, Comer DM, Ragni MV. Prevalence and risk factors associated with hypertension in hemophilia: cross-sectional analysis of a national discharge register. Clin Appl Thromb Hemost 2017;23:871–5.

[47] Takeyama M, Nogami K, Shima M. A new parameter in the thrombin generation assay, mean velocity to peak thrombin, reflects factor VIII activity in patients with haemophilia A. Haemophilia 2016;22:e435–93.

[48] Peyvandi F, Oldenburg J, Friedman KD. A critical appraisal of one-stage and chromogenic assays of factor VIII activity. J Thromb Haemost 2016;14:248–61.

[49] Kitchen S, Blakemore J, Friedman KD, Hart DP, Ko RH, Perry D, Platton S, Tan-Castillo D, Young G, Luddington RJ. A computer-based model to assess costs associated with the use of factor VIII and factor IX one-stage and chromogenic activity assays. J Thromb Haemost 2016;14:757–64.

[50] Tackaberry ES, Ganz PR, Rock G. Measurement of human factor VIII by avidin-biotin dot immunobinding ELISAs. J Immunol Methods 1987;99:59–66.

[51] MyBioSource ELISA Kit, www.MyBioSouce.com.

[52] van Moort I, Joosten M, de Maat MPM, Leebeek FWG, Crossen MH. Pitfalls in the diagnosis of hemophilia severity: what to do? Pediatr Blood Cancer 2017;64(4).

[53] Ames PRJ, Graf M, Archer J, Scarpato N, Iannaccone L. Prolonged activated partial thromboplastin time: difficulties in discriminating coexistent factor VIII inhibitor and lupus anticoagulant. Clin Appl Thromb Hemost 2015;21:149–54.

[54] Miesbach W, Alesci S, Krekeler S, Seifried E. Age-dependent increase of FVIII:C in mild haemophilia A. Haemophilia 2009;15:1022–6.

[55] Haberichter SL, Christopherson RA, Flood VH, Gill JC, Friedman KD, Montgomery RR. In: Von Willebrand factor (VWF) propeptide and factor VIII (FVIII) levels identify the contribution of decreased synthesis and/or increased clearance mechanisms in the

pathogenesis of type 1 Von Willebrand Disease (VWD) in the Zimmerman Program. Abs 874. ASH; 2016.

[56] Schneppenheim R, Budde U, Krey S, Drewke E, Bergmann F, Lechler E, Oldenburg J, Schwaab R. Results of a screening for von Willebrand disease type 2N in patients with suspected haemophilia A or von Willebrand disease type 1. Thromb Haemost 1996;76:598–602.

[57] Viswabandya A, Baidya S, Nair SC, Lakshmi KM, Mathews V, George B, Chandy M, Srivastava A. Clinical manifestations of combined factor V and VIII deficiency: a series of 37 cases from a single center in India. Am J Hematol 2010;85:538–9.

[58] Peyvandi F, Menegatti M. Treatment of rare factor deficiencies in 2016. Hematol Am Soc Hematol Edu Program 2016;2016:663–9.

[59] Zhang B, McGee B, Yamaoka JS, Gulglielmone H, Downes KA, Minoldo S, Jarchum G, Peyvandi F, de Bosch NB, Ruiz-Saez A, Chatelain B, Olpinski M, Bockenstedt P, Sperl W, Kaufman RJ, Nichols WC, Tuddenham EG, Ginsburg D. Combined deficiency of factor V and factor VIII is due to mutations in either LMAN1 or MCFD2. Blood 2006;107:1903–7.

[60] Spreafico M, Peyvandi F. Combined factor V and factor VIII deficiency. Semin Thromb Hemost 2009;35(4):390–9.

[61] Asselta R, Peyvandi F. Factor V deficiency. Semin Thromb Hemost 2009;35:382–9.

Recommended Reading

[1] Ratnoff OD. Why do people bleed? In: Wintrobe MM, editor. Blood, pure and eloquent. New York: McGraw-Hill, Inc; 1980. p. 625–31 [chapter 18].

[2] Mazepa MA, Monahan PE, Baker JR, Riske BK, Soucie JM. Men with severe hemophilia in the United States: birth cohort analysis of a large national database. Blood 2016;127:3073–81.

[3] Blobel CP, Haxaire C, Kalliolias GD, DiCarlo E, Salmon J, Srivastava A. Blood-induced arthropathy in hemophilia: mechanisms and heterogeneity. Semin Thromb Hemost 2015;41:832–7.

[4] Konkle BA, Kessler C, Aledort L, Andersen J, Fogarty P, Kouides P, Quon D, Ragni M, Zakarija A, Ewenstein B. Emerging clinical concerns in the ageing haemophilia patient. Haemophilia 2009;15:1197–209.

[5] van Moort I, Joosten M, de Maat MPM, Leebeek FWG, Crossen MH. Pitfalls in the diagnosis of hemophilia severity: what to do? Pediatr Blood Cancer 2017;64(4).

FVIII Concentrates

The earliest published account of transfusion therapy for hemophilia appeared in the *Lancet* of 1840 [1]. A British surgeon recounted persistent bleeding following strabismus surgery in a boy with hemophilia. To stop the hemorrhaging, the doctor opened a vein in the arm of the patient's sister and directed the stream of blood into a funnel attached to a syringe and needle. The needle was inserted into the patient's vein and the blood entering the syringe was injected directly into the vein. Bleeding ceased and the patient made a satisfactory recovery. Despite this promising beginning, transfusion did not become a popular therapeutic option because blood was difficult to collect, store, and infuse, and adverse reactions were common.

The discovery that blood clotting could be prevented by sodium citrate led to the preparation of plasma, which was much easier to store and infuse. Fresh frozen plasma came into widespread use in the late 1930s; typical doses were 250–500 mL. While these amounts might control bleeding in a small child, they were not ineffective for larger individuals. This is because the FVIII content of plasma is only 0.6 U/mL [2]; giving sufficient plasma to raise FVIII concentrations to more than 20% of normal would result in fluid overload.[1] Because of this limitation, workers attempted to concentrate the FVIII that was in plasma.

THE FIRST FVIII CONCENTRATES

In 1954, British researchers reported the preparation of a bovine antihemophilic globulin (AHG) concentrate for the treatment of individuals with hemophilia. They used this material to control bleeding associated with

[1]The plasma volume of a 70-kg individual is approximately 3000 mL. The maximum volume of plasma that can be safely infused is 1500 mL, providing 900 U of FVIII (0.6 U/mL × 1500 mL = 900 U). Because the final plasma volume will be 4500 mL (1500 + 3000), the FVIII will only be increased by 20% (900 U in 4500 mL = 0.2 U/mL).

dental extractions and subsequently, other surgical procedures [3,4]. Although the bovine concentrate was life saving for patients with severe hemorrhages, only one course of treatment could be given because anaphylaxis was possible with a second exposure. Another disadvantage of bovine FVIII was that it agglutinated human platelets, inducing thrombocytopenia in the recipients.

In the United States, the Merck Sharp and Dohme Company marketed a product for the treatment of hemophilia that was prepared from an ethanol fraction of plasma. This fibrinogen-rich concentrate provided sufficient FVIII for the "supportive management" of a few hemophiliacs undergoing appendectomy and dental extractions [5]. In 1965, investigators from the University of North Carolina and Hyland Laboratories described the preparation of a human FVIII concentrate using a glycine precipitation step. A syringe-full of this product could increase FVIII to normal levels in patients with severe hemophilia and was suitable for the control of surgical bleeding [6,7]. In addition, large amounts of the glycine-precipitated concentrate could partially neutralize antibodies to FVIII. Other methods for concentrating FVIII from plasma were also under investigation at this time, but none had the ease and simplicity of cold- or cryoprecipitation [8].

Cryoprecipitate

When plasma is rapidly frozen and then allowed to gradually thaw, a precipitate (cryoprecipitate) forms that is rich in FVIII, fibrinogen, and macroglobulins. Pool and Shannon [9] described a practical method for extracting this material from a unit of donated blood and storing the cryoprecipitate in a plastic bag. The bag contained about half the FVIII activity of the original 200 mL of plasma, or approximately 70–100 U. The cryoprecipitate was dissolved in a small amount of warm plasma and the contents of several bags were pooled and refrozen. When needed for treatment, the material was thawed and infused intravenously. Although the yield in vitro was less than observed with other methods of FVIII preparation, the overall recovery of FVIII activity was comparable, and provided adequate hemostasis for hemophiliacs undergoing surgery [10]. Cryoprecipitate could be prepared locally by blood banks and hospitals, and individual units pooled and stored frozen until needed. However, the potency of the material being infused into the bleeding patient was unknown because it depended on the levels of FVIII present in the individual donors. Thawing of the bags of cryoprecipitate often required an hour or more, and the insoluble fibrin that formed during thawing would readily occlude intravenous filters, lines, and needles. In addition, allergic reactions consisting of chills, fever, and bronchospasm were not uncommon, and hemolysis might occur if large amounts of cryoprecipitate containing isoantibodies to blood group A or AB antigens were infused into recipients with these blood groups [11]. Nevertheless, the development of cryoprecipitate therapy by Doctor Judith Pool was a seminal event in the history of hemophilia because it introduced a practical and effective treatment that achieved worldwide application.

Plasma-Derived Concentrates

Cryoprecipitate became the source of human FVIII for the large-scale production of purified concentrates for clinical use. In 1971, the American National Red Cross Research Laboratory and the New York University Medical Center described the production of a high-potency FVIII concentrate using cryoethanol precipitation, extraction with tris buffer, and fractionation with polyethylene glycol [12]. Almost 5 L of blood were required to produce 250 U of FVIII [13]. This method was commercialized by Hyland Laboratories, Costa Mesa, California, to produce Hemofil, Antihemophilic Factor (Human) Method Four. A few years later (1974), another commercial concentrate (Koate from Cutter Laboratories, Berkeley, California) became available [14]; it also used cryoprecipitate as the starting material [15].

To obtain the large volumes of plasma required for manufacturing concentrate, companies established blood donation centers. The donated blood was centrifuged to separate the plasma from the red blood cells; reinfusion of the cells enabled the donor to be safely bled as often as once a week. The plasma donations were pooled and transported in tanker trucks to the plant manufacturing the FVIII concentrate. Donors received payment for each unit of their plasma. Although this arrangement provided manufacturers with large volumes of plasma, the downside was the potential for disseminating infectious agents throughout the material.

A major advantage of the commercial concentrates was that a therapeutic dose of FVIII could be administered with a single 30-mL syringe; this made possible the home treatment of hemophilia [16]. Patients and families were taught how to store and reconstitute the concentrates, and learned how to perform venipunctures and intravenous infusions. Home treatment and self-infusion were great improvements over the previous emergency room and hospital management of hemophilia [17]. The individual with hemophilia could institute treatment as soon as he was aware of a hemorrhage, without having to travel to a medical facility and wait for an infusion; importantly, bleeding was quickly controlled, tissue destruction was limited, and an early return to full activity became feasible. Self-infusion fostered independence and enabled hemophiliacs to travel, attend camp or boarding school, and enroll in distant colleges. By the end of the 1970s, most people with hemophilia were being treated with commercial FVIII concentrates.

There were problems associated with the use of these concentrates. Some of the products were found to be contaminated with metal, plastic, and rubber particles [18]. In addition, amorphous microaggregates of IgG, fibrinoprotein, and cold-insoluble globulin capable of passing through the 170-µm filters used for administering the concentrates were occasionally identified [19]. Trapping of these aggregates in the lungs might have accounted for the significant decreases in pulmonary-diffusing capacity noted by some [20] but not all investigators [21], and could explain the primary pulmonary hypertension reported during the 1970s and 1980s in a few hemophiliacs infusing concentrates for ≥ 10 years [22]. Other untoward reactions, such as visual disturbances, headache, dyspnea, bronchospasm, and hypotension, could have been due to the rapid infusion of products containing particulate material [18].

A major disadvantage of the early concentrates was the transmission of viral contaminants. Concentrates manufactured in the late 1970s and early 1980s were prepared from plasma that was usually obtained from paid donors. Some donors wanted the money to support a drug habit; others had contracted viral infections because they shared needles. As described previously, the manufacture of clotting factor concentrate required the pooling of several hundred plasma donations; viruses lurking in a single donor's plasma could spread throughout the entire pool and contaminate several batches of product. A retrospective examination using the polymerase chain reaction (PCR) identified hepatitis C RNA sequences in nearly all the samples of concentrates prepared from paid donors [23]. An epidemiologic study in the early 1980s showed that as many as 27% of all hemophiliacs in Australia had evidence of hepatitis C [24], and infection with hepatitis B virus was also a regular occurrence [25]. An evaluation of liver biopsies from 115 hemophilic patients living in the United States and Europe revealed cirrhosis in 15% and chronic active hepatitis in 7% [26]. Outbreaks of hepatitis A following treatment with FVIII concentrates continued to be reported into the 1990s [27,28].

Another virus that was being transmitted to hemophiliacs in the early 1980s was HIV, the cause of the acquired immunodeficiency syndrome (AIDS) [29]. Epidemiological studies linked patients with AIDS to blood and blood products prepared from donors who subsequently developed AIDS, eventually leading manufacturers to recall clotting factor concentrates made from the plasma of HIV-infected donors. When a laboratory test for the virus became available in 1984, the development of procedures for inactivating the virus in blood products became possible. Cutter Laboratories discovered that pasteurization, the process of heating protein solutions to a specific temperature for a predetermined length of time, was effective in eliminating HIV from clotting factor concentrates while preserving FVIII activity. By July 1985, all concentrates were heat-treated, and there were no further HIV transmissions. However, some 50% of the hemophiliacs that had been exposed to unheated FVIII concentrates between 1980 and 1985 were reported to be HIV-antibody positive [30].

Following the HIV epidemic, manufacturers adopted the method of monoclonal antibody affinity chromatography to increase the purity of their FVIII concentrates, and in the late 1980s two products became available: Monoclate from Armour [31] and Hemofil-M from Hyland-Baxter [32]. The specific activity of these products exceeded 3000 U/mg of protein prior to stabilization with human albumin. Viral inactivation of Monoclate was accomplished initially by dry heat and subsequently by pasteurization, while a solvent/detergent mixture was used for Hemofil-M. In vivo activity recovery was close to values predicted from the labeled unitage when assessed by one-stage FVIII assays [33]. The FVIII in Koate-DVI was purified 300–1000 times from plasma, heated at 80°C for 72h, and treated with a solvent/detergent mixture [34]. Concentrates approved for the treatment of both hemophilia and Von Willebrand Disease, Humate-P (ZLB Behring) and Alphanate SD (Grifols), also were treated with solvent/detergents and pasteurized or treated with dry heat.

RECOMBINANT CONCENTRATES

Despite the purity of the concentrates and steps taken to inactivate viruses, of concern was the possibility that FVIII derived from human plasma might transmit infectious agents such as parvovirus B19, hepatitis A, and the prions associated with variant Creutzfeldt-Jakob disease (CJD) [35]. The development and availability of concentrates prepared by recombinant DNA technology greatly allayed these fears. Recombinant FVIII was manufactured by inserting the cloned gene into Chinese Hamster Ovary (CHO) or Baby Hamster Kidney (BHK) cells and using immunoaffinity chromatography to recover the FVIII from the cell culture medium. Human albumin was added to stabilize the recombinant protein. Two recombinant products were marketed in the early 1990s: Recombinate from Genetics Institute and Baxter Healthcare [36], and Kogenate, developed by Genentech and Cutter Biologics [37]. They were comparable with regard to physical properties and pharmacokinetics [38], and were safe and effective in previously untreated patients with hemophilia A [39,40].

During this same period, it was discovered that B-domain-depleted FVIII retained the same procoagulant activity as the native molecule but was expressed at 10- to 20-fold greater levels by CHO cells [41]. Furthermore, B-domain-depleted FVIII did not require the addition of albumin for stabilization and when given to patients with severe hemophilia, the levels of FVIII achieved were closely related to the dose administered [42]. A study that compared its pharmacokinetics with those of full-length FVIII showed that they were bioequivalent [43]. B-domain-depleted FVIII was approved by the FDA in 2000 and is marketed as Refacto by Wyeth Pharmaceuticals.

Concentrates Free of Extraneous Human and Animal Proteins

By the turn of the century, most individuals with hemophilia were using recombinant products. However, continuing concern about possible viral contamination led to the introduction of several products that minimized or completely avoided the use of human and animal proteins in their manufacture. Kogenate FS and the same product under the name Helixate FS were formulated with sucrose. The cell culture medium for the BHK cells that produced the Kogenate FS contained Human Plasma Protein Solution that was fractionated to remove contaminants, and the purification process included virus inactivation using a solvent/detergent mixture; these and other steps were shown to greatly reduce the infectivity of material spiked with a surrogate CJD agent [44]. The final product was shown to be safe and effective in a clinical trial of patients with severe hemophilia [45]. The FVIII in Advate was stabilized by trehalose and Von Willebrand Factor (VWF), which was coexpressed by the CHO cells [46]. The FVIII was purified using a series of chromatographic steps and viral inactivation with a solvent/detergent treatment. The Wyeth product described previously as ReFacto was reformulated to be free of materials derived from human or animal sources; the manufacturing process included affinity chromatography, solvent/detergent treatment, and a virus-retaining nanofiltration step [47]. The final product was approved in 2008 and marketed as Xyntha Antihemophilic Factor (Recombinant), Plasma/Albumin-Free. More recent entries into the market are Novoeight from Novo Nordisk A/S, approved in 2013, and Kovaltry, Antihemophilic Factor (Recombinant), approved in 2016. Novoeight is a partially B-domain-deleted FVIII analogue, prepared using CHO cells and purified by a series of chromatography steps [48]. No additives of human or animal origin are used in the cell culture, purification, and formulation. Viral clearance is accomplished using solvent/detergent treatment and nanofiltration. In a clinical trial, the percentages of excellent or good hemostatic outcomes were 89.7% for prophylaxis and 96.6% for on-demand treatment [49]. Kovaltry has been modified from Kogenate FS, but is formulated without the addition of human- or animal-derived raw materials [50]. The manufacturing process includes detergent and nanofiltration steps for viral inactivation and removal. Kovaltry was found to be safe and effective for the prevention of bleeding in previously treated patients with severe hemophilia who were followed-up for 1 year [51].

Longer-Acting Concentrates

Although most commercial FVIII concentrates manufactured during the first decade of the 21st century were considered safe, they had to be infused frequently because of the short in vivo half-life ($T/2$) of FVIII: 13.0 ± 3.1 h for ReFacto and 13.6 ± 3.8 h for Advate [30]. Scientists therefore focused their research on modifying recombinant FVIII to prolong its $T/2$, and arrived at several effective methods. These included fusion of FVIII with the Fc domain of immunoglobulin-G1 (IgG1), conjugation with polyethylene glycol (PEG), and synthesis of a truncated, single-chain recombinant molecule with increased affinity for VWF. Table 6.1 lists FVIII concentrates that have been modified to prolong their half-lives.

Fc-fusion had been used to prolong the $T/2$ of proteins as disparate as albumin, etanercept, and romiplostim [52]. The Fc-protein complex enters cells and binds to the neonatal Fc-receptor (Rn); the FcRn-protein is then released back into the circulation, delaying clearance and extending the half-life of the protein. Researchers at the University of North Carolina and Biogen Idec Hemophilia in Waltham, MA, created a coagulant complex by fusing a single B-domain-depleted recombinant FVIII molecule to the Fc domain of IgG1(rFVIIIFc) [53]. Pharmacokinetic studies showed that rFVIIIFc had a longer elimination $T/2$ (1.54–1.70-fold), lower clearance (1.49–1.56-fold), and longer time to 1% FVIII activity above baseline (1.5–1.68-fold) relative to unmodified FVIII [54]. Subsequent clinical trials showed low bleeding rates with extended-interval dosing (every 3–5 days) and infrequent adverse effects [55,56]. The product, marketed as Eloctate, was approved in 2014; extension studies of the original phase 3 trials noted sustained low annualized bleeding rates (ABR) in pediatric, adolescent, and adult patients with severe hemophilia, as well as improvements in quality of life and target joint resolution [57].

Long-chain, chemically activated polyethylene glycol (PEG) is covalently attached to proteins to decrease their clearance and prolong their $T/2$. Mei et al. [58] conjugated PEG to exposed cysteines on

TABLE 6.1   Engineered Factor VIII Products With Longer Half-Lives

Engineered product	Half-life (h)
rFVIII-Fc (Eloctate)	19.0
PEG-FVIII (Adynovate)	14.3
GlycoPEG-FVIII(N8-GP)	18.4
PEG-FVIII (Bay 94-9027)	19.0
Single-chain FVIII (Afstyla)	14.5

Data from Pierce, G. PEN, May 2016.

B-domain-deleted FVIII, and found that the molecule retained functional activity and VWF binding. An extended $T/2$ and lack of immunogenicity was reported in a phase 1 trial of this preparation (Bay 94-9027) in patients with severe hemophilia, and a phase 3 trial showed that doses tailored to individual ABR given twice weekly were effective for prophylaxis and treatment [59,60]. This product is under consideration for FDA approval. Another concentrate, a PEGylated full-length recombinant FVIII, has been developed using the Advate manufacturing process and a novel PEG reagent [61]. Preclinical studies in a murine model demonstrated prevention of joint bleeding for 48 h as compared to 24 h for Advate [62]. In a phase 2/3 open-label study of 137 patients with hemophilia, control of bleeding was rated excellent or good in 92% of episodes and prophylaxis resulted in a greater than 50% reduction in ABR [63]. This product was approved in 2015, and is marketed as Adynovate [Antihemophilic Factor (Recombinant), PEGylated] by Shire Pharmaceuticals. A final product in this category is N8-GP, Novo Nordisk A/S. It is prepared by using sialidase and a sialic acid-PEG reagent to couple PEG to an O-linked glycan in the FVIII B-domain of turoctocog alfa [64]. When activated by thrombin, the B-domain and associated PEG are released, generating activated FVIII with the same structure and specific activity as native FVIII. In previously treated hemophiliacs, the $T/2$ was 19 h, and the time from dosing to a plasma FVIII level of 0.01 U/mL was 6.5 days [65]. For prophylaxis, N8-GP was given every fourth day to 175 patients with severe hemophilia; the median ABR was 1.33, no bleeding occurred in 40%, and the agent was well tolerated [66]. Approval is pending for N8-GP.

FVIII is released from endothelial cells as a two-chain molecule that is stabilized by binding to VWF. To increase FVIII stability and affinity for VWF, investigators synthesized a truncated, single-chain, recombinant FVIII with a modified VWF-binding site [67]. The mean $T/2$ in patients with hemophilia was 14.5 h, as compared to 13.3 h for Advate, and the area under the disappearance curve (AUC) was 35% larger [68]. In an open-label, phase 1/3 study of 173 patients, control of bleeding was rated excellent or good in 72.2%, and hemostasis for 16 surgical procedures was rated as excellent in all but one patient [69]. However, in this larger trial, the product was not long acting; the $T/2$ was only 12.9 h and prophylaxis was given no less than two times per week. The concentrate was approved in 2016 and marketed as Afstyla by CSL Behring.

A Concentrate Prepared From a Human Cell Line

All of the recombinant FVIII molecules described thus far have been manufactured using cultured hamster cell lines that express N-glycolylneuraminic acid or other glycoforms absent from FVIII derived from human cell lines [70]. NUWIQ, Antihemophilic Factor (Recombinant), is a B-domain deleted FVIII produced in genetically modified human embryonic kidney 293F cells with no animal- or human-derived materials added during production [71]. It has been found to be efficacious for prophylaxis, treatment of hemorrhages, and control of surgical bleeding in patients with severe hemophilia A [72]. A study in previously untreated patients (PUPS) reported that the cumulative incidence of high-titer inhibitors was only 12.8%, considerably less than the 25%–30% incidence historically observed with other recombinant FVIII concentrates [73]. Randomized, blinded clinical trials are needed to establish whether it is less antigenic than hamster-cell-derived recombinants. NUWIQ was approved in 2015 and is distributed by Octapharma USA, Inc.

Table 6.2 shows the evolution of concentrates over a span of more than 60 years.

TABLE 6.2   Evolution of FVIII Concentrates Over the Decades From 1950–2010

Decade	Development	Products
1950	Animal FVIII; Cohn Fraction 1	Bovine AHG[a], Fibro-AHG[a]
1960	Cryoprecipitate	N/A[b]
1970	Commercial concentrates	Hemofil; Koate
1980	Viral inactivation	Koate-DVI, Humate-P, Alphanate-SD
	Monoclonal-antibody purified	Hemofil-M, Monoclate-P
1990	Recombinants	Recombinate, Kogenate
2000	B-domain depleted	ReFacto
	Sucrose stabilized	Kogenate-FS, Helixate-FS
	Trehalose stabilized	Advate
	Plasma/albumin-free	Xyntha
2010	No additives of human or animal origin	Novoeight, Kovaltry
	Longer-acting products	Eloctate, Adynovate, NUWIQ, Afstyla (approval pending: Bay-9027, N8-GP)
	Recombinant porcine FVIII	OBIZUR

[a] AHG, *antihemophilic globulin.*
[b] N/A, *not applicable.*

ADVERSE EVENTS WITH CONCENTRATES

Current FVIII concentrates are associated with few adverse events with the exception of alloantibody (inhibitor) formation. Inhibitors appear in up to 30% of hemophiliacs after their initial exposure to concentrates and render these individuals resistant to further therapy with FVIII products. A full discussion of alloantibodies is presented in Chapter 8. Other adverse events are hypersensitivity or allergic reactions, headache, injection site reactions, and infections of central venous catheters used for infusions. Hypersensitivity reactions are usually due to the presence of trace amounts of mouse or hamster protein in the products, and have been reported in fewer than 1% of clinical trial subjects. Another potential adverse event is thrombosis; the risk of thrombus formation is always present when coagulation defects are corrected, especially in patients undergoing orthopedic surgery. For example, Ritchie et al. [74] describe a 67-year-old man with hemophilia who developed extensive deep vein thrombosis following a proximal tibial osteotomy; hemostasis had been secured with FVIII concentrate, and levels had been maintained between 0.5 and 1.0 U/mL. Based on anecdotal experience, it seems reasonable to recommend prophylactic doses of anticoagulants for patients at risk for thrombosis, especially in those receiving intensive replacement therapy for major orthopedic procedures [75].

IMPROVING CONCENTRATES

Efforts to reduce the antigenicity of concentrates and decrease inhibitor development have included increasing the affinity of FVIII for VWF (Afstyla), generating activated FVIII with the same structure as the native molecule (N8-GP), and producing recombinant FVIII in human rather than hamster cell lines (NUWIQ). While these measures might be partially successful, it seems doubtful that they will have a major impact because even plasma-derived concentrates are associated with inhibitor formation. Although the possibility of developing a nonantigenic FVIII concentrate seems remote, there are other products that can control bleeding in patients resistant to current FVIII concentrates. One of these is a partially B-domain-depleted recombinant porcine FVIII(r-pFVIII) expressed by BHK cells in a protein-free medium [76]. It is not inactivated by inhibitors because the amino acid sequence of the A2 and C2 domains, the target of most alloantibodies, is different from that of human FVIII. Available as OBIZUR, it was approved in 2014 for the treatment of patients with acquired hemophilia A, but also appears to be effective in patients with congenital hemophilia A and inhibitors [77] (Chapter 8). However, resistance develops with repeated exposure to this product. OBIZUR is produced by Baxter Healthcare and marketed by Shire.

All approved FVIII concentrates must be infused intravenously; other routes of administration are ineffective. If repeated doses are anticipated, most patients will need an intravenous access device or port, and these require expert placement and regular maintenance. Their use is often complicated by infection, leakage, and vaso-occlusion. Although prescribing a long-acting concentrate might decrease prophylactic dosing from every other day to every 3–5 days, there will still be a need for ready venous access. A major research priority is the development of concentrates that are effective by other than the intravenous route. Therapeutic alternatives to clotting factor concentrates are discussed in Chapter 7.

Although there have been major improvements in FVIII concentrates since the 1960s, they have been accompanied by progressive increases in product prices. The financial burden of hemophilia treatment is enormous; annual prophylaxis for an adult patient can easily top $200,000. Making treatment regimens more affordable will improve the quality of life of hemophiliacs, and there will be long-term benefits in terms of decreased disability and greater productivity.

SUMMARY

The concept of using blood products to control bleeding in people with hemophilia is almost two centuries old. In more recent times, transfusions of whole blood have been replaced by plasma, cryoprecipitate, plasma-derived FVIII concentrates, and currently, recombinant concentrates. Products have been improved by using affinity chromatography purification, viral inactivation steps, elimination of extraneous human proteins, and most recently, modifications to prolong their duration of action. The principal drawbacks of the currently available concentrates are stimulation of inhibitor (antibody) formation, requirement for intravenous infusion, and cost. Current research is focused on the development of less antigenic products and new delivery systems.

References

[1] Lane S. Haemorrhagic diathesis-successful transfusion of blood. Lancet 1840;1:185–8.
[2] Biggs R. The amount of blood required annually to make concentrate to treat patients with haemophilia A and B. In: Biggs R, editor. The treatment of haemophilia A and B and Von Willebrand disease. Oxford: Blackwell Scientific Publications; 1978. p. 101. [chapter 4].
[3] Macfarlane RG, Biggs R, Bidwell E. Bovine antihaemophilic globulin in the treatment of haemophilia. Lancet 1954;266:1316–9.
[4] Macfarlane RG, Mallam PC, Witts LJ, Bidwell E, Biggs R, Fraenkel GJ, Honey GE, Taylor KB. Surgery in haemophilia. The use of animal antihaemophilic globulin and human plasma in thirteen cases. Angio 1957;273:251–9.

[5] McMillan CW, Diamond LK, Surgenor DM. Treatment of classic hemophilia: the use of fibrinogen rich in factor VIII for hemorrhage and for surgery. N Engl J Med 1961;265:277–83.

[6] Webster WP, Roberts HR, Thelin GM, Wagner RH, Brinkhous KM. Clinical use of a new glycine-precipitated antihemophilic fraction. Am J Med Sci 1965;250:643–50.

[7] Brinkhous KM, Shanbrom E, Roberts HR, Webster WP, Fekete L, Wagner RH. A new high-potency glycine-precipitated antihemophilic factor (AHF) concentrate: treatment of classical hemophilia and hemophilia with inhibitors. JAMA 1968;205:613–7.

[8] Pool JG, Hershgold EJ, Pappenhagen AR. High-potency antihaemophilic factor concentrate prepared from cryoglobulin precipitate. Nature (London) 1964;203:312.

[9] Pool JG, Shannon AE. Production of high-potency concentrates of antihemophilic globulin in a closed-bag system. N Engl J Med 1965;273:1443–7.

[10] Cooke JV, Holland PV, Shulman NR. Cryoprecipitate concentrates of FVIII for surgery in hemophiliacs. Ann Intern Med 1968;68:39–47.

[11] Seeler RA. Hemolysis due to anti-A and anti-B in factor VIII preparations. Arch Intern Med 1972;130:101–3.

[12] Newman J, Johnson AJ, Karpatkin MH, Puszkin S. Methods for the production of clinically effective intermediate- and high-purity factor-VIII concentrates. Br J Haemoatol 1971;21:1–20.

[13] Medical World News; November 19, 1971. p. 64–71.

[14] Britton M, Harrison J, Abildgaard CF. Early treatment of hemophilic hemarthroses with minimal dose of new factor VIII concentrate. J Pediatr 1974;85:245–7.

[15] Hershgold EJ, Pool JG, Pappenhagen AR. The potent anti-hemophilic globulin concentrate derived from a cold insoluble fraction of human plasma: characterization and further data on preparation and clinical trial. J Lab Clin Med 1966;67:23–32.

[16] Rabiner SF, Telfer MC. Home transfusion for patients with hemophilia A. N Engl J Med 1970;283:1011–5.

[17] Britten AFH. A little freedom for the hemophiliac. N Engl J Med 1970;283:1051–2.

[18] Couper IA, McAdam JH, Mackenzie MJ, Davidson JF. Contamination of factor-VIII concentrates with metal particles. Lancet 1974;ii:1515.

[19] Eyster ME, Nau MF. Particulate material in antihemophilic factor (AHF) concentrates. Transfusion 1978;18:576–81.

[20] Boese EC, Tantum KR, Eyster ME. Pulmonary function abnormalities after infusion of antihemophilic factor (AJF) concentrates. Am J Med 1979;67:474–6.

[21] Chediak J, Chausow A, Solarski A, Telfer MC. Pulmonary function in hemophiliac patients treated with commercial factor VIII concentrates. Am J Med 1984;77:293–6.

[22] Goldsmith GH, Baily RG, Brettler DB, Davidson Jr. WR, Ballard JO, Driscol TE, Greenberg JM, Kasper CK, Levine PH, Ratnoff OD. Primary pulmonary hypertension in patients with classic hemophilia. Ann Intern Med 1988;108:797–9.

[23] Garson JA, Preston FE, Makris M, Tuke P, Ring C, Machin SJ, Tedder RS. Detection by PCR of hepatitis C virus in factor VIII concentrates. Lancet 1990;1473. i.

[24] Rickard KA, Dority P, Campbell J, Batey RG, Johnson S, Hodgson J. Hepatitis and haemophilia therapy in Australia. Lancet 1982;ii:146–8.

[25] Mannucci PM, Zanetti AR, Colombo M, the Study Group of the Fondazione dell'Emofilia. Prospective study of hepatitis after factor VIII concentrate exposed to hot vapour. Br J Haematol 1988;68:427–30.

[26] Aledort LM, Levine PH, Hilgartner M, Blatt P, Spero JA, Goldberg JD, Bianchi L, Desmet V, Scheuer P, Popper H, Berk PD. A study of liver biopsies and liver disease among hemophiliacs. Blood 1985;66:367–72.

[27] Mannucci PM. Outbreak of hepatitis A among Italian patients with haemophilia. Lancet 1992;i:819.

[28] Robinson SM, Schwinn H, Smith A, Shouval D, Gerlich WH, on behalf of the Working Group on Hepatitis A and Clotting Factors. Clotting factors and hepatitis A. Lancet 1992;340:1465–6.

[29] Green D. Linked by blood: hemophilia and AIDS. London: Elsevier; 2016. p. 147.

[30] Ragni MV, Winkelstein A, Kingsley L, Spero JA, Lewis JH. 1986 update on HIV seroprevalence, seroconversion, AIDS incidence, and immunologic correlates of HIV infection in patients with hemophilia A and B. Blood 1987;70:786–90.

[31] Schreiber AB. The preclinical characterization of monoclate factor VIII:C: antihemophilic factor (human). Semin Hematol 1988;25(2 Suppl. 1):27–32.

[32] Addiego Jr JE, Gomperts E, Liu S-L, Bailey P, Courter SG, Lee ML, Neslund GG, Kingdon HS, Griffith MJ. Treatment of hemophilia A with a highly purified factor VIII concentrate prepared by anti-FVIIIc immunoaffinity chromatography. Thromb Haemost 1992;67:19–27.

[33] Kasper CK, Kim HC, Gomperts ED, Smith KJ, Salzman PM, Tipping D, Miller R, Montgomery RM. In vivo recovery and survival of monoclonal-antibody-purified factor VIII concentrates. Thromb Haemost 1991;66:730–3.

[34] See Koate-DVI package insert.

[35] Kessler CM. Advances in the treatment of hemophilia. Clin Adv Hematol Oncol 2008;6:184–7.

[36] White II GC, McMilllan CW, Kingdon HS, Shoemaker CB. Use of recombinant antihemophilic factor in the treatment of two patients with classic hemophilia. N Engl J Med 1989;320:166–70.

[37] Schwartz RS, Abildgaard CF, Aledort LM, Arkin S, Bloom AL, Brackmann HH, Brettler DB, Fukui H, Hilgartner MW, Inwood MJ, Kasper CK, Kernoff PBA, Levine PH, Lusher JM, Mannucci PM, Scharrer I, MacKenzie MA, Pancham N, Kuo HS, Allred RU, the Recombinant Factor VIII Study Group. Human recombinant DNA-derived antihemophilic factor (factor VIII) in the treatment of hemophilia A. N Engl J Med 1990;323:1800–5.

[38] Kogenate and Recombinate: Phase III. Drug Profiles. 1993;3(1):6–17.

[39] Bray GL, Gomperts ED, Courter S, Gruppo R, Gordon EM, Manco-Johnson M, Shapiro A, Scheibel E, White GIII, Lee M, the Recombinate Study Group. A multicenter study of recombinant factor VIII (Recombinate): safety, efficacy and inhibitor risk in previously untreated patients with hemophilia A. Blood 1994;83:2428–35.

[40] Lusher JM, Arkin S, Abildgaard CF, Schwartz RS. Recombinant factor VIII for the treatment of previously untreated patients with hemophilia A. Safety, efficacy, and development of inhibitors. Kogenate Previously Untreated Patient Study Group. N Engl J Med 1993;328:453–9.

[41] Pittman DD, Alderman EM, Tomkinson KN, Wang JH, Giles AR, Kaufman RJ. Biochemical, immunological, and in vivo functional characterization of B-domain-deleted factor VIII. Blood 1993;81:2925–35.

[42] Fijnvandraat K, Berntorp E, ten Cate JW, Johnsson H, Peters M, Savidge G, Tengborn L, Spira J, Stahl C. Recombinant, B-domain deleted factor VIII (r-VIII SQ): pharmacokinetics and initial safety aspects in hemophilia A patients. Thromb Haemost 1997;77:298–302.

[43] Di Paola J, Smith MP, Klamroth R, Mannucci PM, Kollmer C, Feingold J, Kessler C, Pollmann H, Morfini M, Udata C, Rothschild C, Hermans C, Janco R. Refacto® and Advate®: a single-dose, randomized, two-period crossover pharmacokinetics study in subjects with haemophilia A. Haemophilia 2007;13:124–30.

[44] See Kogenate package insert.

[45] Manco-Johnson MJ, Kempton CL, Reding MT, Lissitchkov T, Goranov S, Gercheva L, Rusen L, Ghinea M, Uscatescu V, Rescia V, Hong W. Randomized, controlled, parallel-group trial of routine prophylaxis vs. on-demand treatment with sucrose-

formulated recombinant factor VIII in adults with severe hemophilia A (SPINART). J Thromb Haemost 2013;11:1119–27.

[46] See Advate package insert.

[47] See Xyntha package insert.

[48] See Novoeight package insert.

[49] Janic D, Matytsina I, Misgav M, Oldenburg J, Ozelo M, Recht M, Korsholm L, Savic A, Santagostino E. Safety and efficacy of turoctocog alfa in prevention and on-demand treatment of bleeding episodes in patients with hemophilia A. Poster presentation at Am Soc Hem 2016, San Diego, December 3–6; 2016.

[50] See Kovaltry package insert.

[51] Saxena K, Lalezari S, Oldenburg J, Tsenekidou-Stoeter D, Beckmann H, Yoon M, Maas Enriquez M. Efficacy and safety of Bay 81-8973, a full-length recombinant factor VIII: results from the LEOPOLD I trial. Haemophilia 2016;22:706–12.

[52] Makris M. Longer FVIII: the 4th generation. Blood 2012;119:2972–3.

[53] Dumont JA, Liu T, Low SC, Zhang X, Kamphaus G, Sakorafas P, Fraley C, Drager D, Reidy T, McCue J, Franck HWG, Merricks EP, Nichols TC, Bitonti AJ, Pierce GF, Jiang H. Prolonged activity of a recombinant factor VIII-Fc fusion protein in hemophilia A mice and dogs. Blood 2012;119:3024–30.

[54] Powell JS, Josephson NC, Quon D, Ragni MV, Cheng G, Li E, Jiang H, LI L, Dumont JA, Goyal J, Zhang X, Sommer J, McCue J, Barbetti M, Luk A, Pierce GF. Safety and prolonged activity of recombinant factor VIII FC fusion protein in hemophilia A patients. Blood 2012;119:3031–7.

[55] Mahlangu J, Powell JS, Ragni MV, Chowdary P, Josephson NC, Pabinger I, Hanabusa H, Gupta N, Kulkarni R, Fogarty P, Perry D, Shapiro A, Pasi KJ, Apte S, Nestorov I, Jiang H, Li S, Neelakantan S, Cristiano LM, Goyal J, Sommer JM, Dumont JA, Dodd N, Nugent K, Vigliani G, Luk A, Brennan A, Pierce GF. Phase 3 study of recombinant factor VIII Fc fusion protein in severe hemophilia A. Blood 2014;123:317–25.

[56] Shapiro AD, Ragni MV, Kulkarni R, Oldenberg J, Srivastava A, Quon DV, Pasi KJ, Hanabusa H, Pabinger I, Mahlangu J, Fogarty P, Lillicrap D, Kulke S, Potts J, Neelakantan S, Nestorov I, Li S, Dumont JA, Jiang H, Brennan A, Pierce GF. Recombinant factor VIII Fc fusion protein: extended-interval dosing maintains low bleeding rates and correlates with von Willebrand factor levels. J Thromb Haemost 2014;12:1788–800.

[57] Wang M, Pasi KJ, Pabinger I, Kerlin BA, Kulkarni R, Nolan B, Liesner R, Brown SA, Hanabusa H, Tsao E, Winding B, Lethagen S, Jain N. Long-term efficacy and quality of life with recombinant factor VIII Fc fusion protein (rFVIIIFc) prophylaxis in pediatric, adolescent, and adult subjects with target joints and severe hemophilia A. Poster presentation at Am Soc Hem 2016, San Diego, December 3-6; 2016.

[58] Mei B, Pan C, Jiang H, Tjandra H, Strauss J, Chen Y, Liu T, Zhang X, Severs J, Newgren J, Chen J, Gu J-M, Subramanyam B, Fournel MA, Pierce GF, Murphy JE. Rational design of a fully active, long-acting PEGylated factor VIII for hemophilia A treatment. Blood 2010;116:270–9.

[59] Coyle TE, Reding MT, Lin JC, Michaels LA, Shah A, Powell J. Phase 1 study of BAY 94-9027, a PEGylated B-domain-deleted recombinant factor VIII with an extended half-life, in subjects with hemophilia A. J Thromb Haemost 2014;12:488–96.

[60] Reding MT, Ng H, Poulsen LH, Eyster ME, Pabinger I, Shin HJ, Walsch R, Lederman M, Wang M, Hardke M, Michaels LA. Safety and efficacy of Bay 94-9027, a prolonged half-life factor VIII. J Thromb Haemost 2017;15:411–9.

[61] Turecek PL, Bossard MJ, Graninger M, Gritsch H, Höllriegl W, Kaliwoda M, Matthiessen P, Mitterer A, Muchitsch EM, Purtscher M, Rottensteiner H, Schiviz A, Schrenk G, Siekmann J, Varadi K, Riley T, Ehrlich HJ, Schwarz HP, Scheiflinger F. Bax 855, a PEGylated rFVIII product with prolonged half-life. Development, functional and structural characterization. Hamostaseologie 2012;32(Suppl. 1):S29–38.

[62] Valentino LA, Cong L, Enockson C, Song X, Scheiflinger F, Muchitsch EM, Turecek PL, Hakobyan N. The biological efficacy profile of Bax 855, a PEGylated recombinant factor VIII molecule. Haemophilia 2015;21:58–63.

[63] Konkle BA, Stasyshyn j O, Chowdary P, Bevan DH, Mant T, Shima M, Engl W, Dyck-Jones J, Fuerlinger M, Patrone L, Ewenstein B, Abbuehl B. Pegylated, full-length, recombinant factor VIII for prophylactic and on-demand treatment of severe hemophilia A. Blood 2015;126:1078–85.

[64] Steinicke HR, Kjalke M, Karpf DM, Balling KW, Johansen PB, Elm T, Ovilsen K, Moller F, Holmberg HL, Gudme CN, Persson E, Hilden I, Pelzer H, Rahbek-Nielsen H, Jespersgaard C, Bogsnes A, Pedersen AA, Kristensen AK, Peschke B, Kappers W, Rode F, Thim L, Tranholm M, Ezban M, Olsen EHN, Bjorn SE. A novel B-domain O-glycoPEGylated FVIII (N8-GP) demonstrates full efficacy and prolonged effect in hemophilic mice models. Blood 2013;121:2108–16.

[65] Tiede A, Brand B, Fischer R, Kavakli K, Lentz SR, Matsushita T, Rea C, Knobe K, Viuff D. Enhancing the pharmacokinetic properties of recombinant factor VIII: first-in-human trial of glycoPEGlylated recombinant factor VIII in patients with hemophilia A. J Thromb Haemost 2013;11:670–8.

[66] Giangrande P, Andreeva T, Chowdary P, Ehrenforth S, Hanabusa H, Leebeek FW, Lentz SR, Nemes L, Poulsen LH, Santagostino E, You CW, Clausen WH, Jonsson PG, Oldenburg J. Clinical evaluation of glycoPEGylated recombinant FVIII: efficacy and safety in severe haemophilia A. Thromb Haemost 2016. https://doi.org/10.1160/TH16-06-0444.

[67] Schulte S. Pioneering designs for recombinant coagulation factors. Thromb Res 2011;128 (Suppl. 1):S9–S12.

[68] Klamroth R, Simpson M, von Depka-Prondzinski M, Gill JC, Morfini M, Powell JS, Santagostino E, Davis J, Huth-Kuhne A, Leissinger C, Neumeister P, Bensen-Kennedy D, Feussner A, Limsakun T, Zhou M, Veldman A, St. Ledger K, Blackman N, Pabinger I. Comparative pharmacokinetics of rVIII-single chain and octocog alfa (Advate®) in patients with severe haemophilia A. Haemophilia 2016;1–9.

[69] Mahlangu J, Kuliczkowski K, Karim FA, Stasyshyn O, Kosinova MV, Lepatan LM, Skotnicki A, Boggio LN, Klamroth R, Oldenburg J, Hellmann A, Santagostino E, Baker RI, Fischer K, Gill JC, P'Ng S, Chowdary P, Escobar MA, Khayat CD, Rusen L, Bensen-Kennedy D, Blackman N, Linsakun T, Veldman A, St Ledger K, Pabinger I. Efficacy and safety of rVIII-single chain: results of a phase 1/3 multicenter clinical trial in severe hemophilia A. Blood 2016;128:630–7.

[70] Kannicht C, Ramstrom M, Kohla G, Tiemeyer M, Casademunt E, Walter O, Sandberg H. Characterisation of the post-translational modifications of a novel, human cell line-derived recombinant human factor VIII. Thromb Res 2013;131:78–88.

[71] See NUWIQ package insert.

[72] Lissitchkov T, Hampton K, von Depka M, Hay C, Rangarajan S, Tuddenham E, Holstein K, Huth-Kühne A, Pabinger I, Knaub S, Bichler J, Oldenburg J. Novel, human cell line-derived recombinant factor VIII (human-cl rhFVIII; Nuwiq®) in adults with severe haemophilia A: efficacy and safety. Haemophilia 2015. https://doi.org/10.1111/hae.12793.

[73] Liesner R, Abashidze M, Aleinikova O, Altisent C, Belletrutti MJ, Borel-Derlon A, Carcao M, Chabost H, Chan A, Dubey L, Ducore JM, Abubacker FN, Gattens M, Gruel Y, Kavardakova N, Khorassani M, Klukowska A, Konigs C, Lambsert T, Lohade S, Sigaud M, Turea V, Wu JK, Vdovin V. Inhibitor development in previously untreated patients with severe hemophilia A treated with Nuwiq, a new generation recombinant FVIII of human origin. Oral presentation at Am Soc Hem 2016, San Diego, December 3–6; 2016.

[74] Ritchie B, Woodman RC, Poon M-C. Deep venous thrombosis in hemophilia A. Am J Med 1992;93:699–700.

[75] Rodriguez-Merchan EC. Thromboprophylaxis in haemophilia patients undergoing orthopaedic surgery. Blood Coagul Fibrinolysis 2014;25:300–2.

[76] Gomperts E. Recombinant B domain deleted porcine factor VIII for the treatment of bleeding episodes in adults with acquired hemophilia A. Expert Rev Hematol 2015;8:427–32.

[77] Mahlangu JN, Andreeva TA, Macfarlane DE, Walsh C, Key NS. Recombinant B-domain-deleted porcine sequence factor VIII (r-pFVIII) for the treatment of bleeding in patients with congenital haemophilia A and inhibitors. Haemophilia 2016. https://doi.org/10.1111/hae.13108.

7

General Management of Hemophilia

The principal aim of treatment is to assist patients in achieving lifestyles that meet their goals and expectations. This requires a regimen that prevents bleeding but also addresses issues related to patient self-esteem and sense of place in the community. Hemorrhages can be prevented by the administration of sufficient clotting factor concentrate to increase FVIII to hemostatic levels until the risk of recurrence has subsided. More difficult is providing care that helps the patient to accept and learn to live with his vulnerabilities. In this chapter, the control of bleeding is described, followed by a discussion of the management of other aspects of hemophilia care, and concludes with future approaches to treatment.

MANAGEMENT OF ACUTE HEMORRHAGES

The essential first steps are:

- Establish whether the patient has hemophilia A, hemophilia B, or other bleeding disorder. Patients with hemophilia B are clinically indistinguishable from those with hemophilia A and lack FIX rather than FVIII.
- Determine if an inhibitor is present. A satisfactory response to recent treatment or a negative inhibitor assay within the past 6 months helps exclude the presence of an inhibitor, but in case of uncertainty, perform an inhibitor assay.
- Infuse concentrate before sending a bleeding patient for imaging or other studies. Prompt treatment limits the extent of bleeding and helps ameliorate pain.
- Administer analgesics for pain relief. These should be given orally or intravenously, but never intramuscularly (provokes muscle hematoma). Appropriate agents for mild-to-moderate pain are

97

acetaminophen with either codeine or hydrocodone, and for more severe pain, hydromorphone. Avoid ketorolac, aspirin, and nonsteroidal antiinflammatory drugs such as naproxen.

The next sections describe the use of FVIII concentrates for acute bleeding, perioperative management, and long-term prophylaxis.

On-Demand Treatment

On-demand treatment refers to the episodic infusion of FVIII concentrates to control acute bleeding. The concentrates might be infused in either a physician's office, hospital emergency room, or at home if the hemophiliac and/or his family have been instructed in the technique of factor infusion. The product used for treatment is selected by the treating physician or treatment facility based on considerations of efficacy, safety, availability, and cost. The doses needed to control bleeding are based on body weight (in kg), the severity and location of the hemorrhage, and previous responses to FVIII infusions. The rise in FVIII activity that will result from a given dose depends on the patient's plasma volume, which is the body weight in kg times 41 mL [1]; a 70-kg patient has a plasma volume of approximately 2870 mL. Because the normal concentration of FVIII is 1 IU (International Unit)/mL, it is necessary to infuse at least 2870 IU or 41 IU/kg to raise the level of FVIII from <0.01 to 1 IU (<1%–100%). The theoretical increase in FVIII per IU per kg is 2.44% ($100 \div 2870 \times 70$), but because only ~80% of the dose is recovered in vivo, two times the total dose divided by body weight will approximate the expected rise in FVIII. Because the half-life of infused FVIII is 12 h, doses are repeated at 12-h intervals to make certain that plasma levels never fall below half of the peak values. Dosing duration depends on the likelihood that bleeding will resume; for example, giving one or two doses is usually sufficient for the treatment of a spontaneous hemarthrosis, while a gastrointestinal or intracranial hemorrhage might require several days of therapy to ensure that bleeding will not recur.

The doses of FVIII concentrate needed to control bleeding range from 15 to 50 IU/kg, depending on whether the hemorrhage is minor, moderate, or major. **Minor hemorrhages** such as persistent epistaxis or bleeding from the tongue usually respond to a single dose of concentrate; for oral bleeding or dental extractions, an initial dose of 20–30 IU/kg is often supplemented with tranexamic acid mouthwash (4.8%). Tranexamic acid inhibits fibrinolysis, preventing the dissolution of clots in tooth sockets or on tongue lesions and the resumption of bleeding. **Moderate hemorrhages** include hemarthroses and muscle hematomas; if treated early, a dose of 20–30 IU/kg rapidly relieves pain and diminishes swelling. Other measures, such as ice, rest, and analgesics, contribute to symptomatic relief. **Major hemorrhages** are gastrointestinal tract, intracerebral, and other life-threatening bleeding events, and are treated with doses of up to 50 IU/kg to achieve FVIII levels of 1.0 IU/mL or 100%. Levels are maintained above 0.5 IU/mL until bleeding has completely resolved and lesions have healed. The doses recommended by manufacturers for the treatment of hemarthroses and major hemorrhages vary by product, and are listed in Table 7.1.

Bleeding in patients with mild hemophilia (FVIII >5%) is often amenable to treatment with desmopressin (1-deamino-8-D-arginine vasopressin, DDAVP). Desmopressin can be administered by intravenous, subcutaneous, or intranasal routes, and stimulates the release of FVIII from endothelial cells. Doses of 0.3 µg/kg achieve peak plasma concentrations within 30–60 min of intravenous infusion or subcutaneous injection, raising FVIII levels two- to fourfold. If bleeding recurs, additional doses can be given daily for up to 4 days, but the response becomes less vigorous with repeated dosing (tachyphylaxis). Hemostatic efficiency of outpatient

TABLE 7.1 Recommended Doses of FVIII Products Approved or Pending Approval

Product	Manufacturers' suggested dose (IU/kg)		
	On-demand treatment		Prophylaxis
	Hemarthroses	Major bleeds	
STANDARD CONCENTRATES			
Advate	15–30 q12–24 h	30–50 q8–24 h	≥25 3–4 ×/wk
Kogenate-FS	15–30 q12–24 h	40–50 q8–12 h	25 q48 h
Kovaltry	15–30 q12–24 h	30–50 q8–24 h	20–40 2–3 ×/wk
Novoeight	15–30 q12–24 h	30–50 q8–24 h	20–50 3 ×/wk
Xyntha	15–30 q12–24 h	30–50 q8–24 h	30±5 3 ×/wk
LONGER-ACTING CONCENTRATES			
Adynovate	15–30 q12–24 h	30–50 q8–24 h	40–50 2 ×/wk
Afstyla	15–30 q12–24 h	30–50 q8–24 h	20–50 2–3 ×/wk[a]
Bay 94-9027	14–62 (mean, 34); ≤2 doses in 48 h		30–60 1–2 ×/wk
Eloctate	20–30 q24–48 h	40–50 q12–24 h	50 q3–5 days[a]
N8-GP	25–75: ≤2 doses in 48 h		50 q4 days
NUWIQ	15–30 q12–24 h	30–50 q8–24 h	30–40 q48h[a]

[a] *Afstyla for children <12: 30–50 IU/kg 2–3 ×/wk; Eloctate for children <6: 50 IU/kg 2 ×/wk; Nuwiq for children 2–11: 30–50 IU/kg q48 h or 3 ×/wk.*

desmopressin treatment is reported as good to excellent in >90% of bleeding episodes [2]. Adverse events are usually mild and consist of facial flushing and tachycardia, but hyponatremia and seizures can occur if fluid intake is not restricted; very rarely, myocardial infarction and stroke have been reported.

PERIOPERATIVE MANAGEMENT

Invasive procedures can be safely accomplished even in patients with severe hemophilia, providing the following conditions are met:

- There are no inhibitory antibodies; an inhibitor assay should be performed during the interval between the patient's last dose of FVIII and the date of the planned procedure
- Previous treatment of the patient has identified the dose of FVIII that will increase levels to $1\,IU/mL$ (100% of normal)
- Intramuscular injections will be avoided
- The use of drugs that impair platelet function, such as aspirin, is curtailed

For surgery or other invasive procedures, concentrates can be given by a continuous infusion, which has the advantage of providing a constant level of FVIII using less product than repeated bolus injections [3,4]. About 30–60 min preoperatively, a bolus dose calculated to raise the FVIII level to ~100% is given, followed by a continuous infusion of the factor to maintain hemostatic levels until wound healing has progressed to the point that bleeding is unlikely [5]. Typically, infusion pumps are used to deliver up to $3\,IU/h/kg$ from bags containing concentrations of FVIII ranging from 100 to $400\,IU/mL$ [6]. The infusion rate depends on the clearance of FVIII, which is affected by a variety of patient characteristics as well as the particular concentrate being infused. To ensure that dosing is appropriate, FVIII levels are assayed 4–6 h after starting the infusion and then daily, and the rate adjusted based on the plasma concentration. To avoid the potential complication of postoperative venous thrombosis, the levels of FVIII should not exceed $1.5\,IU/mL$ (150%) during concentrate infusion [7]. Tranexamic acid is often coadministered to lower bleeding risks in patients having joint replacement or other major procedures, and does not appear to increase the frequency of thrombotic events [5,8].

Although concentrate administration is central to the surgical management of most hemophiliacs, desmopressin can occasionally achieve hemostatic FVIII levels in patients with mild hemophilia. If a patient's baseline levels are $0.10–0.15\,IU/mL$, an infusion of desmopressin is expected to increase FVIII concentrations to $0.30–0.50\,IU/mL$, levels that are considered adequate for minor interventions [9]. However, the responsiveness to desmopressin must be confirmed by giving a challenge dose prior to performing invasive procedures.

PROPHYLAXIS

Regular infusions of clotting factor to maintain normal hemostasis became feasible with the availability of FVIII concentrates in the late 1950s, and the first prophylactic regimens were described by Nilsson in Sweden and Biggs and Rizza in England [10,11]. To prevent bleeding in a target joint, the UK authors gave daily infusions of FVIII accompanied by progressively more vigorous physiotherapy for 1–4 weeks, followed by infusions three times per week for several months. Only modest doses were required because FVIII levels of only 0.01–0.04 IU/mL (1%–4%) appeared to protect patients with moderately severe hemophilia from developing severe hemorrhages [12]. In 1970, Kasper et al. [13] reported that daily infusions of 500 U of FVIII reduced the incidence and severity of bleeding, eliminated hospital admissions, and restored regular school or work attendance. By 1979, patients and families were being taught to infuse concentrates at home, greatly facilitating prophylactic treatment and enabling hemophiliacs to lead nearly normal lives [14]. The prevention of bleeding had become central to the management of hemophilia.

As compared to on-demand treatment, prophylaxis decreases the frequency of joint and other hemorrhages and can prevent joint damage. In a randomized trial, boys started on prophylaxis prior to 30 months of age were more likely to have a normal joint structure and fewer total bleeds by age 6 than those receiving episodic care [15]. In addition, for every year that prophylaxis is delayed after the first joint hemorrhage, there is an 8% increase in the extent of joint damage [16]. Hemophiliacs ranging in age from 7 to 59 have a dramatic decrease in median annualized bleeding rates as well as an improved quality of life after switching from on-demand to prophylaxis [17]. A prospective, cross-sectional analysis of 6126 patients followed for 12 years, from 1999 to 2010, showed that the use of prophylaxis increased from 31% to 59%; by 2010, 75% of severe hemophiliacs under age 20 were receiving prophylaxis [18]. Declines were noted in the mean rate per 6 months of hemarthroses, from 3.03 to 2.36; overall bleeding, from 4.91 to 4.07; and target joint bleeding, from 0.80 to 0.16. Prophylaxis predicted decreased bleeding ($P < .001$), and prophylaxis begun before age 4 and nonobesity predicted preservation of joint motion ($P < .001$).

Primary prophylaxis refers to the administration of FVIII before joint damage has occurred, **secondary prophylaxis** is initiated after there is a steady pattern of recurrent joint hemorrhages, and **tertiary prophylaxis** starts after joint disease is already established. Primary prophylaxis has prevented bleeding and the development of joint disease, secondary prophylaxis has slowed the progression of joint impairment, and tertiary prophylaxis has decreased the frequency of bleeding. All three types of prophylaxis decrease absences from school and work, enable participation in sports and other activities, and improve the quality of life [19].

The decision to implement primary prophylaxis is usually made jointly by the child's parents and the Hemophilia Treatment Center or pediatric hematologist, and is generally recommended when the diagnosis is severe hemophilia (FVIII <0.02 IU/mL or 2%), the child has had one or more hemarthroses, there is a caregiver with the skills needed for intravenous infusions, and the family has insurance and other resources needed to obtain the therapeutic materials. Although FVIII concentrates can be infused through a small butterfly needle inserted into a peripheral vein, repeated infusions in a small child often require a central venous access device (CVAD) such as a tunneled external catheter or implanted port [20,21]. Prior to the insertion of these devices, sufficient concentrate must be infused to achieve fully hemostatic plasma concentrations (>0.5 IU/mL), and following placement a continuous infusion is delivered to maintain levels of 0.3–0.7 IU/mL for 1–5 days. Accessing CVADs requires strict adherence to sterile technique, and gentle flushing of the catheter or port with saline after each use. Because these devices are foreign bodies, antibiotics must be given whenever patients undergo dental or other invasive procedures. There are fewer infections with ports than with external devices and the overall infection rate is 0.66 per 1000 catheter-days [22]. CVADs should be removed as soon as a peripheral venous access is available for the prophylactic infusion of concentrates.

Effective replacement therapy requires FVIII levels that are sufficient to prevent bleeding at all times during the dosing interval. Doses must be individualized because there are a variety of factors that affect the pharmacokinetics of FVIII. For example, the half-life of FVIII is shorter in children than adults and increases continuously with age [23]. A shorter FVIII half-life is reported in individuals with blood group O than in those with blood group A (15.3 h vs 19.7 h; $P = .003$) [24]. And VWF levels, which are lower in those with blood group O, correlate with the length of FVIII dosing intervals [25]; studies of extended half-life products show favorable pharmacokinetics in hemophiliacs with higher VWF levels [26]. Dosing intensity and frequency is also influenced by mutations in other genes that affect hemostasis, such as factor V Leiden; the patient's engagement in sports and other activities; and residual damage from previous hemorrhages [27]. In addition, nonneutralizing FVIII-specific IgG might have developed after earlier exposures to clotting factor concentrates; titers of this IgG ≥40 were recorded in 9 of 42 adults with moderate-to-severe hemophilia and were associated with a significant decrease in FVIII half-life [28]. The low recovery of transfused FVIII in some hemophiliacs without demonstrable inhibitors [29] could be due to other factors, perhaps increased inactivation or clearance of FVIII/VWF.

Because of the difficulty in predicting the dose and dosing interval necessary to achieve effective prophylaxis, an empirical approach is to give a test dose and assay FVIII levels daily with the goal of identifying the dose that maintains FVIII levels above 0.1 IU/mL for the longest interval, because studies have shown that bleeding episodes become more frequent with time spent below this level [30]. Typical regimens are infusions of concentrate in doses of 25–40 IU/kg every other day or 3 days per week; when lower doses are given, there are more hemarthroses and greater loss of function [31]. The use of longer-acting concentrates has decreased the dosing frequency in many, but not all patients. Table 7.1 lists several longer-acting concentrates currently approved or seeking approval and the manufacturers' suggested starting doses; however, it should be recognized that these doses might require modification based on individual patient characteristics. Furthermore, because the extended half-life concentrates have only recently been introduced into clinical practice, there might be delayed or late adverse effects. Chiefly of concern is the possible loss of efficacy with long-term usage, either because of changes in the way the products are metabolized or the formation of inhibitory antibodies.

LIFESTYLE MEASURES

Because of their bleeding tendency, boys with hemophilia are at risk for serious hemorrhages when participating in the rough and tumble that is typical of early childhood. Prior to 1958, severe and prolonged bleeding regularly occurred after sports and other injuries, resulting in joint deformities and muscle atrophy. Physical limitations to their overall activity levels were reported by 68.8% of severe hemophiliacs, and 6.9% missed >10 days of school or work. These figures have declined to 14.9% and 5.6%, respectively, for the cohort of hemophiliacs born during the era of 1983–92, when modern treatment became available.

Today, parents are encouraged to permit their hemophilic sons to engage in a broad range of recreational activities. With adequate prophylaxis, very few pursuits are off-limits for hemophiliacs; those that are forbidden are contact sports such as boxing and wrestling, and games that use projectiles. Recommended sports include swimming, table tennis, golf, bowling, and cycling among others. Properly managed physical exercise and participation in appropriate sports builds muscle strength and aids in preventing joint deterioration [32,33]. A randomized controlled trial in 64 hemophiliacs (59 were severe) demonstrated that a home-based exercise program using strength training devices could improve endurance, subjective physical performance, and general health perception [34]. Another study showed that adult hemophiliacs engaging in more physical activities and sports had a better quality of life than those who were more sedentary [35].

Imposing restrictions on participation in sports and other activities makes children more vulnerable to bullying, alienation, and isolation by their peers. Boys with hemophilia often become reluctant to attend school or engage in scouting or other after-school activities because they fear taunting and ridicule. This affects their socialization and promotes a self-image of weakness and inferiority. Appropriate care and counseling can alter these attitudes. For example, a teenager was quoted as saying "Hemophilia doesn't have to mean not being physically active. As long as you take the proper precautions, keep up-to-date with your treatments, and prepare for the possibilities, you can get out there and push hard in whatever you're doing. There's a lot you can do to set up the environment to your advantage" [36].

Pain is a fact of life for most people with hemophilia; in childhood, pain accompanies trauma-induced hemorrhages and the frequent venipunctures needed to deliver clotting concentrates. During adolescence, athletic activities induce painful bleeds in muscles and joints, and in adulthood, chronic arthritis becomes a source of constant discomfort. Pain management in people with bleeding disorders is challenging because many common analgesics can provoke hemorrhage by impairing platelet function or inducing gastric irritation. More than 40 years ago, the famous coagulation specialist, Armand J. Quick, admonished physicians to avoid giving aspirin to hemophiliacs because it aggravated their bleeding [37]. Today, most individuals rely on acetaminophen for mild pain [38]; nonsteroidal antiinflammatory drugs (NSAIDs), either ibuprofen [39] or selective COX-2 inhibitors such as celecoxib [40], for moderate pain; and opioids such as tramadol for more severe pain [41,42]. Each of these agents has drawbacks: acetaminophen has limited efficacy and a narrow dosing range in individuals with liver impairment, NSAIDS are known to have cardiovascular risks [43], and opioids are subject to abuse and addiction. All medications should be delivered by the oral route unless intravenous infusions are required for very severe pain; intramuscular injections are avoided because of the risk of inciting muscle hemorrhage. A man with chronic hemophilic arthropathy provides this advice: "Once you accept that you will have to deal with pain, life will become a little easier for you. Try alternative therapies and distractions before turning to narcotics" [44]. The several strategies adopted by 381 patients to allay chronic pain included the use of acetaminophen (55%), NSAIDs (49%), ice (33%), and rest (33%) [45]. Opioids were taken infrequently with the exception that hydrocodone-acetaminophen was used by 30% for acute or chronic pain. Other options for chronic pain management are transcutaneous electrical nerve stimulation, massage therapy, biofeedback, yoga, and mindfulness training; some or all of these might be available at comprehensive pain management centers [46].

Problems with adjustment to the vagaries of hemophilia can be unwittingly exacerbated by family members. Mothers are understandably concerned about the potential for trauma-induced hemorrhages and become overly protective. For example, one mother of the author's acquaintance insisted that her son wear a helmet whenever he went outdoors although he had never had a head injury. This immediately identified him to other children as being strange and potentially vulnerable. Boys with hemophilia are often home-schooled because of parental fears of hemorrhages when they are away from home. Siblings can become jealous of the extra attention paid to their hemophilic brother, and fathers might have difficulty forming close relationships with a son who is always under his mother's protective influence.

Depression is often present in adults with hemophilia, and is more likely in individuals with both acute and chronic pain [44]. Symptoms are moderate to severe in the majority, and are significantly associated with a lack of social support and unemployment [47]. Screening for depression should be incorporated into hemophilia care, and appropriate treatment and social support offered to affected individuals and family members.

In this current era of automation, higher education is recommended for all young people and is particularly important for individuals with hemophilia, whose job opportunities might be limited by physical impairment. Prior to the use of prophylaxis, school absenteeism was a major problem for hemophiliacs, and their performance in achievement tests was subpar [48]. More recently, 5.6% of severe hemophiliacs born between 1983 and 1992 reported missing >10 days per year from school or work; in addition, high-school graduation rates were lower than in the general US population [49]. Strategies to improve employment among young adults with hemophilia include raising awareness of career resources, engagement during the teenage years to facilitate career exploration, increasing access to career counselors, and encouraging the hemophilia community to host regional job fairs and provision of mentorship [50].

Individuals with hemophilia and their families also experience considerable economic distress. The bleeding episodes in infants and toddlers usually require frequent visits to emergency facilities and the attention of medical experts. The costs for these visits, hospitalizations, clotting factor concentrates and other drugs rapidly escalate and can quickly exceed insurance limits. For example, the costs of concentrates alone can range from $150,000 to $300,000 per year. In addition, missing work to care for an ill child can threaten a parent's job security. Almost all families of people with hemophilia are under constant financial duress. A recent survey reported that 29% of respondents had been denied insurance coverage, and another 15% had financial hardship [51]. Although 60% had commercial insurance, coverage often came with stipulations that patients use a pharmacy not meeting their needs, required a lengthy prior authorization process, or insisted that participants fail on a product before being allowed to select one of their own choosing. These and other issues resulted in most respondents experiencing delays in receiving products and services. Some have suggested that the high cost of medical care is due to high-cost patients (hemophiliacs), but others have argued that systemic overuse of services and overall wasteful spending by the general population account for the runaway expense [52].

Because concentrates are expensive, manufacturers offer patient assistance programs. These provide limited dollars for insurance copays, but patients must have a prescription for the manufacturer's product and commercial insurance. Insurance policies are expensive, even those with high deductibles and copays, and patients still have out-of-pocket expenses for laboratory tests, emergency room visits, inpatient hospital stays, and doctor bills. To help individuals experiencing financial difficulties, patient organizations have established programs (e.g., Hemophilia Federation of America's "Helping Hands") to provide assistance with housing, utility bills, transportation, and other needs.

Comprehensive Care Centers have been established to serve the overall needs of hemophiliacs and their families. These Centers generally include a physician and nurse skilled in hemophilia care, a social worker, a physical therapist, laboratory personnel, and an assortment of other specialists including orthopedists, dentists, and psychologists. Many also offer genetic counseling to provide test interpretation, risk assessment, and education. The center objectives are to prevent or contain bleeding, enhance physical function, and provide counseling for school, employment, and financial problems. Centers ensure that all patients receive the usual childhood vaccinations for infectious diseases as well as vaccinations against hepatitis A and B. Treatment regimens are reviewed, the response to therapeutic agents is evaluated, and assays are performed annually to detect the presence of inhibitors (see Chapter 8). In addition, some centers offer clotting factor concentrates under the 340B Drug Discount Program. This is a US government program that requires drug manufacturers to provide clotting factor concentrates to eligible health care organizations/covered entities at significantly reduced prices. Qualified Comprehensive Care Centers purchase the concentrates for patients, invoice insurers for the products provided to their subscribers, and use the monetary difference between the amounts paid to the manufacturers and the reimbursements from the insurers to support the center staff and activities.

Several other organizations facilitate the care received by hemophiliacs and their families. Home Health Care companies deliver the concentrates and materials required for intravenous infusions; for patients and families unable to self-infuse, the services of skilled nurses might also be supplied. They work in coordination with the prescribing medical provider, and are tasked with obtaining approval if product substitutions are necessary, ensuring there is an adequate inventory of therapeutic materials, and submitting regular statements listing home products and supplies issued to the patient.

There are also local and national organizations that provide support, education, and advocacy on behalf of people with bleeding disorders. These include the National Hemophilia Foundation, Hemophilia Federation of America, and Committee of Ten Thousand. Some of these groups sponsor summer camps for boys with hemophilia; these camps are a superb resource, teaching campers how to self-infuse concentrates, participate in safe sports and activities, and become independent. The boys see role models in older boys that have successfully coped with hemophilia, and they become more accepting of their limitations. As one mother said when her son returned from camp: "I sent a boy to camp, I got back a man!" [53].

The lifespan of people with hemophilia has progressively lengthened with each new advance in medical therapy. As a consequence, common age-related illnesses such as hypertension, cardiovascular disorders (CVD), renal disease, bone disorders, and cancer are being observed more frequently [54,55]. Therefore, comprehensive health evaluations should be performed at least yearly as men with hemophilia age. In addition to determinations of the annual bleeding rate, recording joint function, and testing the blood for inhibitory FVIII antibodies, these regularly scheduled healthcare visits should include advice about diet, exercise, and smoking avoidance, as well as measurements of blood pressure, body mass index, waist circumference, blood glucose and lipids. Consultations with medical specialists, nutritionists, and physiotherapists are obtained when indicated, and appropriate preventive measures and medications are prescribed.

Prophylaxis with low-dose aspirin (81 mg daily) appears to be safe in middle-aged and older men with mild hemophilia; if the development of acute ischemic events requires more aggressive antithrombotic therapy, patients should receive concomitant FVIII infusions. To avoid peaks of procoagulant activity, concentrates should be administered by continuous infusion to provide FVIII levels ranging between 30% and 60%. It is also recommended that access catheters be placed in the radial rather than the femoral position, stent selection be made in consultation with a cardiologist, and bioprosthetic valves used in place of mechanical valves to avoid requirements for dual antiplatelet therapy and prolonged anticoagulation [56]. This topic is discussed further in Chapter 13.

Many older hemophiliacs suffer the ravages of chronic arthropathy due to previous joint hemorrhages, and are at risk for developing osteoporosis and fractures. Their diets should include adequate amounts of calcium, and periodic measurements of vitamin D and bone density [54]. Older hemophiliacs often have residual damage of the liver and immune system from transfusion-transmitted viruses such as hepatitis C (HCV) and HIV. Viral loads should be assessed regularly and treatment implemented when necessary. There are currently more than a dozen FDA-approved drugs for the treatment of HCV, and several potent antiretroviral combination therapies for HIV. In addition, all hemophiliacs with chronic HCV infection should have regularly scheduled ultrasound evaluations for hepatocellular carcinoma. For cancer patients requiring chemotherapy, FVIII replacement is given prior to biopsies or other invasive procedures, as well as during episodes of drug-induced thrombocytopenia.

The challenges in providing care for the older hemophiliac are illustrated by the following case vignette. A man born in the 1960s with severe hemophilia required repeated infusions of FVIII concentrates for recurrent hemarthroses of the elbows and knees. In the 1970s, he developed hepatitis due to HCV infection, and in the 1980s became infected with HIV. The latter was controlled with antiretroviral therapy, but the liver disease progressed to fulminant hepatic failure requiring liver transplantation. Rapid progression of HCV hepatitis to cirrhosis and liver failure occurs often in patients coinfected with HCV and HIV [56]. After the transplant, his factor VIII levels rose to about 20%, but he continued to have joint and muscle hemorrhages requiring daily infusions of concentrate. The persistent bleeding might have been due to severe thrombocytopenia, which was attributed to HIV infection and improved with corticosteroids. However, because of severe mobility impairment due to chronic arthropathy and osteoporosis exacerbated by the steroid therapy, he developed compression fractures of the spine and had a long bone fracture after a fall. Hemophiliacs in general have reduced bone mineral density and increased fracture rates [57]. This patient became bed-ridden and eventually died of pneumonia.

The NHF-McMaster Guideline on Care Models for Haemophilia Management evaluated best practices for hemophilia care delivery and developed evidence-based recommendations [58]. They suggest using an integrated care model that includes a hematologist, specialized hemophilia nurse, physical therapist, social worker, and continuous availability of a specialized coagulation laboratory. They recommended conducting studies to determine the optimal structure of the integrated care delivery model for hemophilia. All individuals with hemophilia deserve comprehensive care to assist them in meeting the many and varied challenges of this disorder. HIV infection and severe liver disease account for most of the mortality in men with hemophilia, but those who receive medical care at Hemophilia Treatment Centers have better survival [59].

FUTURE APPROACHES

Novel Bypassing Agents

Several novel FVIII-bypassing agents are on the horizon. Emicizumab (Hemlibra) is a humanized bispecific antibody that can replace FVIII in the tenase complex [60]. This agent, when given subcutaneously weekly to 18 severe hemophiliacs, significantly decreased their annualized bleeding rates. The drug was well tolerated and antibodies to emicizumab did not develop. The agent is approved for prophylactic use in patients with inhibitors and is discussed in Chapter 8.

Concizumab (NovoNordisk) is another monoclonal antibody that improves hemostasis in individuals with hemophilia. It targets the tissue factor pathway inhibitor (TFPI) and is given subcutaneously. A phase 1 study in healthy volunteers and individuals with hemophilia showed a favorable safety profile and concentration-dependent procoagulant effects, supporting further study of this agent [61]. Fucoidans are sulfated polysaccharides of botanical origin that inhibit TFPI [62]. A candidate drug, AV513, was given by mouth to hemophilic dogs and improved hemostasis, but no further studies have been reported.

Another agent under study is fitusiran (Alnylam Pharmaceuticals), an RNA interference therapeutic directed against antithrombin. This drug reduced antithrombin levels and improved thrombin generation in a non-human primate model of hemophilia A with anti-FVIII inhibitors [63]. Sixteen patients without inhibitors enrolled in a phase 1/2 trial received fitusiran once monthly subcutaneously [64]. Antithrombin levels were decreased by 80%, the median annualized bleeding rate was reduced to 1, and 69% of participants had no spontaneous bleeding. The agent was generally well tolerated, but a recent report of a fatality due to cerebral venous sinus thrombosis suggests that further assessment of this drug's safety, especially in people with preexisting vascular disease, will be required.

Gene Therapy

The most exciting new treatment for hemophilia is gene therapy. The feasibility of transferring the FVIII gene to a cellular target was demonstrated in 1990 by Israel and Kaufman [65]. A review published in 1994 noted that modified retroviruses or adenoviruses could deliver FVIII to hepatocytes, muscle cells, endothelial cells, keratinocytes, and fibroblasts [66]. Successful gene therapy for canine hemophilia A was reported in 1996, although the correction was short lived because of the development of antibodies to the human FVIII [67]. A unique approach was described in 2006; human skin fibroblasts were obtained by skin biopsy, transfected by electroporation with a factor VIII plasmid, and injected into the patient's omentum [68]. Raised levels of FVIII persisted for several months. In another study, blood outgrowth endothelial cells were cultured from peripheral blood and transfected with a nonviral plasmid carrying complementary DNA for B-domain deleted factor VIII [69]. These cells were infused intravenously into immunodeficient mice and generated therapeutic levels of human FVIII. Other constructs, using different viral vectors and bioengineered FVIII variants, have been described and used to transduce hematopoietic stem cells. Culture and infusion of these cells into appropriately conditioned mice achieved therapeutic FVIII levels [70].

Adenoviruses have often been used for gene therapy because the viral genetic material can be replaced with the transgene of interest, and the viral capsid will direct the cargo to the hepatocytes where it can be integrated into the host genome. Adeno-associated virus (AAV) is favored because it is nonpathogenic and has a minimal risk of insertional mutagenesis [71]. To accommodate the large FVIII molecule within the capsid, expression cassettes have been prepared replacing the entire B-domain with a small spacer and using modified liver-specific promoters; with these constructs, FVIII expression levels above 1.0U/mL could be achieved in macaque monkeys [72]. Small human trials have been initiated using adenoviral vectors (AAV), and demonstrate that gene therapy can raise FVIII levels and decrease bleeding rates. Data from two trials have recently been reported:

Bmn 270 is a factor VIII-AAV construct given in a single intravenous dose. Fifteen patients without detectable preexisting immunity to the AAV capsid or significant liver dysfunction received up to 6×10^{13}/kg doses of the construct [73,74]. Elevations in ALT and AST occurred in up to 73% of recipients but resolved with corticosteroids. Factor VIII levels rose steadily toward the lower end of the normal range by 1 year postgene therapy, and were 0.90 IU/mL at 78 weeks. When factor VIII levels rose to >0.05 IU/mL, bleeding episodes stopped and annualized factor VIII usage was zero. No patient developed inhibitors. BMN 270 has been granted orphan drug status by the FDA.

Spk-8011 is a recombinant AAV vector that encodes a B-domain-deleted human *F8* gene. The construct was infused at a dose of 5×10^{11}/kg in two patients; both achieved factor VIII levels of 0.1–1.2 IU/mL, which have persisted for up to 40 weeks [75]. The next two patients received higher doses (1×10^{12}/kg) but required corticosteroids to control liver inflammation and only achieved levels of 0.08–0.1 IU/mL, suggesting that the smallest possible dose should be used to minimize the immune response to the capsid [71]. Nevertheless, no patient had spontaneous bleeding events following vector infusion and there has been a 98% reduction in factor use to date, with total factor VIII savings of 306,676 IU.

Gene therapy is in the early stages of development, but these preliminary data suggest a promising future. Adverse events are increases in transaminase levels preventable by corticosteroids, and random insertions, which thus far appear benign. Limitations of this approach are the development of high-titer neutralizing antibodies after the first exposure to the vector, precluding repeat dosing [71], and the presence of a strong immune response against adenoviral transgene products in between 30% and 60% of potential recipients, excluding these individuals from gene therapy using this vector.

Lentivirus-mediated platelet-derived FVIII gene therapy was described in 2007, and was accomplished by transducing bone marrow cells with a lentivirus-based gene transfer cassette encoding the platelet-specific integrin αIIb gene promoter [76]. This had the advantage of synthesizing FVIII in the protective environment of the platelet, shielded from inactivation by inhibitor antibodies. In addition, thrombin generation was accelerated more in transgenic mice expressing high levels of platelet FVIII than in mice given FVIII replacement therapy [77]. With continuing research, levels of platelet FVIII that are 30-fold higher than required for hemostasis in hemophilia have been achieved and are not associated with a thrombotic tendency [78]. However, because clots prepared with platelet FVIII differ in structure and stability from those formed with plasma FVIII, some investigators are modifying the FVIII used for the platelet-based gene therapy to provide enhanced hemostasis [79,80]. Most of the reported studies have been in mice; its adaption to other animal models is anticipated.

Gene editing is a promising approach for the cure of hemophilia. CRISPR (clustered, regularly interspaced, short palindromic repeats) uses a specific RNA to guide an endonuclease, Cas9, to create a double-strand break in the DNA. During the repair process, a gene can either be removed (knockout) or added (knock-in) [81]. For example, patient hematopoietic stem cells or induced pluripotent stem cells (iPSCs) can be exposed to CRISPR-Cas9 to remove the defective gene and then transduced with a lentivirus-FVIII vector. Cells with the corrected gene are expanded ex vivo and infused intravenously, with the potential to engraft as self-renewing populations [82]. Although the use of iPSCs is attractive, the numerous mutations that occur during expansion of these cells are a major concern; for example, mutations occurring in *TP53* might predispose to future cancers [83]. The development of tumors might not become apparent until therapeutic levels of the cell products are achieved [84].

There appears a strong possibility that gene therapy will be approved for the treatment of hemophilia in the near future, raising questions about who will be offered this procedure and whether eligible patients will be able to afford it. Treating severely affected young children has the advantage of preventing disability but might have latent effects on growth, aging, and susceptibility to neoplasms. Consequently, many of the current trials are recruiting only teenagers and adults. The cost of gene therapy is also problematic. Currently, manufacturers have set the price of long-acting concentrate higher than that of unmodified concentrate, explaining that the newer concentrates need be administered less often so there is relative price equivalence. Were the same concept applied to gene therapy, which might supplant 10 or more years of concentrate treatment, the cost could be in excess of 5 million dollars. Companies argue that this is justified because the treatment is potentially curative. In response, Aaron Kesselheim, a Professor at Harvard Medical School, is quoted as saying, "When the fire department shows up at a burning house, they don't ask, 'How much is it worth to you to put out the fire?'" [85]. Setting a value on gene therapy will be challenging and is discussed by Orkin and Reilly, two pioneers in this field [86]. They write that the expense of procedures, viral vectors and other modalities, and development costs must all be considered, but might be offset by using some of the economic benefits of the Orphan Drug Act. Other suggestions, such as linking reimbursement to the duration of the desired outcome, exploring new methods for streamlining regulatory processes, and having companies assume the burden of retreatment, will also enhance the accessibility of gene therapy.

SUMMARY

The goals of hemophilia treatment are the prevention and control of hemorrhages as well as assisting affected individuals in developing self-confidence and independence. On-demand treatment refers to the administration of FVIII concentrate to stop ongoing hemorrhage; the dose is calculated based on the location and intensity of bleeding. The control of bleeding in individuals with mild hemophilia, and in those undergoing minor procedures, can often be accomplished with desmopressin, assuming that adequate levels of FVIII are achieved. Patients requiring major surgery are managed with the continuous intravenous infusion of concentrates to maintain levels of FVIII within the normal range until wounds are healed. Prophylaxis is given to prevent bleeding and is indicated for those with severe hemophilia and a history of one or more hemarthroses, and requires that the patient or caregivers have the technical skills to perform intravenous infusions. Individuals and families also need adequate health insurance or other resources to pay for clotting factor concentrates. Successful prophylaxis depends on FVIII levels be sufficiently increased at all times during the dosing interval to prevent bleeding, a condition that might be met through the use of long-acting concentrates. The chief complications of prophylaxis are vascular access device infections and the development of alloantibodies (FVIII inhibitors, Chapter 8).

The activities of children with hemophilia are often restricted because of their bleeding tendency, with negative consequences for socialization and education. Parents are encouraged to have their children attend school regularly, exercise frequently, engage in noncontact sports, and participate in community activities. Pain associated with acute and chronic hemorrhages requires prompt treatment with FVIII replacement therapy and safe analgesics; providing adequate pain relief can avoid the development of depression and drug seeking. As hemophiliacs age, they might experience chronic joint disease and are not immune to other conditions such as diabetes, liver impairment, and cardiovascular disease. Support and resources to meet the life demands of individuals with hemophilia and their families are available from Hemophilia Treatment Centers and a number of patient organizations.

The future for hemophilia treatment appears very promising. A number of unique therapeutic materials unrelated to FVIII are in various phases of development; they include an agent that bypasses FVIII by binding to the tenase complex, an antibody that targets the tissue factor pathway inhibitor, and an RNA therapeutic that decreases antithrombin levels. Research on gene therapy is also well underway, with studies in animal models showing the feasibility of several approaches; a few trials already show promising results in patients with severe hemophilia.

Future research activities might include the following:

- Find a sheltering protein other than VWF that might bring about a clinically meaningful prolongation of FVIII half-life.
- Innovate a method of formulating FVIII concentrates that permits them to be given by mouth or as a long-lasting depot preparation.
- Develop novel economic models and resources to assist patients that require very expensive medications such as clotting factor concentrates.
- Formulate a vector for gene therapy that does not elicit harmful immunologic reactions.

References

[1] Biggs R. Plasma concentrations of factor VIII and factor IX and treatment of patients who do not have antibodies directed against these factors. In: Biggs R, editor. The treatment of haemophilia A and B and von Willebrand disease. Oxford: Blackwell Scientific Publications; 1978. p. 110 [chapter 5].

[2] Leissinger C, Carcao M, Gill JC, Journeycake J, Singleton T, Valentino L. Desmopressin (desmopressin) in the management of patients with congenital bleeding disorders. Haemophilia 2014;20:158–67.

[3] Schulman S, Martinowitz U. Continuous infusion instead of bolus injections of factor concentrate? Haemophilia 1996;2:189–91.

[4] Bidlingmaier C, Deml MM, Kurnik K. Continuous infusion of factor concentrates in children with haemophilia A in comparison with bolus injections. Haemophilia 2006;12:212–7.

[5] Batorova A, Martinowitz U. Continuous infusion of coagulation factors. Haemophilia 2002;8:170–7.

[6] Martinowitz U, Luboshitz J, Bashari D, Ravid B, Gorina E, Regan L, Stass H, Lebetsky A. Stability, efficacy, and safety of continuously infused sucrose-formulated recombinant factor VIII (rFVIII-FS) during surgery in patients with severe haemophilia. Haemophilia 2009;15:676–85.

[7] Coppola A, Franchini M, Makris M, Santagostino E, Di Minno G, Mannucci PM. Thrombotic adverse events to coagulation factor concentrates for treatment of patients with haemophilia and von Willebrand disease: a systematic review of prospective studies. Haemophilia 2012;18:e173–87.

[8] Myles PS, Smith JA, Forbes A, Silbert B, Jayarajah M, Painter T, Cooper DJ, Marasco S, McNeil J, Bussieres JS, McGuinness S, Byrne K, Chan MTV, Landoni G, Wallace S. Tranexamic acid in patients undergoing coronary-artery surgery. N Engl J Med 2017;376:136–48.

[9] Franchini M, Zaffanello M, Lippi G. The use of desmopressin in mild hemophilia A. Blood Coagul Fibrinolysis 2010;21:615–9.

[10] Nilsson IM. Experience with prophylaxis in Sweden. Semin Hematol 1993;30(Suppl. 2):16–9.

[11] Biggs R, Rizza CR. The control of haemostasis in haemophilic patients. In: Biggs R, editor. The treatment of haemophilia A and B and von Willebrand disease. Oxford: Blackwell Scientific Publications; 1978. p. 127 [chapter 6].

[12] Ahlberg A. Haemophilia in Sweden. VII. Incidence, treatment and prophylaxis of arthropathy and other musculoskeletal manifestations of haemophilia A and B. Acta Orthop Scand 1965;77(Suppl):7–80.

[13] Kasper CK, Dietrich SL, Rapaport SI. Hemophilia prophylaxis with factor VIII concentrate. Arch Intern Med 1970;125:1004–9.

[14] Lazerson J. Prophylactic infusion therapy in hemophilia. Hosp Pract 1979;14:49–55.

[15] Manco-Johnson MJ, Abshire TC, Shapiro AD, Riske B, Hacker MR, Kilcoyne R, Ingram JD, Manco-Johnson ML, Funk S, Jacobson L, Valentino LA, Hoots WK, Buchanan GR, DiMichele D, Recht M, Brown D, Leissinger C, Bleak S, Cohen A, Mathew P, Matsunaga A, Medeiros D, Nugent D, Thomas GA, Thompson AA, McRedmond K, Soucie JM, Austin H, Evatt BL. Prophylaxis versus episodic treatment to prevent joint disease in boys with severe hemophilia. N Engl J Med 2007;357:535–44.

[16] Fischer K, van der Bom JG, Mauser-Bunschoten EP, Roosendaal G, Prejs R, de Kleijn P, Grobbee DE, van den Berg M. The effects of postponing prophylactic treatment on long-term outcome in patients with severe hemophilia. Blood 2002;99:2337–41.

[17] Valentino LA, Mamonov V, Hellmann A, Quon DV, Chybicka A, Schroth P, Patrone L, Wong W-Y. A randomized comparison of two prophylaxis regimens and a paired comparison of on-demand and prophylaxis treatments in hemophilia A management. J Thromb Haemost 2012;10:359–67.

[18] Manco-Johnson MJ, Soucie JM, Gill JC, for the Joint Outcomes Committee of the Universal Data Collection, US Hemophilia Treatment Network. Prophylaxis usage, bleeding rates, and joint outcomes of hemophilia 1999 to 2010: a surveillance project. Blood 2017;129:2368–74.

[19] Pipe SW. New therapies for hemophilia. Hematology Am Soc Hematol Educ Program 2016;2016:650–6.

[20] Valentino LA, Kawji M, Grygotis M. Venous access in the management of hemophilia. Blood Rev 2011;25:11–5.

[21] Santagostino E, Mancuso ME. Venous access in haemophilic children: choice and management. Haemophilia 2010;16(Suppl. 1):20–4.

[22] Valentino LA, Ewenstein B, Navickis RJ, Wilkes MM. Central venous access devices in haemophilia. Hemophilia 2004;10:134–46.

[23] Bjorkman S, Oh MS, Spotts G, Schroth P, Fritsch S, Ewenstein BM, Casey K, Fischer K, Blanchette VS, Collins PW. Population pharmacokinetics of recombinant factor VIII: the relationships of pharmacokinetics to age and body weight. Blood 2012;119:612–8.

[24] Vlot AJ, Mauser-Bunschoten EP, Zarkova AG, Haan E, Kruitwagen CLJJ, Sixma JJ, van den Berg HM. The half-life of infused factor VIII is shorter in hemophiliac patients with blood group O than in those with blood group A. Thromb Haemost 2000;83:65–9.

[25] Shapiro AD, Ragni MV, Kulkarni R, Oldenberg J, Srivastava A, Quon DV, Pasi KJ, Hanabusa H, Pabinger I, Mahlangu J, Fogarty P, Lillicrap D, Kulke S, Potts J, Neelakantan S, Nestorov I, Li S, Dumont JA, Jiang H, Brennan A, Pierce GF. Recombinant factor VIII Fc fusion protein: extended-interval dosing maintains low bleeding rates and correlates with von Willebrand factor levels. J Thromb Haemost 2014;12:1788–800.

[26] Klamroth R, Simpson M, von Depka-Prondzinski M, Gill JC, Morfini M, Powell JS, Santagostino E, Davis J, Huth-Kuhne A, Leissinger C, Neumeister P, Bensen-Kennedy D, Feussner A, Limsakun T, Zhou M, Veldman A, St. Ledger K, Blackman N, Pabinger I. Comparative pharmacokinetics of rVIII-single chain and octocog alfa (Advate®) in patients with severe haemophilia A. Haemophilia 2016;22:730–8.

[27] Dunn A. The long and short of it: using the new factor products. Hematology Am Soc Hematol Educ Program 2015;2015:26–32.

[28] Hofbauer CJ, Kepa S, Schemper M, Quehenberger P, Reitter-Pfoetner S, Mannhalter C, Reipert BM, Pabinger I. FVIII-binding IgG modulates FVIII half-life in patients with severe and moderate hemophilia A without inhibitors. Blood 2016;128:293–6.

[29] Mondorf W, Klinge J, Luban NLC, Bray G, Saenko E, Scandella D, the Recombinate Pup Study Group. Low factor VIII recovery in haemophilia A patients without inhibitor titre is not due to the presence of anti-factor VIII antibodies undetectable by the Bethesda assay. Haemophilia 2008;7:13–9.

[30] Collins PW, Blanchette VS, Fischer K, Bjorkman S, Oh M, Fritsch S, Schroth P, Spotts G, Astermark J, Ewenstein B. Breakthrough bleeding in relation to predicted factor VIII levels in patients receiving prophylactic treatment for severe hemophilia A. J Thromb Haemost 2009;7:413–20.

[31] Fischer K, Carlsson KS, Petrini P, Holmstrom M, Ljung R, van den Berg HM, Berntorp E. Intermediate-dose versus high-dose prophylaxis for severe hemophilia: comparing outcome and costs since the 1970s. Blood 2013;122:1129–36.

[32] Lobet S, Lambert C, Hermans C. Stop only advising physical activity in adults with haemophilia…prescribe it now! The role of exercise therapy and nutrition in chronic musculoskeletal diseases. Haemophilia 2016;22:e545–75.

[33] Schafer GS, Valderramas S, Gomes AR, Budib MB, Wolff AL, Ramos AA. Physical exercise, pain and musculoskeletal function in patients with haemophilia: a systematic review. Haemophilia 2016;22:e119–29.

[34] Runkel B, von Mackensen S, Hilberg T. RCT-subjective physical performance and quality of life after a 6-month programmed sports therapy (PST) in patients with haemophilia. Haemophilia 2016;1–8.

[35] von Mackensen S, Harrington C, Tuddenham E, Littley kA, Will A, Fareh M, Hay CR, Khair K. The impact of sport on health status, psychological well-being and physical performance of adults with haemophilia. Haemophilia 2016;22:521–30.

[36] Shinkman S. HFA goes to college. Dateline Fed 2016;50:8.

[37] Quick AJ. No aspirin for hemophiliacs. N Engl J Med 1971;284:218.

[38] Ameer B, Greenblatt DJ. Acetaminophen. Ann Intern Med 1977;87:202–9.

[39] Inwood MJ, Killackey B, Startup SJ. The use and safety of ibuprofen in the hemophiliac. Blood 1983;61:709–11.

[40] Rodriguez-Merchan EC, de la Corte-Rodriguez H, Jimenez-Yuste V. Efficacy of celecoxib in the treatment of joint pain caused by advanced haemophilic arthropathy in adult patients with haemophilia A. Haemophilia 2014;20:e222–42.

[41] Holstein K, Klamroth R, Richards M, Carvalho M, Perez-Garrido R, Gingeri A. Pain management in patients with haemophilia: a European survey. Haemophilia 2012;18:743–52.

[42] Humphries TJ, Kessler CM. Managing chronic pain in adults with haemophilia: current status and call to action. Haemophilia 2015;21:41–51.

[43] Boban A, Lambert C, Hermans C. Is the cardiovascular toxicity of NSAIDS and COX-2 selective inhibitors underestimated in patients with haemophilia? Crit Rev Oncol Hematol 2016;100:25–31.

[44] Tignor D. How I cope with pain. Dateline Fed 2016;(Fall):19–21.

[45] Witkop M, Neff A, Buckner TW, Wang M, Batt K, Kessler CM, Quon D, Boggio L, Recht M, Baumann K, Gut RZ, Cooper DL, Kempton CL. Self-reported prevalence, description and management of pain in adults with haemophilia: methods, demographics and results from the Pain, Functional Impairment, and Quality of life (P-FiQ) study. Haemophilia 2017;23:556–65.

[46] Van Horne C. Pain management and opioid addiction: A Q&A with Kim Mauer, MD and Kirsten Langdon, PhD. Dateline Fed 2017;(Fall):16–8.

[47] Iannone M, Pennick L, Tom A, Cui H, Gilbert M, Weihs K, Stopeck AT. Prevalence of depression in adults with haemophilia. Haemophilia 2012;18:868–74.

[48] Woolf A, Rappaport L, Reardon P, Ciborowski J, D'Angelo E, Bessette J. School functioning and disease severity in boys with hemophilia. J Dev Behav Pediatr 1989;10:81–5.

[49] Curtis R, Baker J, Riske B, Ullman M, Niu X, Norton K, Lou M, Nichol MB. Young adults with hemophilia in the U.S.: demographics, comorbidities, and health status. Am J Hematol 2015;90(Suppl. 2):S11–16.

[50] Quon D, Reding M, Guelcher C, Peltier S, Witkop M, Cutter S, Buranahirun C, Molter D, Frey M, Forsyth A, Tran D, Curtis R, Hiura G, Levesque J, de la Riva D, Compton M, Iyer N, Holot N, Cooper DL. Unmet needs in the transition to adulthood: 18- to 30-year-old people with hemophilia. Am J Hematol 2015;90(Suppl. 2):S17–22.

[51] Verb K. Project Calls 2.0. Dateline Fed 2016;48:12–3. 50:14–5.

[52] McWilliams JM, Schwartz AL. Focusing on high-cost patients-the key to addressing high costs? N Engl J Med 2017;376:807–9.

[53] Seeler RA, Ashenhurst JB, Langehennig PL. Behavioral benefits in hemophilia as noted at a special summer camp. Clin Pediatr (Phila) 1977;16:525–9.

[54] Mannucci PM, Schutgens RE, Santagostino E, Mauser-Bunchoten EP. How I treat age-related morbidities in elderly persons with hemophilia. Blood 2009;114:525.

[55] Konkle BA. The aging patient with hemophilia. Am J Hematol 2012;87:S27–32.

[56] Angelini D, Sood SL. Managing older patients with hemophilia. Hematology Am Soc Hematol Educ Program 2015;2015:41–7.

[57] Gay ND, Lee SC, Liel MS, Sochacki P, Recht M, Taylor JA. Increased fracture rates in people with haemophilia: a 10-year single institution retrospective analysis. Br J Haematol 2015;170:584–6.

[58] Pai M, Key NS, Skinner M, Curtis R, Feinstein M, Kessler C, Lane SJ, Makris M, Riker E, Santesso N, Soucie JM, Yeung CHT, Iorio A, Schunemann HJ. NHF-McMaster guideline on care models for haemophilia management. Haemophilia 2016;22(Suppl. 3):6–16.

[59] Soucie JM, Nuss R, Evatt B, Abdelhak A, Cowan L, Hill H, Kolakoski M, Wilber N, the Hemophilia Surveillance System Project Investigators. Mortality among males with hemophilia: relations with source of medical care. Blood 2000;96:437–42.

[60] Shima M, Hanabusa H, Taki M, Matsushita T, Sato T, Fukutake K, Fukazawa N, Yoneyama K, Yoshida H, Nogami K. Factor VIII-mimetic function of humanized bispecifc antibody in hemophilia A. N Engl J Med 2016;374:2044–53.

[61] Chowdary P, Lethagen S, Friedrich U, Brand B, Hay C, Abdul Karim F, Klamroth R, Knoebl P, Laffan M, Mahlangu J, Miesbach W, Dalsgaard Nielsen J, Martin-Salces M, Angchaisuksiri P. Safety and pharmacokinetics of anti-TFPI antibody (concizumab) in

healthy volunteers and patients with hemophilia: a randomized first human dose trial. J Thromb Haemost 2015;13:743–54.

[62] Prasad S, Lillicrap D, Labelle A, Knappe S, Keller T, Burnett E, Powell S, Johnson KW. Efficacy and safety of a new-class hemostatic drug candidate, AV513, in dogs with hemophilia A. Blood 2008;111:672–9.

[63] Sehgal A, Barros S, Ivanciu L, Cooley B, Qin J, Racie T, Hettinger J, Carioto M, Jiang Y, Brodsky J, Prabhala H, Zhang X, Attarwala H, Hutabarat R, Forst D, Milstein S, Charisse K, Kuchimanchi S, Maier MA, Nechev L, Kandasamy P, Kel'in AV, Nair JK, Rajeev KG, Manoharan M, Meyers R, Sorensen B, Simon AR, Dargaud Y, Negrier C, Camire RM, Akinc A. An RNAi therapeutic targeting antithrombin to rebalance the coagulation system and promote hemostasis in hemophilia. Nat Med 2015;21:492–7.

[64] Ragni MV, Georgiev P, Mant T, Creagh MD, Lissitchkov T, Bevan TD, Austin S, Hay CR, Hegemann I, Kazmi R, Chowdary P, Rangarajan S, Soh C-H, Akinc A, Partisano AM, Sorenson B, Pasi KJ. Fitusiran, an investigational RNAi therapeutic targeting antithrombin for the treatment of hemophilia: updated results from a phase 1 and phase 1/2 extension study in patients without inhibitors. Blood 2016;128:2572 [abstract].

[65] Israel DI, Kaufman RJ. Retroviral-mediated transfer and amplification of a functional human factor VIII gene. Blood 1990;75:1074–80.

[66] Lozier JN, Brinkhous KM. Gene therapy and the hemophilias. JAMA 1994;271:47–51.

[67] Connelly S, Mount J, Mauser Am Gardner JM, Kaleko M, McClelland A, Lothrop Jr CD. Complete short-term correction of canine hemophilia A by in vivo gene therapy. Blood 1996;88:3846–53.

[68] Roth DA, Tawa Jr NE, O'Brien JM, Treco DA, Selden RF, Factor VIII Transkaryotic Therapy Study Group. Nonviral transfer of the gene encoding coagulation factor VIII in patients with severe hemophilia A. N Engl J Med 2001;344:1735–42.

[69] Lin Y, Chang L, Solovey A, Healey JF, Lollar P, Hebbel RP. Use of blood outgrowth endothelial cells for gene therapy for hemophilia A. Blood 2002;99:457–62.

[70] Ramezani A, Hawley RG. Correction of murine hemophilia A following nonmyeloablative transplantation of hematopoietic stem cells engineered to encode an enhanced human factor VIII variant using a safety-augmented retroviral vector. Blood 2009;114:526–34.

[71] George LA. Hemophilia gene therapy comes of age. Blood Adv 2017;1:2591–9.

[72] McIntosh J, Lenting PJ, Rosales C, Lee D, Rabbanian S, Raj D, Patel N, Tuddenham EG, Christophe OD, McVey JH, Waddington S, Nienhuis AW, Gray JT, Fagone P, Mingozzi F, Zhou SZ, High KA, Cancio M, Ng CY, Zhou J, Morton CL, Davidoff AM, Nathwani AC. Therapeutic levels of FVIII following a single peripheral vein administration of rAAV vector encoding a novel human factor VIII variant. Blood 2013;121:3335–44.

[73] Rangarajan S, Walsh L, Lester W, Perry D, Madan B, Laffan M, Yu H, Vettermann C, Pierce GF, Wong WY, Pasi KJ. AAV5-factor VIII gene transfer in severe hemophilia A. N Engl J Med 2017;377:2519–30.

[74] Pasi KJ, Rangarajan S, Kim B, Lester W, Perry D, Madan B, Tavakkoli F, Yang K, Pierce GF, Wong WY. Achievement of normal circulating factor VIII activity following Bmn 270 AAV5-FVIII gene transfer: interim, long-term efficacy and safety results from a phase 1/2 study in patients with severe hemophilia A. Blood 2017;130:603 [abstract].

[75] George LA, Ragni MV, Samelson-Jones BJ, Cuker A, Runoski AR, Cole G, Wright F, Chen Y, Hui DJ, Wachtel K, Takefman D, Couto LB, Reape KZ, Carr ME, Anguela XM, High KA. Spk-8011: preliminary results from a phase 1/2 dose escalation trial of an investigational AAV-mediated gene therapy for hemophilia A. Blood 2017;130:604 [abstract].

[76] Shi Q, Wilcox DA, Fahs SA, Fang J, Johnson BD, Du LM, Desai D, Montgomery RR. Lentivirus-mediated platelet-derived factor VIII gene therapy in murine haemophilia A. J Thromb Haemost 2007;5:352–61.
[77] Baumgartner CK, Zhang G, Kuether EL, Weiler H, Shi Q, Montgomery RR. Comparison of platelet-derived and plasma factor VIII efficacy using a novel native whole blood thrombin generation assay. J Thromb Haemost 2015;13:2210–9.
[78] Baumgartner CK, Mattson JG, Weiler H, Shi Q, Montgomery RR. Targeting factor VIII expression to platelets for hemophilia A gene therapy does not induce an apparent thrombotic risk in mice. J Thromb Haemost 2017;15:98–109.
[79] Neyman M, Gewirtz J, Poncz M. Analysis of the spatial and temporal characteristics of platelet-delivered factor VIII-based clots. Blood 2008;112:1101–8.
[80] Greene TK, Lyde RB, Bailey SC, Lambert MP, Zhai L, Sabatino DE, Camire RM, Arruda VR, Poncz M. Apoptotic effects of platelet factor VIII on megakaryopoiesis: implications for a modified human FVIII for platelet-based gene therapy. J Thromb Haemost 2014;12(12):2102.
[81] Doudna JA, Charpentier E. Genome editing. The new frontier of genome engineering with CRISPR-Cas9. Science 2014;346.
[82] Monahan PE. Emerging genetic and pharmacologic therapies for controlling hemostasis: beyond recombinant clotting factors. Hematology Am Soc Hematol Educ Program 2015;2015:33–9.
[83] Trounson A. Potential pitfall of pluripotent stem cells. N Engl J Med 2017;377:490–1.
[84] Papapetrou EP. Induced pluripotent stem cells, past and future. Science 2016;353:991–3.
[85] Kolata G. New gene-therapy treatments will carry whopping price tags. In: The New York Times; Sept 11. 2017.
[86] Orkin SH, Reilly P. Paying for future success in gene therapy. Science 2016;352:1059–61.

Recommended Reading

[1] Leissinger C, Carcao M, Gill JC, Journeycake J, Singleton T, Valentino L. Desmopressin (desmopressin) in the management of patients with congenital bleeding disorders. Haemophilia 2014;20:158–67.
[2] Manco-Johnson MJ, Soucie JM, Gill JC, for the Joint Outcomes Committee of the Universal Data Collection, US Hemophilia Treatment Network. Prophylaxis usage, bleeding rates, and joint outcomes of hemophilia 1999 to 2010: a surveillance project. Blood 2017;129:2368–74.
[3] Pipe SW. New therapies for hemophilia. Hematology Am Soc Hematol Educ Program 2016;2016:650–6.
[4] Dunn A. The long and short of it: using the new factor products. Hematology Am Soc Hematol Educ Program 2015;2015:26–32.
[5] Angelini D, Sood SL. Managing older patients with hemophilia. Hematology Am Soc Hematol Educ Program 2015;2015:41–7.
[6] Pai M, Key NS, Skinner M, Curtis R, Feinstein M, Kessler C, Lane SJ, Makris M, Riker E, Santesso N, Soucie JM, Yeung CHT, Iorio A, Schunemann HJ. NHF-McMaster guideline on care models for haemophilia management. Haemophilia 2016;22(Suppl 3):6–16.
[7] Doudna JA, Charpentier E. Genome editing. The new frontier of genome engineering with CRISPR-Cas9. Science 2014;346.

8

Antibodies to FVIII

FVIII is a strong immunogen. Nonneutralizing antibodies are immuno-globulin G (IgG) molecules that bind FVIII but do not affect the levels of the clotting factor. They are found in approximately 2%–3% of normal individuals and a higher percentage (7.6%) of those with severe hemophilia prior to any exposure to FVIII concentrates; their incidence increases to 36% after exposure to FVIII concentrates [1,2]. The presence of nonneutralizing antibodies enhances the likelihood that a hemophiliac will acquire an alloantibody (inhibitor) that inactivates FVIII. Inhibitors develop significantly more often in previously untreated patients (PUPs) with nonneutralizing antibodies than in those without these anti-bodies (hazard ratio, 2.74; 95% confidence interval, 1.23–6.12) [1], and they are eventually observed in as many as a third of all hemophiliacs. Inhib-itors frequently appear after only a few infusions of FVIII concentrate, and the risk of antibody development is present even in those who have been inhibitor-free for decades. This accounts for the recommendation that test-ing for inhibitors be conducted annually in all hemophiliacs receiving clot-ting factor infusions.

Antibodies that inactivate FVIII are not confined to individuals with hemophilia; they occur, albeit infrequently, in people without prior clot-ting defects or a history of exposure to blood or blood products. They are typical autoantibodies, and differ from alloantibodies in the speed of their reaction with FVIII (reaction kinetics), the epitopes they target, and most importantly, in their response to therapeutic agents [3]. Alloan-tibodies and autoantibodies will be discussed separately in this chapter.

ALLOANTIBODIES

The majority of alloantibodies have IgG1 and IgG4 heavy chains and kappa light chains [4], and most are directed against epitopes located in the A2, C1, and C2 domains of FVIII [5]. In a study of 115 patients, the fre-quency of neutralizing antibodies to each domain was 23% to A2, 78% to

C1, and 68% to C2 [6]. Their high affinity for FVIII epitopes interferes with the assembly of a complex composed of factors VIII, IX, phospholipid, and calcium, the tenase complex (see Fig. 2.1). Antibody binding to FVIII is time-, temperature-, and pH-dependent: it occurs maximally at 37°C and neutral pH. The loss of FVIII activity over time is linear when plotted on log-log paper, consistent with type 1 reaction kinetics [7]. The concentration of the antibody can be assessed by incubating the patient's plasma with FVIII and recording the residual FVIII activity after a fixed interval. The Bethesda assay uses pooled normal plasma as the source of FVIII, incubation at 37°C for 2h, and a residual FVIII activity of 50% to define 1U of inhibitor [8]. Several modifications of the assay have improved its accuracy and reproducibility; alternatives to clotting-based methods are chromogenic assays, enzyme-linked immunosorbent assays (ELISA), and fluorescence-based immunoassays (FLI) [9].

The ability of these assays to accurately measure inhibitor concentrations can be compromised by the presence of FVIII in the patient sample. This occurs in patients with low titer inhibitors receiving prophylactic therapy or undergoing immune tolerance induction (ITI), and can result in a false-negative inhibitor test. This problem can be circumvented by heating the patient's plasma, usually to 56°C for 30min, followed by centrifugation [10]. Heating might improve inhibitor detection by dissociating bound FVIII molecules, but there is a risk of overdetection or false-positive results [11]. Currently, the Center for Disease Control and Prevention (CDC) defines a positive inhibitor titer as $\geq$0.5 Nijmegen-Bethesda Units, and recommends confirmation of low titer inhibitors by reassay and verification by a second method, such as chromogenic or immunologic assay.

Risk Factors for Alloantibody Development (Table 8.1)

Certain characteristics of the patient and of the FVIII product appear to predispose to antibody development. Foremost among the patient features is the individual's genetic composition. Antibody formation shows a familial predilection; having a sibling with an inhibitor greatly increases the patient's odds of developing an antibody [12]. Hemophiliacs with nonsense mutations or large deletions in the FVIII gene have a 7–10 times higher inhibitor prevalence as compared to those with milder gene defects such as missense mutations [13]. Intron-22 inversions are common in severe hemophiliacs and are associated with a modest increase in inhibitor risk, but the risk is significantly lower than in those with large deletions (odds ratio = 3.6) or nonsense mutations (odds ratio = 1.4) [14]. This unexpected occurrence has been explained by the observation that patients with intron-22 inversions express the complete FVIII amino acid sequence intracellularly, which likely mitigates the immune response to therapeutic FVIII [15].

TABLE 8.1 Risk Factors for the Development of FVIII Antibodies

---

Alloantibodies

Patient factors

Siblings with inhibitors

Genetic factors: *Factor VIII gene:* large deletions, nonsense mutations, exon 21; *Other genes:* IL-10, TGF-β, TNF-α, CTLA-4

Black race

Danger signals

Product features

Second vs Third generation recombinants

Recombinant vs Plasma-derived FVIII

Chemically modified to prolong half-life?

Autoantibodies

Elderly without prior coagulopathy

Peripartum

Autoimmune disorders (SLE, rheumatoid arthritis, pemphigus)

Neoplasms (lymphomas, macroglobulinemia)

Drug reactions (penicillin, interferon)

Astermark et al. [16] have recorded the prevalence of the various categories of mutations in 833 patients with inhibitors. Inversions were observed in 48.4%, large deletions, 4.7%; nonsense mutations, 7.4%; small deletions/insertions (indels), 10.4%; splice-site mutations, 1.6%; and missense mutations, 12.2%. Some of the more common missense mutations are Tyr2105 > Cys, Arg2150 > His, Arg2163 > His, Trp2229 > Cys, and Pro2300 > Leu. Another common mutation, Arg593 > Cys, is often found in patients with mild hemophilia; such patients should not be exposed to exogenous FVIII treatment because they readily form alloantibodies.

Mutations in more than a dozen immune response genes have been shown to increase the risk or protect against alloantibody formation, and include the genes for interleukins, transforming growth factor β, tumor necrosis factor α, and cytotoxic T-lymphocyte antigen-4 [17]. Studies of these and other immune response genes have recorded 53 single-nucleotide polymorphisms that predict inhibitor status, attesting to the complexity of the immune response and the involved pathways [12]. Although there is evidence that hemophiliacs of African and Latino ancestry have a higher prevalence of inhibitors than those of European ancestry [18,19], the hypothesis [20] that the alloantibody formation is due mainly to a mismatch between the patient's minority-race FVIII molecules and the infused multiracial FVIII molecules is still unconfirmed.

Alloantibodies most often appear within the first 20 FVIII-exposure days, but they can develop at any time during a patient's life. They are more likely to occur during intensive FVIII replacement therapy and in association with surgery, trauma, or infection. A study of previously untreated children showed that high doses of FVIII given for bleeding or surgery doubled inhibitor incidence as compared to prophylactic therapy [21]. Even adults with multiple lifetime exposures to blood products can develop antibodies during periods of intensive FVIII replacement. One of the author's patients was an elderly man with severe hemophilia who had been exposed to plasma, cryoprecipitate, and concentrates many times in the past; yearly testing invariably showed the expected treatment-induced increases in FVIII levels. At age 66, he had an uneventful transurethral resection of the prostate. Bleeding was well controlled with therapeutic doses of concentrate infused during surgery and continued into the immediate postoperative period. However, hematuria developed on day 5 and worsened despite continuous FVIII therapy, and an inhibitor was detected. Alternative hemostatic agents were given with control of bleeding, and without further exposure to FVIII concentrates over the next few months, the titer of the inhibitor gradually declined.

The development of FVIII antibodies during stressful periods has been attributed to "danger signals." According to the danger hypothesis, antigen-presenting cells (APCs) become activated and initiate inflammation when there is tissue destruction [22]. For example, a joint hemorrhage in a hemophiliac might release uric acid, heat shock proteins, cytokines, and DNA from the injured tissues; these products of inflammation exhibit damage-associated molecular patterns (DAMPs) [23]. DAMPs induce the expression of costimulatory molecules in APCs while they are ingesting red cells, cellular debris, and transfused FVIII. The APCs, now bearing major histocompatibility complex (MHC)-bound FVIII and expressing costimulatory molecules, migrate to regional lymph nodes and encounter T-cells. They activate the T-cells, which then proliferate, release proinflammatory cytokines, and stimulate the production of alloantibodies by B-cells that have been primed by exposure to FVIII. Thus, an immune system that is activated by cellular damage and exposed to a foreign antigen (FVIII) generates antibodies to FVIII. This accounts for the recommendation that patients with systemic inflammatory reactions in the absence of major bleeding should not be exposed to FVIII products [13].

Qualitative changes in FVIII during manufacture can result in alloantibody formation. In 1993, the treatment of hemophiliacs with a concentrate called Factor VIII CPS-P was associated with a 4.5-fold increase in inhibitor development, leading to the withdrawal of this product from clinical use [24]. All current FVIII concentrates are either plasma-derived or prepared from recombinant DNA; the latter are produced in baby hamster kidney cells (second generation) or Chinese hamster ovary cells (third generation). An observational study found that second-generation recombinants (Kogenate-FS, Helixate-FS) were 60% more likely to give rise to inhibitors than third-generation products (Advate) [25]. Further study using a mouse model confirmed that second-generation products were significantly more immunogenic, perhaps because they had a lower proportion of high-mannose glycans and more sialic acid and fusosylated glycans [26].

Suspicion that recombinant concentrates are more immunogenic than plasma-derived concentrates is supported by a randomized, controlled trial (SIPPET) that has compared plasma-derived with recombinant concentrates. The incidence of inhibitors in 251 previously untreated patients was found to be 87% higher with recombinants, even when second-generation products were excluded from the analysis [27]. Inhibitors developed in 37.3% of patients receiving recombinants compared to 23.2% for those treated with plasma-derived products, and the hazard ratio for high-titer inhibitors was 2.59 (95% CI, 1.11–6.00). Furthermore, inhibitors appeared earlier and the titers were higher with recombinant than plasma-derived concentrates [28]. These data confirm that FVIII alloantibodies are common, and that the risk of inhibitor development is greater with recombinant products than plasma-derived concentrates.

It might be anticipated that modifications of the methods for producing recombinant FVIII might affect its immunogenicity. Deletion of the B-domain is used to increase the expression of FVIII by cultured cells, but this step does not appear to increase alloantibody formation; furthermore, a study in neonatal mice showed no immunologic differences between wild-type and B-domain deleted FVIII [29]. In addition, modifications such as PEGylation and Fc-fusion have not been shown to increase the incidence of inhibitors in clinical trials of previously treated patients (Table 8.2). Studies of these long-acting concentrates in previously untreated patients (PUPs) have not yet reported, but the incidence of inhibitors after exposure to Nuwiq, a FVIII expressed by human cell lines, was 20.8% [31], quite similar to the 23.9% reported with an older recombinant [30].

Management of Hemophiliacs Developing Alloantibodies

The appearance of an alloantibody is a major adverse event, rendering the patient unresponsive to FVIII replacement therapy and at risk of uncontrolled bleeding. Alternative methods of achieving hemostasis

TABLE 8.2 Inhibitor Incidence in Previously Treated Patients (PTP) and Previously Untreated Patients (PUP) Exposed to Modified FVIII Concentrates

FVIII concentrate	Modification	Patient population	Inhibitor incidence
Recombinate	Unmodified	PUP	17/71; 23.9% [30]
Nuwiq	Human cell line	PUP	15/66; 20.8% [31]
		PTP	0/54 [32]
Afstyla	Single chain	PTP	0/173 [33]
N8-GP	GlycoPEGylated	PTP	1/175; 0.6% [34]
Adynovate	PEGylated	PTP	0/120 [35]
BAY 94-9027	B-domain-deleted; PEGylated	PTP	0/132 [36]
Eloctate	Fc fusion protein	PTP	0/165 [37]

currently rely on bypassing agents or porcine FVIII. Bypassing agents activate coagulation by pathways independent of FVIII, and currently include Factor Eight Inhibitor Bypassing Activity (FEIBA) and recombinant FVIIa (rVIIa, NovoSeven RT). FEIBA contains activated FVII, nonactivated factors II, IX, and X, and small amounts of FVIII antigen and traces of kinins [38]. It is prepared from human plasma; the final product is nanofiltered and vapor heated to remove viruses. When administered in daily doses of 50–100 U/kg, it shortens the activated partial thromboplastin time of inhibitor-containing hemophilic plasma and restores hemostasis in most patients with inhibitors. FEIBA in doses of 85 U/kg every other day can be used to prevent bleeding [39]. Because it promotes thrombin generation, FEIBA increases the risk of thromboembolic events especially when daily doses exceed 200 U/kg or patients have risk factors for thrombosis. Other adverse effects are hypersensitivity reactions and increases in inhibitor titers (amnamesis) due to the presence of FVIII antigen in the product.

Recombinant human FVII is synthesized by cultured baby hamster kidney cells and catalytically converted into the active two-chain form, rFVIIa [40]. The chromatographic purification process removes contaminating viruses and no human proteins are used in the production or formulation of the product. When administered intravenously, rFVIIa forms a complex with tissue factor that activates FX on the surface of activated platelets at the site of injury, where the FXa might be protected from inactivation by clotting inhibitors. NovoSeven RT is approved for the treatment of bleeding episodes and the preoperative management of hemophiliacs with FVIII inhibitors, and is given in doses of 90 μg/kg.

Because of its short half-life, the doses must be repeated as often as every 2–4 h until satisfactory hemostasis is achieved. In clinical trials, hypersensitivity reactions and thromboembolism were the most common serious adverse events. A Cochrane Database Systemic Review that examined published studies comparing plasma-derived products such as FEIBA and prothrombin complex concentrate with rFVIIa found that the agents had similar hemostatic efficacy and thromboembolic risks, but the authors suggested that additional adequately powered, randomized controlled trials were needed [41].

Antihemophilic factor (Recombinant, Porcine Sequence), available as Obizur, was described in Chapter 6. It is a B-domain deleted form of porcine FVIII that generally has a low crossreactivity with alloantibodies against human FVIII. In a phase II study of hemophiliacs with nonlife, nonlimb threatening hemorrhages and inhibitor titers of ≤15.7 B.U. toward human and ≤6 B.U. against porcine FVIII, doses of 50 U/kg every 6 h controlled all 25 bleeding episodes in nine patients [42]. An advantage of this agent is that the levels of FVIII during treatment can be monitored using assays for porcine FVIII, and treatment discontinued when patients are no longer responsive. Most individuals will have an increase in their porcine FVIII inhibitor titers following a course of treatment, and a second course of therapy should not be prescribed because of the risk of severe hypersensitivity reactions and anaphylaxis with re-exposure to the product.

Alloantibody formation must be anticipated in every hemophiliac exposed to FVIII concentrates; responses to infusions must be carefully observed, and posttreatment FVIII levels measured whenever the therapeutic response is less than expected. If an inhibitor is detected, measures to induce tolerance to FVIII should be implemented if at all possible. Candidates for immune tolerance induction (ITI) are individuals with severe hemophilia A and inhibitor titers >5 B.U., confirmed on more than one repeat measurement [43]. Other candidates for ITI are hemophiliacs with inhibitor titers of <5 B.U. persisting for >6 months and poorly controlled bleeding despite increased FVIII doses or if bypassing agents are required to achieve hemostasis. ITI is initiated immediately in patients with inhibitor titers of 5–10 B.U., but is delayed in those with higher titers until the titer is ≤10 B.U. Patients with higher titer inhibitors might be selected for ITI if they have serious or life-threatening bleeding or are being considered for bypassing agent prophylaxis because of frequent hemorrhages.

For tolerance induction, FVIII concentrate is given to children <8 years of age in daily doses of 200 I.U./kg, and in most older children and adults in doses of 50–100 I.U./kg three times per week. Either plasma-derived or recombinant products have been used, but recently Pipe [44] summarized experimental evidence that the long-acting product, rFVIIIFc, might be particularly effective for tolerance induction, even in those individuals resistant to other concentrates. Indeed, a preliminary study reported that tolerance was achieved in three inhibitor patients in only 4–12 weeks; one of them had failed another concentrate [45]. A clinical trial to evaluate rFVIIIFc for tolerance induction is in progress.

In hemophiliacs undergoing ITI, inhibitor titers are measured monthly; tolerance has been achieved when inhibitor titers are persistently negative (<0.6 B.U.), FVIII recovery is ≥66%, and FVIII half-life is ≥6 h. The success of tolerance induction is predicted by younger age, inhibitor titer <10 B.U., historical peak titer <200 B.U., peak of <100 B.U. after the start of ITI, and low-risk FVIII genotype [46]. Tolerization to FVIII requires rigorous adherence to dosing schedules and often takes many months; once patients have attained tolerance, they start prophylaxis with FVIII concentrates. Hemorrhages occurring before tolerance has been achieved are usually managed with bypassing agents such as FEIBA and rFVIIa. Past efforts to replace rFVIIa with a longer-acting analog were unsuccessful because the new agent was found to be immunogenic [47]. However, one or two doses of a mixture of plasma-derived FVIIa and FX (MC710) have been shown to be safe and effective in a phase III clinical trial of patients with inhibitors [48]. However, studies in more individuals will be needed to assess its thrombotic risks.

ITI is expensive, approaching $1 million for small children and considerable more for larger children and adults. A Markov decision analysis model that examined expected clinical outcomes and expenses over the lifetime of a 5-year-old with a high titer inhibitor showed an increase in life expectancy of 4.6 years and a reduction in costs by approximately $1.7 million [49]. The authors comment that expensive treatments can be cost effective when analyzed from a long-term societal perspective. It is anticipated that further improvements in ITI will be associated with less onerous, more cost-efficient regimens.

An alternative might be enhancing the ability of T-regulatory cells (Tregs) to suppress T- and B-cell responses to FVIII. Investigators report the generation of a FVIII-specific chimeric antigen receptor (CAR) specific for the A2 domain of FVIII. These ANS8 CAR Tregs were able to suppress the proliferation of FVIII-specific T-effector cells with specificity for a different FVIII domain in vitro [50]. In addition, the cells were able to suppress the recall antibody response of previously immunized murine splenocytes in vitro and in vivo, suggesting that this approach might be capable of eliciting tolerance in patients with inhibitors.

A unique hemostatic agent developed for the treatment of patients with hemophilia, including those with inhibitors, is emicizumab (ACE-910), a humanized bispecific antibody that can replace FVIII in the tenase complex [51]. This agent, when given subcutaneously weekly to 18 severe hemophiliacs, significantly decreased their annualized bleeding rates; furthermore, 8–11 patients with inhibitors had no bleeding. The drug was well tolerated and antibodies to emicizumab did not develop.

Emicizumab was approved by the FDA in November, 2017; it is given subcutaneously in doses of 3 mg/kg weekly for 4 weeks, followed by 1.5 mg/kg weekly thereafter. It has the potential to replace the bypassing agents currently used to control bleeding in inhibitor patients.

Another potential bypassing agent is a serine protease inhibitor (serpin) of activated protein C (APC). Under normal circumstances, APC partially lyses activated factors V and VIII, preventing excessive thrombin generation. A common mutation in FV, factor V Leiden, increases the resistance of FV to proteolysis by APC, and if present in people with hemophilia, appears to reduce the severity of their bleeding. Investigators have mutated endogenous serpins, PCI and α_1-antitrypsin, to enhance their ability to target and inhibit APC [52]. Studies in mice with hemophilia B showed that the engineered inhibitors promoted thrombin generation and controlled bleeding after a tail clip injury.

AUTOANTIBODIES

Autoantibodies directed against native FVIII develop in only 1–2 per million individuals per year, but have a serious impact on morbidity, mortality, and cost of therapy. These pathogenic antibodies appear to result from the breakdown in tolerance that normally protects FVIII from inactivation. Affected individuals, who usually have no previous history of abnormal bleeding, spontaneously develop extensive ecchymoses and painful hematomas, as well as limb- or life-threatening hemorrhages after trivial or minor trauma. Most patients with acquired hemophilia are elderly, but occasionally autoantibodies develop in young women in association with pregnancy [53], and in patients with rheumatoid arthritis, psoriasis, pemphigus, and systemic lupus erythematosus (SLE). They also occur after exposure to drugs such as penicillin and interferon, and in patients with lymphoma, macroglobulinemia, and myeloma (Table 8.1).

Autoantibodies differ from alloantibodies in a number of major respects [2]. Autoantibodies preferentially bind to epitopes in the C2 domain of FVIII; a study of 63 patients showed antibodies to C2 in 81%, A2 in 52%, and C1 in 57% [6]. In addition to the difference in target epitopes, alloantibodies inactivate FVIII more rapidly than alloantibodies. Although low-titer alloantibodies can be saturated with FVIII in vitro and high doses of concentrate can achieve hemostatic levels in some patients in vivo, autoantibodies are difficult to saturate and FVIII concentrate therapy is almost always ineffective. After exposure to FVIII, anamnesis occurs regularly in patients with alloantibodies but rarely in those with autoantibodies. Finally, immunosuppressive drugs eliminate FVIII inhibitors much more often in patients with autoantibodies than in those with alloantibodies.

The diagnosis of acquired hemophilia is suspected when excessive or unusual bleeding occurs in association with a prolonged activated partial thromboplastin time (aPTT) and normal prothrombin time. In addition, the aPTT of normal plasma is prolonged when mixed with the patient's plasma. The Bethesda-Nijmegen Assay is used to confirm the diagnosis, but because the inactivation of FVIII by autoantibodies is nonlinear, the method usually underestimates the concentration of the inhibitor and frequently fails to predict the response to treatment [54]. In addition, autoantibodies are often accompanied by considerable residual FVIII. As noted previously, residual FVIII in the plasma samples can result in false-negative inhibitor assays. In one study, negative assays in six patients suspected of having FVIII autoantibodies revealed low-titer inhibitors when the plasma samples were heated at 58°C for 90 min [55]. These samples, when unheated, tested positive for FVIII autoantibodies using an enzyme-linked immunoadsorbent (ELISA) assay, showing that this method is less affected by residual FVIII activity.

The etiology of the autoantibody is generally obscure, and investigations searching for underlying neoplasms or other conditions rarely seem warranted. Occasionally, nonspecific lupus anticoagulants prolong the aPTT and 1-stage factor assays display decreased levels of FVIII and other clotting factors, but if two-stage assays are available, they will reveal that the levels of FVIII are actually increased. Furthermore, patients with SLE usually have a variety of serologic markers of the disease and are more likely to have thrombosis than bleeding.

The control of hemorrhage in patients with acquired hemophilia frequently constitutes a medical emergency. Whether bleeding is emanating from a dental extraction, an intramuscular injection, a simple fall, or a major surgical procedure, the hemorrhage will be persistent, extensive, and often sufficiently severe to produce hypotension and shock. Transfusions of blood, plasma, and FVIII concentrates are relatively ineffective, even though clotting assays demonstrate FVIII activity in the patient's plasma. Furthermore, antibodies thought to be weak based on the results of inhibitor assays are nonetheless capable of inactivating far more FVIII in vivo than suspected. Treatment with human FVIII, even in large doses, usually fails to control hemorrhages, but recombinant porcine FVIII (r-pFVIII, Obizur, Baxter Healthcare) or bypassing agents can be life saving. R-pFVIII is given in a dose of 200 U/kg and titrated to maintain target FVIII levels until satisfactory hemostasis has been achieved. A clinical trial in 28 patients with acquired hemophilia showed positive responses 24 h after initiation of therapy, and ultimately control of bleeding in 24 of the patients [56]. Alternatively, bypassing agents such as FEIBA might be given in doses of 50 U/kg, not to exceed 200 U/kg daily; or NovoSeven RT, 90 μg/kg, repeated every 2–4 h depending on the clinical situation. Patients must be closely monitored, lest excessive dosing of these bypassing agents promote thrombotic events. Agents currently under evaluation that were mentioned previously, such as MC710 and emicizumab, might provide greater flexibility and safety for the control of bleeding in patients with autoantibodies as well as those with alloantibodies. Occasionally, individuals that require surgical exploration because of massive bleeding can be managed with very large doses of human or porcine FVIII concentrate combined with therapeutic apheresis to diminish the concentration of antibody.

In view of the limb- and life-threatening nature of acquired hemophilia, every effort should be made to eliminate the FVIII autoantibodies. Corticosteroids have been the mainstay of treatment for many years; doses of 1 mg/kg daily for up to 3 weeks suppress or eliminate inhibitors in nearly half of all patients [57]. Patients resistant to or intolerant of corticosteroids often respond to modest doses of cyclophosphamide, given orally or intravenously. Another agent frequently effective in patients with autoimmune diseases is rituximab, a monoclonal antibody directed against the B-lymphocyte CD20 surface antigen. When given in doses of $375\,mg/m^2$, but sometimes as low as $100\,mg/m^2$, rituximab was noted to eliminate FVIII autoantibodies in 77% of 160 patients [58]. However, this experience was mostly anecdotal, and a Cochrane Database Systemic Review in 2016 was unable to locate any randomized trials of rituximab in this disorder [59]. For patients with autoantibodies recurring after second trials of steroids and rituximab, alternative monoclonal antibodies or immunosuppressive drugs that have been effective in other autoimmune diseases, such as azathioprine and mycophenolate mofetil, might be considered [60]. Treatment with these agents is occasionally complicated by cytopenias and serious infections, especially in the elderly.

The use of harsh immunosuppressive agents might be avoided if more specific therapies for autoimmunity were available. It is known that regulatory T-cells (Tregs) are critical for the maintenance of peripheral tolerance. Recent work suggests that in autoimmune disorders such as type 1 diabetes, certain of these Tregs malfunction and lose their ability to control Th17 cell development [61]. Expansion of Th17 cells shifts the Th17/Treg balance and promotes inflammation and autoimmunity in disorders such as rheumatoid arthritis, SLE, Sjogren syndrome, and psoriasis, as well as type 1 diabetes [62]. Drugs that inhibit Th17 are being developed to correct the Th17/Treg imbalance and restore immune tolerance [63]. If individuals with acquired hemophilia are discovered to have decreased Tregs and increased Th17 cells, pharmacological inhibition of Th17 will offer another approach to the management of this serious disorder.

Acquired hemophilia is a very expensive disorder. Recurrent, severe hemorrhages result in frequent, prolonged hospitalizations. Surgical procedures are often required to maintain the airway in patients with subglottic hemorrhages, prevent paralysis due to nerve compression secondary to compartment syndromes, and relieve intracranial pressure associated with intracranial hematomas. Control of bleeding requires large doses of very expensive clotting factor concentrates, and costly immunosuppressive drugs are needed for the elimination of the autoantibody. Treatments costing several million dollars are not unusual; they exhaust the resources of most patients and exceed the limits of most private insurance plans. Reimbursements by Medicare and Medicaid are almost always insufficient to cover drug and concentrate costs, adversely impacting the bottom line of hospitals and physician practices. As a consequence, most hospitals strongly discourage physicians from admitting patients with this disorder and suggest that they be transferred to an Academic Medical Center. Unfortunately, by the time patients reach these centers, hemorrhages have often progressed and problems have multiplied. Even though the last decade has seen a great improvement in therapy for acquired hemophilia, the full benefits will not be realized until the cost issue is resolved.

References

[1] Cannavo A, Valsecchi C, Garagiola I, Roberta P, Mannucci PM, Rosendaal FR, Peyrandi F. Nonneutralizing antibodies against factor VIII and risk of inhibitor development in severe hemophilia A. Blood 2017;129:1245–50.
[2] Hofbauer CJ, Kepa S, Schemper M, Quehenberger P, Reitter-Pfoertner S, Mannhalter C, Reipert BM, Pabinger I. FVIII-binding IgG modulates FVIII half-life in patients with severe and moderate hemophilia A without inhibitors. Blood 2016;128:293–6.
[3] Green D. Factor VIII inhibitors: a 50-year perspective. Haemophilia 2011;17:831–8.
[4] Whelan SFJ, Hofbauer CJ, Horling FM, Allacher P, Wolfsegger MJ, Oldenburg J, Male C, Windyga J, Tiede A, Schwarz HP, Scheiflinger F, Reipert BM. Distinct characteristics of antibody responses against factor VIII in healthy individuals and in different cohorts of hemophilia A patients. Blood 2013;121:1039–48.
[5] Prescott R, Nakai H, Saenko EL, Scharrer I, Nilsson IM, Humphries JE, Hurst D, Bray G, Scandella D. The inhibitor antibody response is more complex in hemophilia A patients than in most nonhemophiliacs with factor VIII autoantibodies. Blood 1997;89:3663–71.
[6] Kahle J, Orlowski A, Stichel D, Healey JF, Parker ET, Jacquemin M, Krause M, Tiede A, Schwabe D, Lollar P, Königs C. Frequency and epitope specificity of anti-factor VIII C1 antibodies in acquired and congenital hemophilia A. Blood 2017, https://doi.org/10.1182/blood-2016-11-751347.
[7] Biggs R, Bidwell E. A method for the study of antihaemophilic globulin inhibitors with reference to six cases. Br J Haematol 1959;5:379–95.
[8] Kasper CK, Aledort LM, Counts RB, et al. A more uniform measurement of factor VIII inhibitors. Thromb Diath Haemorrh 1975;34:869–72.
[9] Favaloro EJ, Verbruggen B, Miller CH. Laboratory testing for factor inhibitors. Haemophilia 2014;20(Suppl. 4):94–8.
[10] Miller CH, Platt SJ, Rice AS, Kelly F, Soucie JM, Hemophilia Inhibitor Research Study Investigators. Validation of Nijmegen-Bethesda assay modifications to allow inhibitor measurement during replacement therapy and facilitate inhibitor surveillance. J Thromb Haemost 2012;10:1055–612.

[11] Miller CH. Improving the performance of factor VIII inhibitor tests in hemophilia A. Thromb Res 2015;136:1047–8.

[12] Astermark J, Berntorp E, White GC, Kroner BL, MIBS Study Group. The Malmo International Brother Study (MIBS): further support for genetic predisposition to inhibitor development in hemophilia patients. Haemophilia 2001;7:267–72.

[13] Oldenburg J, El-Maarri O, Schwaab R. Haemophilia 2002;8(Suppl. 2):23–9.

[14] Gouw SC, van den Berg HM, Oldenburg J, Astermark J, de Groot PG, Margaglione M, Thompson AR, van Weerde W, Boekhorst J, Miller CH, le Cessie S, van der born JG. F8 gene mutation type and inhibitor development in patients with severe hemophilia A: systematic review and meta-analysis. Blood 2012;119:2922–34.

[15] Sauna ZE, Lozier JN, Kasper CK, Yanover C, Nichols T, Howard TE. The intron-22-inverted F8 locus permits factor VIII synthesis: explanation for low inhibitor risk and a role for pharmacogenomics. Blood 2015;125:223–8.

[16] Astermark J, Donfield SM, Gomperts ED, Schwarz J, Menius ED, Pavlova A, Oldenburg J, Kessing B, DiMichele DM, Shapiro AD, Winkler CA, Berntorp E, Hemophilia Inhibitor Genetics Study (HIGS) Combined Cohort. The polygenic nature of inhibitors in hemophilia A: results from the Hemophilia Inhibitor Genetics Study (HIGS) Combined Cohort. Blood 2013;121:1446–54.

[17] Astermark J. FVIII inhibitors: pathogenesis and avoidance. Blood 2015;125:2045–51.

[18] Aledort LM, Dimichele DM. Inhibitors occur more frequently in African-American and Latino hemophiliacs. Haemophilia 1998;4:68.

[19] Carpenter SL, Soucie JM, Sterner S, Presley R, Hemophilia Treatment Center Network (HTCN) Investigators. Increased prevalence of inhibitors in Hispanic patients with severe haemophilia A enrolled in the Universal Data Collection database. Haemophilia 2012;18:e260–5.

[20] Viel KR, Ameri A, Abshire TC, Iyer RV, Watts RG, Lutcher C, Channell C, Cole SA, Fernstrom KM, Nakaya S, Kasper CK, Thompson AR, Almasy L, Howard TE. Inhibitors of factor VIII in black patients with hemophilia. N Engl J Med 2009;360:1618–27.

[21] Gouw SC, van den Berg HM, Fischer K, Auerswald G, Carcao M, Chalmers E, Chambost H, Kurnik K, Liesner R, Petrini P, Platokouki H, Altisent C, Oldenburg J, Nolan B, Garrido RP, Mancuso ME, Rafowicz A, Williams M, Clausen N, Middelburg RA, Ljung R, van der Bom JG, PedNet and Research of Determinants of INhibitor development (RODIN) Study Group. Intensity of factor VIII treatment and inhibitor development in children with severe hemophilia A: the RODIN study. Blood 2013;16(121):4046–55.

[22] Matzinger P. Tolerance, danger, and the extended family. Annu Rev Immunol 1994;12:991–1045.

[23] Lovgren KM, Sondergaard H, Skov S, Winberg B. Non-genetic risk factors in haemophilia A inhibitor management-the danger theory and the use of animal models. Haemophilia 2016;22:657–66.

[24] Rosendaal FR, Nieuwenhuis HK, van den Berg HM, Heijboer H, Mauser-Bunschoten EP, van der Meer J, Smit C, Strengers PF, Briët E. A sudden increase in factor VIII inhibitor development in multitransfused hemophilia A patients in The Netherlands. Dutch Hemophilia Study Group. Blood 1993;81:2180–6.

[25] Gouw SC, van der Bom JG, Ljung R, Escuriola C, Cid AR, Claeyssens-Donadel S, van Geet C, Kenet G, Mäkipernaa A, Molinari AC, Muntean W, Kobelt R, Rivard G, Santagostino E, Thomas A, van den Berg HM. PedNet and RODIN Study Group. Factor VIII products and inhibitor development in severe hemophilia A. N Engl J Med 2013;368:231–9.

[26] Lai J, Swystun LL, Cartier D, Zhang C, Nesbitt K, Dennis J, Hough C, Lillicrap D. Differential glycosylation between recombinant factor VIII produced in baby hamster kidney and Chinese hamster ovary cells confers differences in immunogenicity in a humanized hemophilia α mouse model. Blood 2016;128:326 (Abstract).

[27] Peyvandi F, Mannucci PM, Garagiola I, El-Beshlawy A, Elalfy M, Ramanan V, Eshghi P, Hanagavadi S, Varadarajan R, Karimi M, Manglani MV, Ross C, Young G, Seth T, Apte S, Nayak DM, Santagostino E, Mancuso ME, Sandoval Gonzalez AC, Mahlangu JN, Bonanad Boix S, Cerqueira M, Ewing NP, Male C, Owaidah T, Soto Arellano V, Kobrinsky NL, Majumdar S, Perez Garrido R, Sachdeva A, Simpson M, Thomas M, Zanon E, Antmen B, Kavakli K, Manco-Johnson MJ, Martinez M, Marzouka E, Mazzucconi MG, Neme D, Palomo Bravo A, Paredes Aguilera R, Prezotti A, Schmitt K, Wicklund BM, Zulfikar B, Rosendaal FR. A randomized trial of factor VIII and neutralizing antibodies in hemophilia A. N Engl J Med 2016;374:2054–64.

[28] Peyvandi F, Cannavo A, Garagiola I, Palla R, Mannucci PM, Rosendaal FR, for the SIPPET Study Group. Timing and severity of inhibitor development in recombinant versus plasma-derived factor VIII concentrates: a SIPPET analysis. J Thromb Haemost 2018;16:39–43.

[29] Pittman DD, Alderman EM, Tomkinson KN, Wang JH, Giles AR, Kaufman RJ. Biochemical, immunological, and in vivo functional characterization of B-domain-deleted factor VIII. Blood 1993;81:2925–35.

[30] Bray GL, Gomperts ED, Courter S, Gruppo R, Gordon EM, Manco-Johnson M, Shapiro A, Scheibel E, White III G, Lee M, the Recombinate Study Group. A multicenter study of recombinant factor VIII (Recombinate): safety, efficacy, and inhibitor risk in previously untreated patients with hemophilia A. Blood 1994;83:2428–35.

[31] Liesner R, Abashidze M, Aleinkova O, Altisent C, Bellerutti MJ, Borel-Derlon A, Carcao M, Chambosi H, Chan A, Dubey L, Ducore JM, Abubacker FN, Gattens M, Gruel Y, Kavardakova N, Khorassani M, Klukowska A, Lambert T, Lohade S, Sigaud M, Tunea V, Wu JK, Vdovin V, Pavlova A, Jansen M, Belyanskaya L, Walter O, Knaub S, Neufeld EJ. Immunogenicity, efficacy and safety of Nuwiq® (human-cl rhFVIII) in previously untreated patients with severe haemophilia A-interim results from the NuProtect Study. Haemophilia 2017;1–10.

[32] Tiede A, Oldenburg J, Lissitchkov T, Knaub S, Bichler J, Manco-Johnson MJ. Prophylaxis vs. on-demand treatment with Nuwiq® (Human-cl rhFVIII) in adults with severe haemophilia A. Haemophilia 2016;22:374–80.

[33] Mahlangu J, Kuliczkowski K, Karim FA, Stasyshyn O, Kosinova MV, Lepatan LM, Skotnicki A, Boggio LN, Klamroth R, Oldenburg J, Hellmann A, Santagostino E, Baker RI, Fischer K, Gill JC, P'Ng S, Chowdary P, Escobar MA, Khayat CD, Rusen L, Bensen-Kennedy D, Blackman N, Limsakun T, Veldman A, St. Ledger K, Pabinger I, and the AFFINITY Investigators, Efficacy and safety of rVIII-SingleChain: results of a phase 1/3 multicenter clinical trial in severe hemophilia A. Blood 2016;128:630–7.

[34] Giangrande P, Andreeva T, Chowdary P, Ehrenforth S, Hanabusa H, Leebeek FW, Lentz SR, Nemes L, Poulsen LH, Santagostino E, You CW, Clausen WH, Jonsson PG, Oldenburg J, and the Pathfinder™ 2 Investigators. Clinical evaluation of glycoPEGylated recombinant FVIII: efficacy and safety in severe haemophilia A. Thromb Haemost 2016.

[35] Konkle BA, Stasyshyn O, Chowdary P, Bevan DN, Mant T, Shima M, Engl W, Dyck-Jones J, Fuerlinger M, Patrone L, Ewenstein B, Abbuehl B. Pegylated, full-length, recombinant factor VIII for prophylactic and on-demand treatment of severe hemophilia A. Blood 2015;126:1078–85.

[36] Reding MT, Ng HJ, Poulsen LH, Eyster ME, Pabinger I, Shin HJ, Walsch R, Lederman M, Wang M, Hardtke M, Michaels LA. Safety and efficacy of BAY 94-9027, a prolonged-half-life factor VIII. J Thromb Haemost 2016.

[37] Mahlangu J, Powell JS, Ragni MV, Chowdary P, Josephson NC, Pabinger I, Hanabusa H, Gupta N, Kulkarni R, Fogarty P, Perry D, Shapiro A, Pasi KJ, Apte S, Nestorov I, Jiang H, Li S, Neelakantan S, Cristiano LM, Goyal J, Sommer JM, Dumont JA, Dodd N, Nugent K, Vigliani G, Luk A, Brennan A, Pierce GF, and the A-LONG Investigators. Phase 3 study of recombinant factor VIII Fc fusion protein in severe hemophilia A. Blood 2014;123:317–25.

[38] FEIBA anti-inhibitor coagulant complex, see FEIBA package insert.

[39] Antunes SV, Tangada S, Stasyshyn O, Mamonov V, Phillips J, Guzman-Becerra N, Grigorian A, Ewenstein B, Wong WY. Randomized comparison of prophylaxis and on-demand regimens with FEIBA NF in the treatment of haemophilia A and B with inhibitors. Haemophilia 2014;20:65–72.

[40] NovoSeven RT, Coagulation factor VIIa (recombinant). See NovoSeven RT package insert.

[41] Matino D, Makris M, Dwan K, D'Amico R, Iorio A. Recombinant factor VIIa concentrate versus plasma-derived concentrates for treating acute bleeding episodes in people with haemophilia and inhibitors. Cochrane Database Syst Rev 2015;16.

[42] Mahlangu JN, Andreeva TA, Macfarlane DE, Walsh C, Key NS. Recombinant B-domain-deleted porcine sequence factor VIII (r-pFVIII) for the treatment of bleeding in patients with congenital haemophilia A and inhibitors. Haemophilia 2016, https://doi.org/10.1111/hae.13108.

[43] Valentino LA, Kempton CL, Kruse-Jarres R, Mathew P, Meeks SL, Reiss UM, on behalf of the International Immune Tolerance Induction Study Investigators. US guidelines for immune tolerance induction in patients with haemophilia A and inhibitors. Haemophilia 2015;21:559–67.

[44] Pipe SW. New therapies for hemophilia. Hematol Am Soc Hematol Edu Program 2016;2016:650–6.

[45] Malec LM, Journeycake J, Ragni MV. Extended half-life factor VIII for immune tolerance induction in haemophilia. Haemophilia 2016;22:e545–75.

[46] Kempton CL, Meeks SL. Toward optimal therapy for inhibitors in hemophilia. Blood 2014;124:3365–72.

[47] Lentz SR, Ehrenforth S, Karim FA, Matsushita T, Weldingh KN, Windyga J, Mahlangu JN, adept™2 Investigators. Recombinant factor VIIa analog in the management of hemophilia with inhibitors: results from a multicenter, randomized, controlled trial of vatreptacog alfa. J Thromb Haemost 2014;12:1244–53.

[48] Shinkoda Y, Shirahata A, Fukutake K, Takamatsu J, Shima M, Hanabusa H, Mugishima H, Takedani H, Kawasugi K, Taki M, Matsushita T, Tawa A, Nogami K, Higasa S, Kosaka Y, Fujii T, Sakai M, Migita M, Uchiba M, Kawakami K, Sameshima K, Ohashi Y, Saito H. A phase III clinical trial of a mixture agent of plasma-derived factor VIIa and factor X (MC710) in haemophilia patients with inhibitors. Haemophilia 2017;23:59–66.

[49] Colowick AB, Bohn RL, Avorn J, Ewenstein BM. Immune tolerance induction in hemophilia patients with inhibitors: costly can be cheaper. Blood 2000;96:1698–702.

[50] Yoon J, Schmidt A, Zhang AH, Konigs C, Kim YC, Scott DW. FVIII-specific human chimeric antigen receptor T-regulatory cells suppress T- and B-cell responses to FVIII. Blood 2017;129:238–45.

[51] Shima M, Hanabusa H, Taki M, Matsushita T, Sato T, Fukutake K, Fukazawa N, Yoneyama K, Yoshida H, Nogami K. Factor VIII-mimetic function of humanized bispecifc antibody in hemophilia A. N Engl J Med 2016;374:2044–53.

[52] Polderdijk SG, Adams TE, Ivanciu L, Camire RM, Baglin TP, Huntington JA. Design and characterization of an APC-specific serpin for the treatment of hemophilia. Blood 2017;129:105–13.

[53] Gibson CJ, Berliner N, Miller AL, Loscalzo J. A bruising loss. N Engl J Med 2016;375:76–81.

[54] Tiede A, Werwitzke S, Scharf RE. Laboratory diagnosis of acquired hemophilia A: limitations, consequences, and challenges. Semin Thromb Haemost 2014;40:803–11.

[55] Batty P, Platton S, Bowles L, Pasi KJ, Hart DP. Pre-analytical heat treatment and a FVIIII ELISA improve factor VIII antibody detection in acquired haemophilia A. Br J Haematol 2014;166:946–57.

[56] Kruse-Jarres R, St-Louis J, Greist A, Shapiro A, Smith H, Chowdary P, Drebes A, Gomberts E, Bourgeois C, Mo M, Novack A, Farin H, Ewenstein B. Efficacy and safety of OBI-1, an antihaemophlic factor VIII (recombinant), porcine sequence, in subjects with acquired haemophilia A. Haemophilia 2015;21:162–70.

[57] Green D, Rademaker AW, Briet E. A prospective, randomized trial of prednisone and cyclophosphamide in the treatment of patients with factor VIII autoantibodies. Thromb Haemost 1993;70:753–7.

[58] D'Arena G, Grandone E, Di Minno MND, Musto P, Di Minno G. The anti-CD20 mono-clonal antibody rituximab to treat acquired haemophilia A. Blood Transfus 2016;14:255–61.

[59] Zeng Y, Zhou R, Duan X, Long D. Rituximab for eradicating inhibitors in people with acquired haemophilia A. Cochrane Database Syst Rev 2016;7.

[60] Gilhus NE. Myasthenia gravis. N Engl J Med 2016;375:2570–81.

[61] Shao S, He F, Yang Y, Yuan G, Zhang M, Yu X. Th17 cells in type 1 diabetes. Cell Immunol 2012;280:16–21.

[62] Singh RP, Hasan S, Sharma S, Nagra S, Yamaguchi DT, Wong DT, Hahn BH, Hossain A. Th17 cells in inflammation and autoimmunity. Autoimmun Rev 2014;13:1174–81.

[63] Fasching P, Stradner M, Graninger W, Dejaco C, Fessler J. Therapeutic potential of targeting the Th17/Treg axis in autoimmune disorders. Molecules 2017;22:134. https://doi.org/10.3390/molecules22010134.

The Von Willebrand Factor

The Von Willebrand factor (VWF) is a multifunctional protein that plays essential roles in hemostasis. There is evidence that VWF evolved in ancestral vertebrates some 500 million years ago; a *VWF* transcript was identified in a fish of ancient origin, the Atlantic hagfish or slimy eel [1]. Although the hagfish protein has a simpler structure than contemporary vertebrate VWF, it nonetheless forms high-molecular-weight multimers and supports botrocetin-induced platelet aggregation. These characteristics suggest that the hagfish protein is probably a true antecedent of modern VWF.

VWF is expressed mainly by endothelial cells (85%) and megakaryocytes (15%). In 1964, Ewald Weibel and George Palade [2] described rod-like organelles in endothelial cells, and, two decades later, Denisa Wagner and colleagues [3] reported that VWF was the major constituent of these organelles, by then designated Weibel-Palade Bodies (WPBs). VWF is synthesized in the endoplasmic reticulum of the endothelial cells and polymerizes in the acidic pH of the Golgi apparatus, forming long chains of multimers called concatamers. It adopts a tubular configuration and is packaged and stored in the WPBs. FVIII is also synthesized by endothelial cells and becomes bound to VWF within the endothelial cell WPBs [4]. Stimulated endothelial cells form secretory bodies that fuse with the cell membrane, releasing uncoiled concatamers that are cleaved by a unique endothelial protease, ADAMTS13 (A Disintegrin And Metalloproteinase with ThromboSpondin type 1 repeats-13), into multimers of various sizes. The multimers circulate in a globular form, but unfold when vascular injury induces vasoconstriction and VWF is subject to elongation forces [5]. The extended VWF molecules adhere to subendothelial connective tissue and provide binding sites for platelets. Traces of thrombin form on the platelet surface and dissociate FVIII from VWF, providing FVIII for the tenase complex and the generation of sufficient thrombin to make a hemostatic plug. The elongated VWF is cleaved by ADAMTS13 into smaller multimers and fragments that circulate and are eventually removed by receptors on endothelial cells and macrophages. The plasma concentration of VWF is $10\,\mu g/mL$, but there is a diurnal variation with the peak levels occurring at noon [6]. In this chapter, the research that contributed to our knowledge about the life cycle of VWF will be described in greater detail and the function of the protein explored.

THE VWF GENE

Investigators described the cloning and characterization of the complementary DNA of VWF in 1985 (reviewed by Ginsburg and Bowie [7]). The human VWF gene is located on the short arm of chromosome 12, locus 12p13.3, and spans 178 kb. It contains 52 exons; 17 exons encode the signal peptide and the propeptide, and 35 exons in approximately 100 kilobases encode the mature subunit of the VWF [8]. The largest exon is #28, which encodes the entire A1 and A2 repeats (Fig. 9.1). There is a second sequence, a pseudogene, on chromosome 22, that has 97% homology with the authentic gene on chromosome 12, and represents a nonprocessed duplication spanning exons 23–24. There are several single-nucleotide polymorphisms (SNPs) that associate with VWF levels; most are intronic and located in two haplotype blocks [9].

THE VWF PROTEIN

VWF is synthesized exclusively in endothelial cells and megakaryocytes. The VWF complex consists of a signal peptide of 22 aminoacids (aa), a propeptide of 741 aa, and a mature protein of 2050 aa [10]. The signal peptide directs the transport of the 220-kDa subunits to the

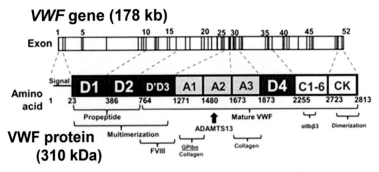

FIG. 9.1   The Von Willebrand Factor (VWF) gene and protein. *Reproduced from Lwystun LL, James PD. Genetic diagnosis in hemophilia and von Willebrand disease. Blood Rev 2017;31:47–56 (Elsevier).*

endoplasmic reticulum, where two pro-VWF monomers form a disulfide-bonded dimer at their carboxyterminal cysteine knot (CK) domains (Fig. 9.1). In endothelial cells, the VWF dimers are transported to the Golgi complex where the acidic pH and Ca^{2+} milieu trigger the formation of dimeric bouquets [11]. These are assembled into a helicoidal structure or tubule, with the D'-D3 domains forming the interior walls and the A1-CK domains protruding outward. The formation of disulfide bonds between D3 domains generates multimers of up to 50 dimers. Tubules containing many such multimers are packed together inside the WPBs. Other proteins in WPBs are P-selectin, endothelin, tissue plasminogen activator, angiopoietin-2, mediators of inflammation, and several proteins involved in the exocytosis of VWF. Also within the Golgi lumen, the VWF propeptide is cleaved from the mature molecule. After separation from the Golgi, WPBs move to the cell periphery where they become anchored to actin fibers.

The VWF propeptide was originally thought to be a second Von Will-ebrand protein and reported to be deficient in patients with severe VWD [12]. It was called vW AgII, but subsequently was found to be identical to the propeptide of VWF [13]. The plasma concentration of the propeptide is 1 µg/mL and it has a half-life of 2–3 h [14]. It is required for the intracellular processing of VWF, catalyzing the multimerization process and directing the multimers to the trans-Golgi network [15]. Upon release from the endothelial cells, the propeptide and VWF dissociate and circulate independently.

FVIII is also synthesized by endothelial cells and binds to the D'-D3 region of VWF (Fig. 9.1), intracellularly within the WPBs, and extracellularly after release from these organelles [16]. The FVIII bound to intracellular VWF in transduced blood outgrowth endothelial cells affects tubular organization and hampers the ability of VWF strings to bind platelets; on the other hand, VWF strings with bound FVIII derived from glomerular microvascular and umbilical vein endothelial cells readily recruit platelets [4].

The VWF molecule undergoes extensive glycosylation, sialylation, and sulfation; 13% of the N-linked glycans and 1% of the O-linked glycans harbor ABO (H) carbohydrate determinants [17]. In 1964, Preston [18] reported that FVIII activity was lower in people with blood group O than in those with other blood groups, and 13 years later, Gill et al. [19] determined that VWF antigen levels were also lowest in those with blood group O (mean, 0.75 IU/mL). The mean concentrations of VWF antigen in groups A, B, and AB were 1.06, 1.17, and 1.23 IU/mL, respectively. These ABO blood groups are found on the glycans of VWF originating from endothelial cells but not from megakaryocytes [20]. Endothelial cells express the A and B transferases that convert the H antigen to the corresponding blood groups, so that H antigen expression per unit of VWF is highest in individuals with blood group O [21]. On the other hand, people who completely lack the H antigen, those with the rare Bombay phenotype, have the lowest level of VWF (0.69 IU/mL) [22].

Toll-like receptors not only play a fundamental role in innate immunity, but also participate in the synthesis and expression of VWF. The intestinal expression of Toll-like receptor 2 (Tlr2) is stimulated by the gut microbiota and promotes the synthesis of VWF by hepatic endothelial cells [23]. In a mouse model, a 30% decrease in VWF levels and reduced thrombus growth was observed in animals raised germ-free or homozygous for Tlr2 deficiency; replenishing the gut flora or administering Tlr2 corrected both VWF deficiency and thrombus growth. Another study reported that Tlr2 activates WPB exocytosis in human aortic endothelial cells [24].

There are three pathways for the release of VWF from the endothelium. *Basal* secretion occurs when the WPB adheres to the endothelial cell's filamentous actin network, regulated by myosin IIa [25]. The WPB then undergoes tethering, docking, priming, and eventually fusion to the cell membrane, releasing its entire contents into the circulation [17]. WPBs can also selectively release their lower-molecular-weight contents by transiently fusing with the membrane [26]. *Regulated* secretion occurs when endothelial cells are stimulated by agonists such as thrombin, fibrin, or the calcium ionophore A23187 [27,28]. This elicits the aggregation of multiple WPBs, fusion into a large secretory vesicle, and release of a massive number of VWF multimers [29]. While most VWF resides in WPBs and is secreted, a smaller amount of low-molecular-weight VWF is not stored but available for *constitutive* release into the subendothelium. VWF is also stored in the α-granules of platelets and released during platelet activation. As compared with plasma VWF, platelet VWF has reduced N-linked sialylation and does not express AB blood group determinants [30].

Upon their release from endothelial cells, VWF multimers become tethered to the endothelial surface and unfold into very long strings [31] or concatamers [32] (chain-like linkages of multimers up to several millimeters long) that can capture circulating platelets and form platelet thrombi. The concatamers are cleaved by ADAMTS13 [33,34]; the cleavage site is located in the A2 domain of VWF, between residues 1605 and 1606 [35] (Fig. 9.1). Cleavage by ADAMTS13 is strongly dependent on shear and platelet binding, but does not require calcium [36]. ADAMTS13 is primarily synthesized by hepatic stellate cells, but also produced by endothelial cells and released directly from the Golgi apparatus [37,38]. Multimer cleavage probably accounts for the mixture of intact multimers and proteolytic fragments found in plasma; these fragments range in size from 0.85 to 12×10^3 kDa [39]. The slower release of ADAMTS13 in comparison with VWF permits accumulation of multimers with continued endothelial cell stimulation. These multimers assume a globular configuration in the circulation, which protects them from further cleavage by ADAMTS13.

When VWF mediates platelet adhesion and aggregation in flowing blood, shear-induced stretching of the multimers promotes cleavage by ADAMTS13 [40]. During inflammation, this proteolysis is limited because oxidants produced by activated neutrophils oxidize methionine residues in the A1A2A3 domains exposed by the shear-stretched conformation of VWF; these oxidized residues resist further cleavage by ADAMTS13 [41]. The susceptibility of VWF to cleavage by ADAMTS13 is also modified by N-linked glycans [42]; terminal sialic acid residues on these glycans enhance the proteolysis of VWF by ADAMTS13 [43]. While platelet VWF has reduced N-linked sialylation and is resistant to ADAMTS13, a greater degree of sialylation and increased proteolysis could account for the lower VWF levels associated with blood group O [44]. On the other hand, the VWF of individuals with the Bombay phenotype lacks ABO determinants and H-antigen; the absence of these structures probably increases VWF cleavage by other proteases such as trypsin and cathepsin B [22].

The observation that pneumoencephalography and other central nervous system stimulation increased FVIII levels suggested that a mediator released from the hypothalamus-pituitary axis might be responsible. This speculation was confirmed when vasopressin, and subsequently its analogue, desmopressin (1-deamino-8-D-arginine vasopressin or DDAVP), was found to consistently raise the levels of FVIII and VWF in healthy persons as well as individuals with hemophilia A and VWD [45,46]. Desmopressin has a direct effect on endothelial cells, binding to the vasopressin V2 receptor and activating cyclic adenosine-monophosphate-mediated signaling [47]. The agent has been extremely valuable in studying the FVIII/VWF complex and in the treatment of hemostatic disorders (Chapter 11).

CLEARANCE OF VWF

The half-life of VWF in patients with severe VWD following infusion of VWF-containing concentrate is 15.2 h [48]. Studies that have examined the half-life of endogenous VWF released by desmopressin have shown a broad range of values, depending on the individual's blood group. The elimination half-life after desmopressin treatment is significantly shorter in O than non-O persons (10 vs 25.5 h, $P < .01$), as is the higher propeptide to antigen ratio (1.6 vs 1.2, $P < .001$), a predictor of increased VWF clearance [49]. In addition, aging and blood type are interrelated; age-related increases in VWF are significantly greater in non-O individuals [50]. Whether the increased half-life of VWF in people with non-O blood group is due solely to less rapid clearance is unsettled [51]. The contributions to the variation in VWF levels are estimated as 4.8% for age, 1.6% for body mass index, and 30% for ABO blood type [9].

When VWF encounters shear stress, the molecule unfolds and is cleared by binding to the low-density lipoprotein receptor-related protein-1 (LRP1) of the macrophage [52–54]. In the absence of such stress, circulating VWF gradually loses sialic acid residues and becomes bound and internalized by macrophages in the liver and spleen [55]. The binding sites include the Ashwell-Morell receptor and MGL, both macrophage-specific galactose-type lectin receptors [56], and the C-type lectin receptor, CLEC4M, a mannose-specific receptor expressed by endothelial cells in the liver and lymph nodes [57]. Polymorphisms in these receptors might affect their binding affinity for VWF and account for some of the variation in VWF levels among individuals. These and other pathways of VWF clearance have been recently reviewed by Denis and Lenting [58].

FUNCTIONS OF VWF

The biologic activities associated with VWF are platelet adhesion, platelet aggregation, stabilization of circulating FVIII, and regulation of angiogenesis. Hemostatic functions are expressed when circulating VWF encounters an area of vasoconstriction associated with vascular injury. The change from shear to elongated blood flow converts VWF from a globular configuration to an extended thread, and exposes binding sites for subendothelial collagen and platelets. Plasma VWF concatamers can contain from 40 to 200 multimers and thread lengths from 3 to 15 μm [59]. Experiments in mice show that hemostasis is mainly supported by endothelial VWF, but platelet VWF can also contribute to the control of bleeding if endothelial VWF is absent [60].

The role of VWF in mediating the adhesion of platelets to the vessel wall was elucidated by Weiss et al. in 1978 [61]. These investigators reported that a plasma fraction containing FVIII activity (and VWF) corrected the impaired platelet adhesion observed when blood from patients with VWD was perfused over de-endothelialized segments of rabbit aorta. It was subsequently shown that VWF bound to the subendothelium at shear rates $>500\,s^{-1}$ [62,63]. High shear rates induce endothelial cells to release ultrahigh-molecular-weight multimers into the extracellular subendothelial matrix [64,65]. These concatamers are anchored to the endothelial cells by P-selectin [66], and the A3 domain (Fig. 9.1) of VWF binds to collagen types I and III [67]. The A1 domain of VWF binds to collagen type VI, the principal collagen component of the subendothelium [68]. When bound to collagen, the globular VWF assumes an extended conformation, exposing the A1 domain and enabling its interaction with the platelet Gp1bα receptor [69]. Platelets roll on the endothelium until they are immobilized by direct platelet-collagen binding mediated by platelet GPVI and integrin

$\alpha_2\beta_1$ [70]. In addition, platelet vimentin engages the A1, A2, and A3 domains of VWF during platelet adhesion at high shear stresses [71].

A second attribute of the VWF is its ability to mediate platelet aggregation. Under shear stress, VWF multimers bind to the platelet GpIbα receptor, promoting an influx of Ca^{2+} and inducing platelet aggregation [72]. VWF released from aggregating platelets, and specifically its C1–6 domain (Fig. 9.1), binds to the platelet $\alpha_{IIb}\beta_3$ receptor [73,74]. To examine the interaction between VWF and platelets in detail, in vitro studies have employed the antibiotic, ristocetin.

Ristocetin was introduced in the late 1950s for the treatment of staphylococcal and enterococcal infections. A common adverse effect, thrombocytopenia, occurred because of the rapid, dose-dependent decrease of circulating platelets that was followed by a swift recovery of the platelet count [75]. The antibiotic was withdrawn from the market, but Australian investigators obtained a quantity of the drug from the manufacturer and found that it aggregated platelets in normal and hemophilic platelet-rich plasma, but not platelets in the plasma of patients with VWD [76,77]. Subsequently, Weiss et al. [78] showed that the impaired ristocetin-induced platelet aggregation was corrected by the FVIII-VWF complex, but was not inhibited by antibodies against FVIII procoagulant activity, and concluded that ristocetin-induced platelet aggregation was a function of VWF, not FVIII. Other workers reported that VWF becomes bound to platelets in the presence of ristocetin [79,80], and it was subsequently shown that the binding site was the platelet GpI complex [81]. Ristocetin binds to the A1 domain of VWF (Fig. 9.1), specifically to sequences adjacent to a loop bounded by two cysteines, and this binding facilitates the attachment of VWF to platelet GpIb, inducing platelet aggregation [82]. The binding of ristocetin introduces a conformational change in VWF that enhances its affinity for platelet GpIb [83]. Botrocetin, a snake venom, also induces attachment of VWF to platelet GpIb VWF, but by binding directly to the A1 loop [84].

VWF also interacts with neutrophils. Neutrophils activated by inflammation release traps (neutrophil extracellular traps or NETS) consisting of DNA fibers, histones, and antimicrobial proteins [85]. Immunochemistry studies showed that VWF binds to NETS [86]; furthermore, in the presence of ristocetin, VWF binds directly to the DNA of NETS [87]. The data are consistent with a role for VWF in leukocyte adhesion and thrombus formation. In addition, VWF is bound by a *Staphylococcus aureus* protein, the von Willebrand-binding protein (vWbp). The binding of VWF enables the microbe to adhere to sites of vascular injury under flow conditions [88]. The vWbp activates prothrombin to generate thrombin and *S. aureus*-fibrin-platelet aggregates [89]. These attributes of *S. aureus* might account for its ability to bind to heart valves and induce endocarditis.

The third biologic activity associated with VWF is FVIII stabilization and delivery of the procoagulant to the platelet plug. Many early workers observed that FVIII coagulant activity coeluted in the void volume of chromatographic columns along with the high-molecular-weight multimers of VWF, but could be dissociated from VWF by adjustment of the ionic strength of the eluting solution [90]. Subsequent studies showed that the binding site for purified FVIII was near the N-terminus of VWF, corresponding to the D'D3 domain (Fig. 9.1) [91]. The light chain of FVIII binds to VWF with high affinity [92], and is essential for the stabilization of FVIII in the circulation. Patients with severe VWD [93] or critical mutations in the VWF D-domain (VWF Normandy) [94] have accelerated clearance of FVIII. In addition, VWF shields FVIII from degradation by LRP1 [95] and endocytosis by dendritic cells [96]. The stoichiometry of the FVIII/VWF complex is approximately 0.5 FVIII per VWF monomer [97]. When in a complex with VWF, FVIII is unable to bind to platelets or enter the tenase complex [98]. FVIII is released from VWF by thrombin, which removes the site (FVIII residue 1689) that binds it to VWF [99].

Lastly, VWF interacts with several factors involved in the regulation of blood vessel growth and development [100]. VWF regulates the storage, release, and synthesis of angiopoietin-2 within endothelial cells. Angiopoietin-2 synergizes with vascular endothelial growth factor-2 signaling to destabilize blood vessels and promote angiogenesis. In addition, VWF promotes vascular maturation by binding to the integrin, αvβ3, on vascular smooth muscle cells. As discussed in the next chapter, absence of VWF is associated with angiodysplasia and disrupted blood vessel formation.

SUMMARY

The gene for Von Willebrand factor (VWF) is located on the short arm of chromosome 12, at locus 12p13.3, and contains 52 exons. VWF is synthesized by endothelial cells and is stored, along with FVIII, in specific organelles called Weibel-Palade bodies (WPBs). Agonists such as fibrin and desmopressin (DDAVP) stimulate WPBs to release medium-sized and ultralarge multimers (concatamers) of VWF, as well as FVIII. Most concatamers become attached to the surface of endothelial cells; under flow conditions, they unravel, bind platelets, and are cleaved by ADAMTS13, an endothelial cell protease. The FVIII/VWF complex that enters the plasma is gradually cleared from the circulation by macrophage and endothelial cell receptors. When vascular integrity is breached, the tissue factor-FVIIa complex generates FXa and small amounts of thrombin; thrombin activates FVIII and releases it from the VWF multimers. Platelets clinging to multimeric VWF threads undergo aggregation and adhesion to subendothelial collagen, and the end result is the formation of a hemostatic plug. Another biologic activity associated with VWF is the regulation of endothelial cell proliferation and migration, as well as vascular maturation.

References

[1] Grant MA, Beeler DL, Spokes KC, Chen J, Dharaneeswaran H, Sciuto TE, Dvorak AM, Interlandi G, Lopez JA, Aird WC. Identification of extant vertebrate *Myxine glutinosa* vWF: evolutionary conservation of primary hemostasis. Blood 2017;130:2548–58.
[2] Weibel ER, Palade GE. New cytoplasmic components in arterial endothelia. J Cell Biol 1964;23:101–12.
[3] Wagner DD, Olmsted JB, Marder VJ. Immunolocalization of von Willebrand protein in Weibel-Palade bodies of human endothelial cells. J Cell Biol 1982;95:355–60.
[4] Turner NA, Moake JL. Factor VIII is synthesized in human endothelial cells, packaged in Weibel-Palade bodies and secreted bound to ULVWF strings. PLoS ONE 2015;10.
[5] De Ceunynck K, De Meyer SF, Vanhoorelbeke K. Unwinding the von Willebrand factor strings puzzle. Blood 2013;121:270–7.
[6] Timm A, Fahrenkrug J, Jorgensen HL, Sennels HP, Goetze JP. Diurnal variation of von Willebrand factor in plasma: the Bispebjerg study of diurnal variations. Eur J Haematol 2014;93:48–53.
[7] Ginsburg D, Bowie EJW. Molecular genetics of von Willebrand disease. Blood 1992;79:2507–19.
[8] Mancuso DJ, Tuley EA, Westfield LA, Worrall NK, Shelton-Inloes BB, Sorace JM, Alevy YG, Sadler JE. Structure of the gene for human von Willebrand factor. J Biol Chem 1989;264:19514–27.
[9] Campos M, Sun W, Yu F, Barbalic M, Tang W, Chambless LE, Wu KK, Ballantyne C, Folsom AR, Boerwinkle E, Dong J-F. Genetic determinants of plasma von Willebrand factor antigen levels: a target gene SNP and haplotype analysis of ARIC cohort. Blood 2011;117:5224–30.
[10] Schneppenheim R, Budde U. von Willebrand factor: the complex molecular genetics of a multidomain and multifunctional protein. J Thromb Haemost 2011;9(Suppl. 1):209–15.
[11] McCormack JJ, Lopes da Silva M, Ferraro F, Patella F, Cutler DF. Weibel-Palade bodies at a glance. J Cell Sci 2017;130:3611–7.
[12] Montgomery RR, Zimmerman TS. von Willebrand's disease antigen II. A new plasma and platelet antigen deficient in severe von Willebrand's disease. J Clin Invest 1978;61:1498–507.
[13] Fay PJ, Kawai Y, Wagner DD, Ginsburg D, Bonthron D, Ohlsson-Wilhelm BM, Chavin SI, Abraham GN, Handin RI, Orkin SH, Montgomery RR, Marder VJ. Propolypeptide of von Willebrand factor circulates in blood and is identical to von Willebrand antigen II. Science 1986;232:995–8.
[14] van Mourik JA, Boertjes R, Huisveld IA, Fijnvandraat K, Pajkrt D, van Genderen PJ, Fijnheer R. von Willebrand factor propeptide in vascular disorders: a tool to distinguish between acute and chronic endothelial cell perturbation. Blood 1999;94:179–85.
[15] Haberichter SL. Von Willebrand factor propeptide: biology and clinical utility. Blood 2015;126:1753–61.
[16] Bouwens EAM, Mourik MJ, van den Biggelaar M, Eikenboom JCJ, Voorberg J, Valentijn KM, Mertens K. Factor VIII alters tubular organization and functional properties of von Willebrand factor stored in Weibel-Palade bodies. Blood 2011;118:5947–56.

[17] Lenting PJ, Christophe OD, Denis CV. Von Willebrand factor biosynthesis, secretion, and clearance: connecting the far ends. Blood 2015;125:2019–28.

[18] Preston AE, Barr A. The plasma concentration of factor VIII in the normal population. Br J Haematol 1964;10:238–45.

[19] Gill JC, Endres-Brooks J, Bauer PJ, Marks Jr. WJ, Montgomery RR. The effect of ABO blood group on the diagnosis of von Willebrand disease. Blood 1987;69:1691–5.

[20] Matsui T, Shimoyama T, Matsumoto M, Fujimura Y, Takemoto Y, Sako M, Hamako J, Titani K. ABO blood group antigens on human plasma von Willebrand factor after ABO-mismatched bone marrow transplantation. Blood 1999;2895–900.

[21] O'Donnell J, Boulton FE, Manning RA, Laffan MA. Amount of H antigen expressed on circulating von Willebrand factor is modified by ABO blood group genotype and is a major determinant of plasma von Willebrand factor antigen levels. Arterioscler Thromb Vasc Biol 2002;22:335–41.

[22] O'Donnell J, McKinnon TAJ, Crawley JTB, Lane DA, Laffan MA. Bombay phenotype is associated with reduced plasma-VWF levels and an increased susceptibility to ADAMTS13 proteolysis. Blood 2005;106:1988–91.

[23] Jäckel S, Kioupsti K, Lillich M, Hendrikx T, Khandagale A, Kollar B, Hörmann N, Reiss C, Subramaniam S, Wilms E, Ebner K, Brühl MV, Rausch P, Baines JF, Haberichter S, Lämmle B, Binder CJ, Jurk K, Ruggeri ZM, Massberg S, Walter U, Ruf W, Reinhardt C. Gut microbiota regulate hepatic von Willebrand factor synthesis and arterial thrombus formation via Toll-like receptor-2. Blood 2017;130:542–53.

[24] Into T, Kanno Y, Dohkan J, Nakashima M, Inomata M, Shibata K, Lowenstein CJ, Matsushita K. Pathogen recognition by Toll-like receptor 2 activates Weibel-Palade body exocytosis in human aortic endothelial cells. J Biol Chem 2007;282:8134–41.

[25] Li P, Wei G, Cao Y, Deng Q, Han X, Huang X, Huo Y, He Y, Chen L, Luo J. Myosin IIa is critical for cAMP-mediated endothelial secretion of von Willebrand factor. Blood 2017; https://doi.org/10.1182/blood-2017-08-802140.

[26] Babich V, Meli A, Knipe L, Dempster JE, Skehel P, Hannah MJ, Carter T. Selective release of molecules from Weibel-Palade bodies during a lingering kiss. Blood 2008;111:5282–90.

[27] Ribes JA, Francis CW, Wagner DD. Fibrin induces release of von Willebrand factor from endothelial cells. J Clin Invest 1987;79:117–23.

[28] Sporn LA, Marder VJ, Wagner DD. Inducible secretion of large, biologically potent von Willebrand factor multimers. Cell 1986;46:185–90.

[29] Valentijn KM, van Driel LF, Mourik MJ, Hendriks G-J, Arends TJ, Koster AJ, Valentijn JA. Multigranular exocytosis of Weibel-Palade bodies in vascular endothelial cells. Blood 2010;116:1807–16.

[30] McGrath RT, van den Biggelaar M, Byrne B, O'Sullivan JM, Rawley O, O'Kennedy R, Voorberg J, Preston RJ, O'Donnell JS. Altered glycosylation of platelet-derived von Willebrand factor confers resistance to ADAMTS13 proteolysis. Blood 2013;122:4107–10.

[31] De Ceunynck K, Rocha S, Feys HB, De Meyer SF, Uji-i H, Deckmyn H, Hofkens J, Vanhoorelbeke K. Local elongation of endothelial cell-anchored von Willebrand factor strings precedes ADAMTS13 protein-mediated proteolysis. J Biol Chem 2011;286:36361–7.

[32] Springer TA. Von Willebrand factor, Jedi knight of the bloodstream. Blood 2014;124:1412–25.

[33] Fujikawa K, Suzuki H, McMullen B, Chung D. Purification of human von Willebrand factor-cleaving protease and its identification as a new member of the metalloproteinase family. Blood 2001;98:1662–6.

[34] Dong JF, Moake JL, Nolasco L, Bernardo A, Arceneaux W, Shrimpton CN, Schade AJ, McIntire LV, Fujikawa K, López JA. ADAMTS-13 rapidly cleaves newly secreted ultra-large von Willebrand actor multimers on the endothelial surface under flowing conditions. Blood 2002;100:4033–9.

[35] Dent JA, Berkowitz SD, Ware J, Kasper CK, Ruggeri ZM. Identification of a cleavage site directing the immunochemical detection of molecular abnormalities in type IIA von Willebrand factor. Proc Natl Acad Sci 1990;87:6306–10.

[36] Gogia S, Kelkar An Zhang C, Dayananda KM, Neelamegham S. Role of calcium in regulating the intra- and extracellular cleavage of von Willebrand factor by the protease ADAMTS13. Blood Adv 2017;1:2063–74.

[37] Turner N, Nolasco L, Tao Z, Dong JF, Moake J. Human endothelial cells synthesize and release ADAMTS-13. J Thromb Haemost 2006;4:1396–404.

[38] Turner NA, Nolasco L, Ruggeri ZM, Moake JL. Endothelial cell ADAMTS13 and VWF: production, release, and VWF string cleavage. Blood 2009;114:5102–11.

[39] Hoyer LW, Shainoff JR. Factor VIII-related protein circulates in normal human plasma as high molecular weight multimers. Blood 1980;55:1056–9.

[40] Muia J, Zhu J, Gupta G, Haberichter SL, Friedman KD, Feys HB, Deforche L, Vanhoorelbeke K, Westfield LA, Roth R, Tolia NH, Heuser JE, Sadler JE. Allosteric activation of ADAMTS13 by von Willebrand factor. Proc Natl Acad Sci U S A 2014;111:18584–9.

[41] Fu X, Chen J, Gallagher R, Zheng Y, Chung DW, Lopez JA. Shear stress-induced unfolding of VWF accelerates oxidation of key methionine residues in the A1A2A3 region. Blood 2011;118:5283–91.

[42] McKinnon TAJ, Chion ACK, Millington AJ, Lane DA, Laffan MA. N-linked glycosylation of VWF modulates its interaction with ADAMTS13. Blood 2008;111:3042–9.

[43] McGrath RT, McKinnon TAJ, Byrne B, O'Kennedy R, Terraube V, McRae E, Preston RJS, Laffan MA, O'Donnell JS. Expression of terminal α2-6-linked sialic acid on von Willebrand factor specifically enhances proteolysis by ADAMTS13. Blood 2010;115:2666–73.

[44] Bowen DJ. Sugar targets VWF for the chop. Blood 2010;115:2565.

[45] Mannucci PM. Desmopressin: an historical introduction. Haemophilia 2008;14(Suppl. 1):1–4.

[46] Mannucci PM, Ruggeri ZM, Pareti FI, Capitanio A. 1-Deamino-8-d-arginine vasopressin: a new pharmacological approach to the management of haemophilia and von Willebrand's diseases. Lancet 1977;i:869–72.

[47] Kaufmann JE, Oksche A, Wollheim CB, Gunther G, Rosenthal W, Vischser UM. Vasopressin-induced von Willebrand factor secretion from endothelial cells involves V2 receptors and cAMP. J Clin Invest 2000;106:107–16.

[48] Dobrkovska A, Krzensk U, Chediak JR. Pharmacokinetics, efficacy and safety of Humate-P in von Willebrand disease. Haemophilia 1998;4(Suppl. 3):33–9.

[49] Gallinaro L, Cattini MG, Sztukowska M, Padrini R, Sartorello F, Pontara E, Bertomoro A, Daidone V, Pagnan A, Casonato A. A shorter von Willebrand factor survival in O blood group subjects explains how ABO determinants influence plasma von Willebrand factor. Blood 2008;111:3540–5.

[50] Albanez S, Ogiwara K, Michels A, Hopman W, Grabell J, James P, Lillicrap D. Aging and ABO blood type influence von Willebrand factor and factor VIII levels through interrelated mechanisms. J Thromb Haemost 2016;14:953–63.

[51] Groeneveld DJ, van Bekkum T, Cheung KL, Dirven RJ, Castaman G, Reitsma PH, van Vlijmen B, Eikenboom J. No evidence for a direct effect of von Willebrand factor's ABH blood group antigens on von Willebrand factor clearance. J Thromb Haemost 2015;13:592–600.

[52] Chion ACK, O'Sullivan JM, Drakeford C, Bergsson G, Dalton N, Aquila S, Ward S, Fallon PG, Brophy TM, Preston RJ, Brady L, Shells O, Laffan M, McKinnon TA, O'Donnell JS. N-linked glycans within the A2 domain of von Willebrand factor modulate macrophage-mediated clearance. Blood 2016;128:1959–68.

[53] Castro-Nunez L, Dienava-Verdoold I, Herczenik E, Mertens K, Meijer AB. Shear stress is required for the endocytic uptake of the factor VIII-von Willebrand factor complex by macrophages. J Thromb Haemost 2012;10:1929–37.

[54] Rastegarlari G, Pegon JN, Casari C, Odouard S, Navarrete A-M, Saint-Liu N, van Vlijmen BJ, Legendre P, Christophe OD, Denis CV, Lenting PJ. Macrophage LRP1 contributes to the clearance of von Willebrand factor. Blood 2012;119:2126–34.

[55] Van Schooten CJ, Shahbazi S, Groot E, Oortwijn BD, van den Berg HM, Denis CV, Lenting PJ. Macrophages contribute to the cellular uptake of von Willebrand factor and factor VIII in vivo. Blood 2008;112:1704–12.

[56] Ward SE, O'Sullivan JM, Drakeford C, Aguila S, Jondle CN, Sharma J, Fallon PG, Brophy TM, Preston RJS, Smyth P, Sheils O, Chion A, O'Donnell JS. A novel role for the macrophage galactose-type lectin receptor in mediating von Willebrand factor clearance. Blood 2018;131:911–6.

[57] Rydz N, Swystun LL, Notley C, Paterson AD, Riches JJ, Sponagle K, Boonyawat B, Montgomery RR, James PD, Lillicrap D. The C-type lectin receptor CLEC4M binds, internalizes, and clears von Willebrand factor and contributes to the variation in plasma von Willebrand factor levels. Blood 2013;121:5228–37.

[58] Denis CV, Lenting PJ. VWF clearance: it's glycomplicated. Blood 2018;131:842–3.

[59] Springer TA. Biology and physics of von Willebrand factor concatamers. J Thromb Haemost 2011;9(Suppl. 1):130–43.

[60] Kanaji S, Fahs SA, Shi Q, Haberichter SL, Montgomery RR. Contribution of platelet vs. endothelial VWF to platelet adhesion and hemostasis. J Thromb Haemost 2012;10:1646–52.

[61] Weiss HJ, Baumgartner HR, Tschopp TB, Turitto VT, Cohen D. Correction by factor VIII of the impaired platelet adhesion to subendothelium in von Willebrand disease. Blood 1978;51:267–79.

[62] Sakariassen KS, Bolhuis PA, Sixma JJ. Human blood platelet adhesion to artery subendothelium is mediated by factor VIII-Von Willebrand factor bound to the subendothelium. Nature 1979;279:636–8.

[63] Stel HV, Sakariassen KS, de Groot PG, van Mourik JA, Sixma JJ. Von Willebrand factor in the vessel wall mediates platelet adherence. Blood 1985;65:85–90.

[64] Galbusera M, Zoja C, Donadelli R, Paris S, Morigi M, Benigni A, Figliuzzi M, Remuzzi G, Remuzzi A. Fluid shear stress modulates von Willebrand factor release from human vascular endothelium. Blood 1997;90:1558–64.

[65] Tannenbaum SH, Rick ME, Shafer B, Gralnick HR. Subendothelial matrix of cultured endothelial cells contains fully processed high molecular weight von Willebrand factor. J Lab Clin Med 1989;113:372–8.

[66] Padilla A, Moake JL, Bernardo A, Ball C, Wang Y, Arya M, Nolasco L, Turner N, Berndt MC, Anvari B, Lopez JA, Dong J-F. P-selectin anchors newly released ultralarge von Willebrand factor multimers to the endothelial cell surface. Blood 2004;103:2150–6.

[67] Cruz MA, Yuan H, Lee JR, Wise RJ, Handin RI. Interaction of the von Willebrand factor (vWF) with collagen. J Biol Chem 1995;270:10822–7.

[68] Rand JH, Wu XX, Potter BJ, Uson RR, Gordon RE. Co-localization of von Willebrand factor and type VI collagen in human vascular subendothelium. Am J Pathol 1993;142:843–50.

[69] Nowak AA, Canis K, Riddell A, Laffan MA, McKinnon TAJ. O-linked glycosylation of von Willebrand factor modulates the interaction with platelet receptor glycoprotein Ib under static and shear stress conditions. Blood 2012;120:214–22.

[70] Shida Y, Rydz N, Stegner D, Brown C, Mewburn J, Sponagle K, Danisment O, Crawford B, Vidal B, Hegadorn CA, Pruss CM, Nieswandt B, Lillicrap D. Analysis of the role of von Willebrand factor, platelet glycoprotein VI-, and α2β1-mediated collagen binding in thrombus formation. Blood 2014;124:1799–807.

[71] Da Q, Behymer M, Correa JI, Vijayan KV, Cruz MA. Platelet adhesion involves a novel interaction between vimentin and von Willebrand factor under high shear stress. Blood 2014;123:2715–21.

[72] Chow TW, Hellums JD, Moake JL, Kroll MH. Shear stress-induced von Willebrand factor binding to platelet glycoprotein Ib initiates calcium influx associated with aggregation. Blood 1992;80:113–20.

[73] Parker RI, Gralnick HR. Identification of platelet glycoprotein IIb/IIIa as the major binding site for released platelet-von Willebrand factor. Blood 1986;68:732–6.

[74] Zhou YF, Eng ET, Zhu J, Lu C, Walz T, Springer TA. Sequence and structure relationships within von Willebrand factor. Blood 2012;120:449–58.

[75] Gangarosa EJ, Johnson TR, Ramos HS. Ristocetin-induced thrombocytopenia: Site and mechanism of action. AMA Arch Intern Med 1960;105:107–13.

[76] Howard MA, Firkin BG. Ristocetin-a new tool in the investigation of platelet aggregation. Thromb Diath Haemorrh 1971;26:362–71.

[77] Howard MA, Sawers RJ, Firkin BG. Ristocetin: a means of differentiating von Willebrand's disease into two groups. Blood 1973;41:687–90.

[78] Weiss HJ, Rogers J, Brand H. Defective ristocetin-induced platelet aggregation in von Willebrand's disease and its correction by factor VIII. J Clin Invest 1973; 52:2697–707.

[79] Green D, Potter EV. Platelet-bound ristocetin aggregation factor in normal subjects and patients with von Willebrand's disease. J Lab Clin Med 1976;87:976–86.

[80] Zucker MB, Kim S-J, McPherson J, Grant RA. Binding of factor VIII to platelets in the presence of ristocetin. Br J Haematol 1977;35:535–49.

[81] Nachman RL, Jaffe EA, Weksler BB. Immunoinhibition of ristocetin-induced platelet aggregation. J Clin Invest 1977;59:143–8.

[82] Flood VH, Gill JC, Morateck PA, Christopherson PA, Friedman KD, Haberichter SL, Branchford BR, Hoffman RG, Abshire TC, JA DP, Hoots WK, Leisssinger C, Lusher JM, Ragni MV, Shapiro AD, Montgomery RR. Common VWF exon 28 polymorphisms in African Americans affecting the VWF activity assay by ristocetin cofactor. Blood 2010;116:280–6.

[83] Kim J, Zhang CZ, Zhang X, Springer TA. A mechanically stabilized receptor-ligand flexbond important in the vasculature. Nature 2010;466(7309):992–5.

[84] Sugimoto M, Mohri H, McClintock RA, Ruggeri ZM. Identification of discontinuous von Willebrand factor sequences involved in complex formation with botrocetin. A model for the regulation of von Willebrand factor binding to platelet glycoprotein Ib. J Biol Chem 1991;266:18172–8.

[85] Brinkmann V, Reichard U, Goosmann C, Fauler B, Uhlemann Y, Weiss DS, Weinrauch Y, Zychlinsky A. Neutrophil extracellular traps kill bacteria. Science 2004;303:1532–5.

[86] Fuchs TA, Brill A, Duerschmied D, Schatzberg D, Monestier M, Myers Jr. DD, Wroblewski SK, Wakefield TW, Hartwig JH, Wagner DD. Extracellular DNA traps promote thrombosis. Proc Natl Acad Sci U S A 2010;107:15880–5.

[87] Grässle S, Huck V, Pappelbaum KI, Gorzelanny C, Aponte-Santamaría C, Baldauf C, Gräter F, Schneppenheim R, Obser T, Schneider SW. von Willebrand factor directly interacts with DNA from neutrophil extracellular traps. Arterioscler Thromb Vasc Biol 2014;34:1382–9.

[88] Claes J, Vanassche T, Peetermans M, Liesenborghs L, Vandenbriele C, Vanhoorelbeke K, Missiakas D, Schneewind O, Hovlaerts MF, Heying R, Verhamme P. Adhesion of *Staphylococcus aureus* to the vessel wall under flow is mediated by von Willebrand factor-binding protein. Blood 2014;124:1669–76.

[89] Kroh HK, Panizzi P, Bock PE. Von Willebrand factor-binding protein is a hysteretic conformational activator of prothrombin. Proc Natl Acad Sci U S A 2009;106:7786–91.

[90] Cooper HA, Wagner RH. The defect in hemophilic and von Willebrand's disease plasmas studied by a recombination technique. J Clin Invest 1974;54:1093–9.

[91] Takahashi Y, Kalafatis M, Girma P-P, Sewerin K, Andersson L-O, Meyer D. Localization of a factor VIII binding domain on a 34 kilodalton fragment of the N-terminal portion of von Willebrand factor. Blood 1987;70:1679–82.

[92] Leyte A, Verbeet MP, Brodniewicz-Proba T, Van Mourik JA, Mertens K. The interaction between human blood-coagulation factor VIII and von Willebrand factor. Characterization of a high-affinity binding site on factor VIII. Biochem J 1989;257:679–83.

[93] Weiss HJ, Sussman II, Hoyer LW. Stabilization of factor VIII in plasma by the von Willebrand factor. Studies on posttransfusion and dissociated factor VIII and in patients with von Willebrand's disease. J Clin Invest 1977;60:390–404.

[94] Gaucher C, Jorieux S, Mercier B, Oufkir D, Mazurier C. The "Normandy" variant of von Willebrand disease: characterization of a point mutation in the von Willebrand factor gene. Blood 1991;77:1937–41.

[95] Lenting PJ, Neels JG, van den Berg BM, Clijsters PP, Meijerman DW, Pannekoek H, van Mourik JA, Mertens K, van Zonneveld AJ. The light chain of factor VIII comprises a binding site for low density lipoprotein receptor-related protein. J Biol Chem 1999;274:23734–9.

[96] Dasgupta S, Repessé Y, Bayry J, Navarrete AM, Wootla B, Delignat S, Irinopoulou T, Kamaté C, Saint-Remy JM, Jacquemin M, Lenting PJ, Borel-Derlon A, Kaveri SV, Lacroix-Desmazes S. VWF protects FVIII from endocytosis by dendritic cells and subsequent presentation to immune effectors. Blood 2007;109:610–2.

[97] Vlot AJ, Koppelman SJ, van den Berg MH, Bouma BN, Sixma JJ. The affinity and stoichiometry of binding of human factor VIII to von Willebrand factor. Blood 1995;85:3150–7.

[98] Nesheim M, Pittman DD, Giles AR, Fass DN, Wang JH, Slonosky D, Kaufman RJ. The effect of plasma von Willebrand factor on the binding of human factor VIII to thrombin-activated human platelets. J Biol Chem 1991;266:17815–20.

[99] Lollar P, Hill-Eubanks DC, Parker CG. Association of the factor VIII light chain with von Willebrand factor. J Biol Chem 1988;263:10451–5.

[100] Randi AM, Laffan MA. Von Willebrand factor and angiogenesis: basic and applied issues. J Thromb Haemost 2017;15:13–20.

Recommended Reading

[1] Turner NA, Nolasco L, Ruggeri ZM, Moake JL. Endothelial cell ADAMTS13 and VWF: production, release, and VWF string cleavage. Blood 2009;114:5102–11.

[2] Springer TA. Von Willebrand factor, Jedi knight of the bloodstream. Blood 2014;124:1412–25.

[3] Lenting PJ, Christophe OD, Denis CV. Von Willebrand factor biosynthesis, secretion, and clearance: connecting the far ends. Blood 2015;125:2019–28.

[4] Turner NA, Moake JL. Factor VIII is synthesized in human endothelial cells, packaged in Weibel-Palade bodies and secreted bound to ULVWF strings. Plos One 2015;10:.

[5] Haberichter SL. Von Willebrand factor propeptide: biology and clinical utility. Blood 2015;126:1753–61.

[6] Chion ACK, O'Sullivan JM, Drakeford C, et al. N-linked glycans within the A2 domain of von Willebrand factor modulate macrophage-mediated clearance. Blood 2016;128:1959–68.

10

Von Willebrand Disease: Classification and Diagnosis

Von Willebrand disease (VWD) is a congenital hemorrhagic disorder due to a deficient or defective protein, the Von Willebrand Factor (VWF). It is characterized by mucosal bleeding, impaired platelet adhesion at sites of vascular injury, and a decrease in FVIII, the latter in common with hemophilia. As noted in Chapter 1, Erik A. Von Willebrand described the disorder in 1926. In 1981, researchers from Scandinavia and the Netherlands reinvestigated the extant members of the original bleeder family reported by Von Willebrand. They observed that 6 of 10 family members had decreased Von Willebrand antigen and ristocetin cofactor, along with low FVIII levels in some cases, thereby confirming the diagnosis of VWD [1].

EPIDEMIOLOGY

Determining the incidence of VWD is difficult because the criteria for the diagnosis have been unclear. If only those with unequivocal, severe disease (VWF level < 0.1 IU/mL) are counted, the incidence is 1.38 per million in the North American population and 1.53 per million in the European population [2]. Using less strict criteria, such as referral for significant bleeding symptoms and mean VWF 0.49 IU/mL, the incidence is 1 in 1000 [3]. The frequency is similar (23–110 per million) if the criterion is a VWF level of ≤ 0.3 IU/mL (with ≥ 0.5 IU/mL in healthy people) [4]. The prevalence is much higher, 0.6%–1.3%, if populations are screened for evidence of VWD. Alternatively, the diagnosis could require demonstration of a relevant mutation; for example, in the genes for VWF or ADAMTS13 (**a d**isintegrin **a**nd **m**etalloproteinase with a **t**hrombospondin type 1 motif, member **13**).

149

Genetic analysis using exome sequencing has revealed a number of common *VWF* coding variants in African-Americans; those in exon 28 (I1380V, N1435S, and D1472H) are associated with a relative decrease in ristocetin cofactor (RCoF) [5,6]. Consequently, African-Americans have lower ratios of RCoF to VWF antigen (RCof:VWFAg) on average than Caucasians; the differences are statistically significant for those with blood group O (0.79 vs 0.97) [7]. In particular, individuals with the Asp1472His variant have lower VWF antigen, RCof, and RCof:VWFAg ratio than those without this variant, but do not have an increase in bleeding events [8]. Therefore, the use of the RCof:VWFAg ratio to diagnose VWD type 2 results in the overdiagnosis of this disorder in African-Americans.

CLASSIFICATION

Early workers recognized that VWD was very heterogeneous, both in its clinical manifestations and laboratory features [9]. Although the disorder was inherited as an autosomal dominant in most patients with classical VWD, recessive and X-linked forms of the disease were also thought to exist [10,11]. Patients with findings consistent with VWD but normal levels of FVIII were said to have "variant VWD" [12], whereas those with low levels of FVIII but normal VWF activity and an autosomal pattern of inheritance were diagnosed as "autosomal hemophilia"; some of the latter were eventually identified as variants of VWD [13]. There were other reports of patients with VWF deficient in carbohydrate content [14]. It was not until improved assays for VWF were introduced and genetic studies became available that a classification that encompassed the many variants of VWD became possible.

Single gene disorders are classified according to whether the relevant protein is decreased, dysfunctional, or completely absent. In people with VWD, VWF is decreased in type 1, has defective hemostatic activity in type 2, and is usually undetectable in type 3. Within types 1 and 2, there are subtypes characterized by specific disease mechanisms and location of mutations, with a greater loss of functional activity than protein in most type 2 variants. The inheritance pattern is autosomal dominant for the majority of VWD types. Type 1 is the most frequently encountered and will be discussed first.

Type 1

Reduced VWF in patients with type 1 disease is due to decreased synthesis, intracellular retention, or increased clearance of the molecule.

In 1990, Ewenstein et al. [15] cultured endothelial cells from the umbilical veins of two neonates with type 1 VWD and reported that the content of VWF messenger RNA was reduced and the secreted factor diminished. More recently, blood outgrowth endothelial cells from patients with type 1 VWD have been examined, and the earlier findings confirmed and extended [16,17]. Defects were observed in processing, storage, and formation of VWF strings, and these alterations could be tied to specific mutations in the A1 and A2 domains of the molecule. Mutations in the A3 and D4 domains have been linked to an inability to form Weibel-Palade bodies [18]. In fact, pathogenic mutations have been identified in 65% of patients with type 1 disease, and 70% of these are missense substitutions affecting VWF trafficking, storage, secretion, and clearance [19]. Sequence mutations occur in over 80% of those with VWF levels of <0.3 IU/mL but only 44% of those with levels 0.3–0.5 IU/mL.

As noted in Chapter 9, VWF is cleaved by ADAMTS13. Increased vulnerability of VWF to cleavage by ADAMTS13 is associated with an exon 28 mutation encoding the amino acid polymorphism, Tyr1584Cys [20]. The susceptibility of VWF to cleavage by ADAMTS13 is also influenced by N-linked glycosylation [21]. Because blood group A and B glycans increase VWF survival, the absence of these glycans from those with blood group O is accompanied by a decreased VWF survival and lower VWF levels [22].

Increased clearance of VWF is also observed in some people with type 1 VWD. Individuals with the **Vicenza** variant of VWD have a lifelong bleeding tendency, decreased VWF activity in plasma but normal levels in platelets, and decreased high-molecular-weight multimers (HMWM) in the circulation [23]. The disorder is associated with an Arg1205His substitution in the VWF D3 domain and enhanced clearance of the mutant protein [24]. The abbreviated VWF survival is mediated by macrophages, most likely in the liver and spleen. Another factor that affects the clearance of VWF is the C-type lectin receptor, CLEC4M [25]. CLEC4M binds to the N-linked glycans of VWF, affecting its plasma levels and clearance. Specific CLEC4M polymorphisms are associated with decreased levels of VWF and have been reported in some patients with type 1 VWD.

Problems in classifying individuals as having type 1VWD arise when the level of VWF is only mildly decreased (0.3–0.5 IU/mL) or was low in the past but is now within the normal range. Patients are often referred for investigation because of a bleeding tendency and VWF levels that are <0.5 IU/mL but >0.3 IU/mL; currently, such patients are classified as having **low VWF** rather than VWD [26]. Lavin et al. [27] surveyed 126 Irish patients from 118 kindreds with low VWF. Table 10.1 lists the characteristics of the study patients.

TABLE 10.1   Characteristics of Patients With Low VWF (0.3–0.5 IU/dL)

Female (89%)
Group O blood type (89%)
Positive bleeding assessment tool (BAT) in 77% of females
Increase in VWF with age
Increased FVIII:VWF
VWF increase >0.5 IU/mL 1 h after desmopressin

Data from Lavin M, Aguila S, Schneppenheim S, Dalton N, Jones KL, O'Sullivan JM, O'Connell NM, Ryan K, White B, Byrne M, Rafferty M, Doyle MM, Nolan M, Preston RJS, Budde U, James P, Di Paola J, O'Donnell JS. Novel insights into the clinical phenotype and pathophysiology underlying low VWF levels. Blood 2017;130:2344–53.

Low VWF was observed mainly in women, perhaps because heavy menstrual bleeding brought them to medical attention, and the bleeding phenotype could not be explained by other clotting factor deficiencies or platelet function disorders. Blood group O was significantly more common in those with low VWF as compared to the general Irish population (89% vs 55%). VWF increased with aging at a rate of 0.02 IU/mL per year in this cohort. Studies of VWF propeptide were consistent with reduced synthesis and/or secretion of VWF and with elevated FVIII:VWF ratios. Finally, treatment with desmopressin produced substantial increases in VWF in all patients, and should be selected for excessive bleeding or invasive procedures (Chapter 11).

The increase in VWF with age is reported in mild type 1 patients; VWF antigen, RCof, and FVIII rise by 0.30, 0.20, and 0.20 IU/mL per decade, respectively [28]. As a consequence, about 20% of type 1 patients will normalize their VWF levels over a 10-year period [29], and one might wonder whether the rise in VWF mitigates the bleeding tendency. One set of investigators reported an age-related amelioration of bleeding symptoms [30], but another group showed that the frequency of bleeding episodes and treatment of VWD were not significantly decreased in patients older or younger than age 65 [31]. Increases in VWF with aging are less likely to occur in people with VWF variants, a positive family history, and non-O blood group [32]. Lastly, exercise increases FVIII/VWF in healthy individuals but does not appear to raise the levels in patients with type 1 disease, although further investigation is needed to confirm this point [33]. Because the levels of FVIII and VWF can be influenced by age, exercise, stress, phase of the menstrual cycle, and pregnancy, these factors should be taken into consideration when deciding whether patients have low VWF or type 1 VWD; genotyping is usually not helpful in this situation [34].

Type 2

There are four major subtypes of type 2 VWD. In contrast to type 1 patients, no age-related increases in VWF levels are observed in patients with type 2 disease, and elderly patients report significantly more bleeding than younger patients ($P = .048$) [29]. A potential explanation is that the elderly have a greater propensity to age-related disorders associated with bleeding such as cancer and falls.

Type 2A occurs in 20%–25% of VWD patients and is characterized by decreased levels of VWF activity and a lack of high-molecular-weight multimers (HMWM) [32]. The deficiency of HMWM is associated with decreased platelet adhesion and a bleeding phenotype, particularly gastrointestinal bleeding [35]. The responsible mechanisms include defective multimerization associated with mutations in the propeptide, D3, or A2 domains; defective dimerization when mutations are in the CK domain; and enhanced proteolysis by ADAMTS13 if mutations are in the A2 domain at the Tyr1605-Met1606 cleavage site [26,36]. Not all variants fit neatly into these categories, but the overall impact is on the steps needed for VWF processing and storage [37].

Although most type 2A variants are inherited as autosomal dominants, VWD2AIIC is characterized by recessive inheritance, low VWF antigen, lack of HMWM, and absence of VWF proteolytic fragments [38]. An unusual variant of VWD2AIIC has autosomal dominant inheritance and increased VWF antigen, and has been designated VWD type IIC **Miami** [39]. Genetic analysis revealed a duplication of exons 9–10, resulting in impairment of multimer assembly, an altered structure of the mature VWF, and reduced clearance in vivo [40].

Type 2B occurs in about 5% of VWD patients and is associated with mutations in the A1 domain of VWF. It features enhanced binding of VWF to platelet GpIbα, resulting in a decrease in circulating HMWM as well as platelet aggregation by unusually low concentrations of ristocetin. Some patients have thrombocytopenia, occasionally provoked by acute stress or pregnancy [41], and associated with a change in the binding conformation of GpIbα [42]. The thrombocytopenia is probably due to spontaneous platelet aggregation and subsequent removal of the VWF/platelet complexes by macrophages [43]. Bleeding tends to be more severe in these thrombocytopenic patients. Exercise decreases the mean platelet count but increases the levels of FVIII/VWF, ADAMTS13, and the percentage of HMWM [33]. The decline in platelet count might be a reflection of the larger amounts of mutant VWF bound to platelets, but the reasons for the increases in ADAMTS13 and HMWM are unclear.

The chief characteristic of type 2M is defective binding of the mutant VWF to collagen or platelet GpIbα. In contrast to types 2A and 2B, the content of HMWM is not diminished. Some mutations in the A1 domain alter

VWF binding to type IV collagen under flow conditions and are found in 27% of patients with type 2M [44]. Type IV collagen is a key component of the basement membrane of the vascular endothelium; patients with type 2M VWD have excessive bleeding as a consequence of impaired VWF-mediated platelet adhesion to this collagen. In addition, mutations in exon 28 affecting the A1 domain can alter the binding of VWF to GpIbα and impair ristocetin, but not botrocetin-mediated platelet aggregation [45,46]. The variable expression of one exon 28 mutant, M1304R, among family members was found to be due to differences in the extent of mutant monomer incorporation in the final multimer structure of VWF [47]. Bleeding tends to be more severe in patients with type 2A than 2M VWD, particularly gastrointestinal hemorrhage [35].

Type 2N VWD is associated with decreased circulating FVIII due to mutations in the binding site within the N-terminal D'D3 domain of VWF, and affected patients are either homozygous for a type 2N mutation or compound heterozygous for a type 1 and type 2N mutation. The most common mutation is R854Q; study of 13 unrelated Italian families led to an estimate that this mutation took place from 10,000 to 40,000 years ago when human populations moved from Africa toward Europe [48]. The incidence of type 2N VWD is unclear because it has occasionally been mistaken for mild hemophilia; in one study, 5 of 14 patients originally presumed to have hemophilia A were found to have type 2N VWD [49]. It is distinguished from hemophilia by the autosomal recessive inheritance pattern, persistent increases in FVIII after infusion of FVIII/VWF concentrate, and decreased binding of FVIII to the mutant VWF [50,51]. Patients have excessive bleeding but normal concentrations of VWF and HMWM. Infusions of desmopressin induce a 2.3-fold increase in VWF but a variable and transient increase in FVIII, and do not improve VWF binding of FVIII [52].

Type 3

Patients with type 3 VWD have extremely low or undetectable levels of VWF and <10% FVIII; they can be distinguished from those with very severe type 1 disease by measurement of the VWF propeptide, which is ≥5 IU/dL in type 1 and very low or absent in type 3 [53]. A majority of the mutations are in the propeptide, and include large deletions and other sequence variations that affect VWF synthesis and secretion [54]. Families display either codominant or recessive inheritance patterns. Some members of the kindreds originally studied by Von Willebrand had VWD type 3 [55]; the original proband, Hjördis, had life-threatening bleeding with minor trauma and exsanguinated at age 13 during her fourth menstrual period [1].

CLINICAL AND LABORATORY EVALUATION

Most patients with VWD have a history of bleeding from the nose and gums as well as extensive bruising with minimal trauma. Heavy menstrual bleeding often contributes to iron deficiency anemia. Also common are persistent vaginal bleeding postpartum, continued oozing after dental extractions, and wound hematomas complicating surgical procedures. Other sites of hemorrhage are the gastrointestinal and urinary tracts, but hemarthroses and large muscle hematomas are mainly limited to patients with severe type 1 and type 3 VWD.

Gastrointestinal bleeding occasionally arises from angiodysplastic lesions [56,57]. These vascular malformations occur most commonly in middle-aged and older patients and are located principally in the cecum and ascending colon, but can be anywhere in the gastrointestinal tract [58]. A survey showed that only those lacking VWF HMWM had angiodysplasia; these were patients with type 2 (2%) and type 3 (4.5%) disease [59]. An investigation of VWD blood outgrowth endothelial cells showed that those from patients with type 1 and 3 disease have impaired migration velocity and enhanced directionality [60]. Type 2A and 2B cells are the most proliferative and many have impaired tubule formation on Matrigel. However, there is significant heterogeneity among individual VWD phenotypes.

The pathophysiology of angiodysplasia is still incompletely understood. VWF-deficient mice have increased vascularization, indicating that VWF plays a role in retarding angiogenesis [61]. In vitro, inhibiting endothelial VWF expression enhances angiogenesis, decreases integrin $\alpha v \beta 3$, increases vascular endothelial growth factor receptor-2 (VEGFR2)-dependent proliferation and migration of endothelial cells, and augments the release of angiopoietin-2 (Ang-2) [62]. Integrin $\alpha v \beta 3$ is required for the binding of VWF to endothelial cells and regulates the sensitivity of these cells to VEGFR2. In the absence of VWF, $\alpha v \beta 3$ is internalized; its loss from the cell surface probably enhances VEGFR2 signaling, contributing to the formation of the immature, fragile blood vessels characteristic of angiodysplasia. Ang-2 is normally present in Weibel-Palade (W-P) bodies; absence of VWF results in the disassembly of W-P bodies and the release of Ang-2. Ang-2 can then bind to the Tie-2 receptor, antagonizing angiopoietin-1 and destabilizing cell-cell junctions, thereby contributing to vessel malformation.

The variability in the hemorrhagic manifestations of VWD has prompted investigators to construct questionnaires that capture and present the patient's bleeding history in a relatively objective fashion. These **bleeding assessment tools** or BATs are scored and can be used to provide a quantitative description of the disease [63]. The International Society of Thrombosis and Haemostasis has developed a detailed BAT [64];

a condensed version is available on the American Society of Hematology Education website at http://www.hematology.org [65]. Points are assigned based on bleeding severity; the normal range is 0–3 for men, 0–5 for women, and 0–2 for children [66]. Bleeding scores are inversely related to laboratory-determined levels of VWF ($P < .001$) and are uniformly increased in patients with severe type 1 disease (VWF antigen $\leq 0.05\,IU/mL$) [19]. However, they do not correlate with VWF antigen levels $>0.05\,IU/mL$, and a low bleeding score usually predicts that routine laboratory tests will be uninformative. Patients with bleeding scores >10 are likely to have frequent, severe hemorrhages and might benefit from prophylactic treatment for VWD [67]. Because VWD is very heterogeneous, several laboratory tests will usually be required to define the disorder; a partial list is shown in Table 10.2. A quantitative estimate of FVIII concentration is required for all individuals suspected of having VWD, using any of several methods (see Chapter 5, Diagnosis of Hemophilia).

Knowing an individual's **blood group** can be helpful in the evaluation of a mild bleeding tendency. Some individuals with blood group O might have quite low levels of VWD; deciding whether they have VWD usually requires the presence of other features, such as excessive bleeding, affected family members, and abnormal laboratory tests.

TABLE 10.2   Tests Used for the Diagnosis of Von Willebrand Disease

Bleeding assessment tool (BAT)
PFA-100 system
FVIII assay (clotting, chromogenic, ELISA)
VWF antigen assay
Ristocetin-induced platelet aggregation
Ristocetin cofactor activity
Botrocetin-induced platelet aggregation
Collagen-binding assay
Modified GpIb-binding assay
VWF propeptide assay
VWF multimer analysis
FVIII-binding assay
DDAVP challenge test
Genotyping

Many patients with VWD have a prolonged skin bleeding time, but this test is invasive and has been supplanted by the platelet function analyzer (PFA-100 System, Siemens Healthineers, Tarrytown, New York). The **PFA-100** is meant to simulate primary hemostasis by measuring the time required (the closing time, in seconds) for a platelet plug to completely occlude a microscopic aperture in a membrane coated with collagen and epinephrine (C-EPI) or collagen and ADP (C-ADP) under controlled high shear stress. Closure times with the C-EPI and C-ADP cartridges are longer (>160 s) in most patients with VWD than in healthy controls [68]. The test differentiates between VWD samples and samples from people without VWD who have recently ingested aspirin; closing times are prolonged with both cartridges in the former, but only with the C-EPI cartridge in the latter. Patient samples must be processed within 4 h of collection and kept at room temperature. Interestingly, the closure time with the C-EPI cartridge in blood group O samples is significantly longer than with other blood groups, and this should be taken into consideration when testing blood group O patients suspected of having VWD [69]. Other factors affecting the closure times are a platelet count <100,000/μL, hematocrit <30%, renal failure, and a high sedimentation rate. The PFA-100 is not sensitive enough to be used as a screening test for low VWF and VWD in adolescents with menorrhagia [70].

VWF antigen has been quantitated by immunologic methods since the 1970s. Electroimmunoassays [71] and solid-phase immunoradiometric assays [72,73] were developed but either had large coefficients of variation or required the use of radioactivity, and there was only modest agreement between laboratories [74]. Most laboratories currently use the enzyme-linked immunosorbent assay (ELISA) and commercial reagents to measure VWF antigen.

VWF function is generally assessed by studying **ristocetin-induced platelet aggregation**, ristocetin cofactor activity, or the binding of VWF to collagen or modified glycoprotein 1b (Gp1b) fragments. The addition of ristocetin, 1.2–1.5 mg/mL, to normal platelet-rich plasma readily induces aggregation; diminished aggregation is observed with most types of VWD. VWD 2B is distinguished by responses to lower concentrations of ristocetin than required for platelet aggregation in normal plasma [75]. **Ristocetin cofactor** activity is measured using either fresh, formalin-fixed, or reconstituted lyophilized platelets [76]. The addition of dilutions of patient plasma and ristocetin results in a family of aggregation curves whose slopes are compared to the calibrator. The assay is relatively insensitive, has a large coefficient of variation, and has been supplanted by fully automated procedures [77]. Although the introduction of ristocetin has been a valuable contribution to the diagnosis of VWD, it has important limitations [78]. False-negative results occur because some ristocetin-based assays are insensitive to moderately reduced levels of VWF;

conversely, false-positive diagnoses occur with mutations at the ristocetin-binding site of VWF. These mutations can produce an abnormal test that does not reflect VWF function under physiologic conditions.

Functional assays have been developed that avoid the use of ristocetin by measuring the **binding of VWF to GpIb** fragments modified by introducing gain-of-function mutations into GpIbα [79,80]. These assays are sensitive to a broad range of VWF levels and closely correlate with ristocetin cofactor assays. The INNOVANCE VWF Ac assay (Siemens, Tarrytown, New York) uses latex particles coated with an antibody against GpIb to which the recombinant modified GpIb is added. The addition of patient plasma induces a VWF-dependent agglutination that is detected turbidimetrically. The ACL AcuStar (Instrumentation Laboratory, Bedford, Massachusetts) is an automated immunoassay analyzer using chemiluminescence that is sensitive to low levels of VWF and has small interassay variability [81]. A VWF:GpIbM assay that has a lower limit of detection of 2 IU/dL and coefficient of variation of 5.6% has an ELISA version available in the United States [82,83].

Botrocetin, a snake venom, agglutinates platelets by binding to platelet glycoprotein 1b in the presence of VWF [45,84]. Attachment is initially to the A1 domain of VWF, followed by binding of the botrocetin-A1 interface to glycoprotein 1bα [85]. As noted previously, certain mutations affecting the A1 domain alter the binding of VWF to GpIbα and impair ristocetin but not botrocetin-induced platelet aggregation [46].

Collagen binding assays use ELISA methodology and either type I, III, IV, or VI collagen; VWF A1 and A3 domains bind to collagen types I and III, while the A1 domain binds to collagen types IV and VI [86]. Patients with VWD types 2A and 2B have defective binding to collagen types I and III, suggesting that the assay is sensitive to the HMWM of VWF [87]. Binding defects to collagen types IV and VI are observed in patients with type 1 and 2M VWD, depending on the specific mutations within the A1 and A3 domains and the type of collagen used by the assay method [44].

The **FVIII binding** assay is performed when the ratio of FVIII to VWF antigen is ≤0.6 and type 2N VWD is suspected. A simplified assay technique using clotting methods was described in 1998 [88]; currently, an ELISA system using commercial reagents is available [89]. A kinetic analysis determined that the time needed for VWF to bind 50% of FVIII is approximately 2 s [90].

The **VWF propeptide** is secreted into the circulation along with the VWF monomer; their respective concentrations are 1 and 10 µg/mL but by definition there is 1 U/mL each of propeptide and VWF, and a ratio at steady state of 1.0 [91]. This 1:1 ratio is increased if the clearance of VWF is accelerated, as occurs in some forms of VWD [92]. Mutant forms of the propeptide or its complete absence from the plasma are associated with failures of multimerization, storage, or secretion of VWF [53].

The propeptide is measured using an ELISA method and an antipropeptide monoclonal antibody [93].

VWF multimer analysis is infrequently performed for diagnostic purposes but is used mainly to investigate the pathophysiology of VWD. The highest-molecular-weight multimers are the most hemostatically active; their reduction in type 2 VWD accounts for the bleeding tendency in affected individuals with normal concentrations of VWF. The multimer content of plasma is examined using sodium dodecyl sulfate agarose gels and manipulating the gel concentration to resolve the multimer patterns. The procedure is difficult, time consuming, and requires considerable expertise.

Desmopressin challenge testing can be used to assist in the diagnosis of some types of VWD. After baseline measurements are obtained, desmopressin in a dose of 0.3 μg/kg in 50-mL normal saline, is infused intravenously over 30 min and the extent of increase in FVIII/VWF is determined at 4 h. A twofold increase has been reported in 95% of patients with low VWF (0.3–0.5 IU/mL) and 86% of those with levels of 0.2–0.3 IU/mL (type 1 VWD) [94]. Patients with abnormalities of the HMWM of VWF have smaller increases in FVIII/VWF than those with normal HMWM ($P = .002$), and most partial and nonresponsive patients have mutations in the A1–A3 domains [95]. The largest increases are observed in type 2N VWD, as well as the shortest postinfusion FVIII/VWF half-life. Multimer analysis after the desmopressin challenge test is effective in distinguishing many of the VWD types [96]. Desmopressin infusion is contraindicated in patients suspected of having type 2B VWD because it induces in vivo platelet aggregation and thrombocytopenia [97].

Genetic testing is not performed routinely but is useful in assessing patients with types 2B and 2N, and for genetic prenatal diagnosis in type 3 disease [34]. Genetic analyses generally use next-generation sequencing to detect single-nucleotide variants. Fig. 10.1 shows the numbers of sequence variations by VWF domain. While most variants in patients with

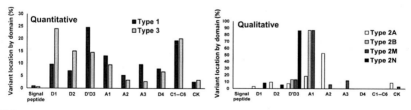

FIG. 10.1   Unique pathogenic Von Willebrand Factor gene variants by protein domain. *Data from: http://www.vwf.group.shef.ac.uk/vwd.html [Accessed June 2016]. Reproduced with permission from Swystun LL, James PD. Genetic diagnosis in hemophilia and von Willebrand disease. Blood Rev 2017;31:47–56.*

type 1 and 3 VWD are located in the C and D domains, type 2 variants are mainly in the A domains, with the exception of type 2N in the D'D3 domain.

Missense mutations constitute the majority of variants in type 1 VWD; they contribute to the impaired intracellular routing, storage, secretion, or faster clearance of the protein [15]. Sequence variation frequency varies with VWF antigen level; it is 82% at levels <0.3 IU/mL and 44% at levels ≥0.3 IU/mL [19]. The more severe phenotypes might be due to incorporation of the mutant VWF subunit into the multimers. On the other hand, some sequence variations observed in people with VWF antigen levels between 0.3 and 0.5 IU/mL might be nonpathologic or neutral. Several of these mutations have been found in individuals without a bleeding phenotype and with normal levels of VWF; in fact, sequence variations are observed in 11.4% of a healthy control population [98]. Type 2 disorders are mainly associated with missense mutations and have autosomal dominant inheritance with the exception of type 2N, which is autosomal recessive. The diagnosis of type 2B can be confirmed by the identification of well-characterized genetic variants. Type 3 VWD is often found in populations with consanguineous marriages; offspring are homozygous or compound heterozygous and most have null alleles. Table 10.3 displays the genetic features of VWD types [99].

DIAGNOSIS

The importance of a thorough history and physical examination cannot be overemphasized. In particular, all of the elements of the bleeding assessment tool (BAT) should be addressed; a web-based version is available at https://bh.rockefeller.edu/ISTH-BATR/. A frequent element of the bleeding history in women with VWD is heavy menstrual bleeding (HMB), defined as a menstrual period lasting >7 days with large clots and a heavy flow (hourly change of pad or tampon) [100]. In addition, a low ferritin and anemia are often present. HMB is reported by 80% of women with VWD [101], and the prevalence of VWD in women with HMB is reported to be 5%–20% [102]. But the diagnosis of VWD might not be appropriate in all these women, and the difficulty of establishing a firm diagnosis is illustrated by the following case history.

A 27-year-old woman had persistent vaginal bleeding after a dilatation and curettage for endometriosis. Despite cautery and packing, she continued hemorrhaging. The past history revealed easy bruising for the last 2 years but no bleeding problems during and after two uneventful pregnancies. Monthly migraine headaches were partially relieved by aspirin. The family history was unremarkable. Laboratory data showed a hemoglobin 11.1 g/dL, hematocrit 32%, and platelet count 170,000/μL.

TABLE 10.3 Genetic Features of VWD Types

VWD type	Prevalence	Mechanism	Location (exon) of mutation	Variant	Inheritance pattern
1	65%–80%	Partial deficiency	1–52, promoter	Point mutations, indels[a]	Autosomal dominant[b]
1C	15% of type 1	Accelerated clearance	18–27, 33–38	Missense	Autosomal dominant
2A	10%–20%	Impaired multimerization	2–28,51,52	Missense	Autosomal dominant
2B	5%–10%	↑binding to GpIb	28	Missense	Autosomal dominant
2M	3%–5%	↓binding to GpIb or collagen	28–32	Missense	Autosomal dominant
2N	2%–5%	↓binding to FVIII	17–27	Missense, null variants	Autosomal recessive homozygous, compound heterozygous
3	1 per million	Severe deficiency	1–52	Point mutations, indels[a]	Autosomal recessive, codominant

[a] Indels = insertions, deletions.
[b] Occasionally recessive or codominant.
Modified from Swystun LL, James PD. Genetic diagnosis in hemophilia and von Willebrand disease. Blood rev 2017;31:47–56.

The aPTT was 49 s (control, 36 s), prothrombin time 14 s (13 s), bleeding time 8 min (4–7 min), FVIII 0.45 IU/mL (0.6–1.6 IU/mL), VWF antigen 0.4 IU/mL (0.5–2.0 IU/mL), and ristocetin cofactor 0.6 IU/mL (0.5–2.0 IU/mL).

This patient has a bleeding tendency of recent onset and laboratory studies that might be consistent with VWD type 1. Additional history is needed that should include a detailed menstrual history and other elements of the BAT including bleeding with oral surgery. Factors that might affect the decision to pursue more sophisticated testing are:

- Was there recent ingestion of aspirin, which prolongs the bleeding time and the PFA-100 closing time with the C-EPI but not with the C-ADP cartridge.
- When the FVIII/VWF was measured in relation to her menstrual cycle, FVIII is significantly lower on days 5–7 of the cycle [103].
- The blood group: group O is associated with a 30% lower level of FVIII/VWF than other blood groups [104]. If she is non-O, a low ristocetin cofactor level would be a strong predictor of VWD [105].
- If she is African-American, common polymorphisms in the *VWF* gene might account for the decreased ristocetin cofactor activity [8].

When she is no longer bleeding and has not taken aspirin or other drugs known to affect platelet function, a PFA-100 should be ordered and the FVIII, VWF antigen, and ristocetin cofactor assays repeated. Values within the normal range exclude the diagnosis of VWD, but if abnormal values persist, further evaluation is appropriate to arrive at a specific diagnosis. Levels of VWF antigen and ristocetin cofactor <30 IU/dL are diagnostic of VWD. If her ristocetin cofactor is modestly reduced to 0.3–0.5 IU/mL and the ratio of FVIII to VWF antigen is >0.6, she falls into the category of **low VWF** [26]. If her VWF levels are not consistent with a diagnosis of VWD, she might still require treatment with desmopressin to raise the VWF concentration prior to surgery or to control bleeding.

Occasional patients with low VWF have bleeding that seems excessive considering that the decrease in VWF might be only modest; these patients often have other hemostatic defects, particularly impaired platelet aggregation with collagen [106]. Patients with type 1 VWD have an increased frequency of the 807C allele of the α_2 gene for the $\alpha_2\beta_1$ collagen receptor; this allele is associated with a decrease in the platelet collagen receptor density [107]. The low receptor density can be recognized by a decrease in collagen-induced platelet aggregation and an increase in the PFA-100 closing time.

Diagnostic Features of VWD Subtypes

Bleeding severity can range from mild to severe in patients with type 1 VWD but is often life threatening in those with type 3 VWD. BAT scores are significantly higher in type 2A than in type 2M ($P < .001$) [35], and bleeding is more severe in type 2B when thrombocytopenia is present. Bleeding events in type 2N mimic those observed in patients with hemophilia [108].

Type 1 VWD: ristocetin cofactor is <0.30 IU/mL or binding to platelet GpIb is decreased, the ratios of ristocetin cofactor to VWF antigen and FVIII to VWF antigen are >0.6, and HMWM are normal. Severe type 1 (VWF antigen <5 IU/dL) is distinguished from type 3 by propeptide levels ≥5 IU/dL. **Type 2 VWD:** all except 2N have ristocetin cofactor <0.3 IU/mL and a decreased ratio (≤0.6) of ristocetin cofactor to VWF antigen. 2A has normal or decreased ristocetin-induced platelet aggregation and loss of HMWM; **2B** has enhanced ristocetin-induced platelet aggregation, loss of HMWM, and occasionally thrombocytopenia; **2M** has normal HMWM and decreased VWF binding to type IV collagen; and **2N** has a decreased ratio (≤0.6) of FVIII to VWF antigen, and decreased binding of FVIII to VWF. **Type 3 VWD:** FVIII is very low, and ristocetin cofactor and VWF antigen are undetectable. VWF propeptide levels <0.05 IU/mL distinguish type 3 from severe type 1. Finally, patients can have comorbidities that alter hemostasis or take medications that affect platelet function. A comprehensive history in such patients is invaluable for achieving an accurate diagnosis.

Gene Mutations in VWD Subtypes

The genetic features of the various types of VWD are displayed in Table 10.3; identifying mutations is useful in confirming the diagnosis, especially in those with type 2 VWD. The value of genetic analysis in type 1 VWD is less clear; two large studies were unable to identify mutations in 27% and 36% of those suspected of having the disorder [109,110]. The failure to detect *VWF gene* mutations does not necessarily negate the diagnosis of VWD because polymorphisms might be present in other genes involved in VWF processing, secretion, or clearance. On the other hand, sequence variations are often not detected in patients with milder phenotypes, and these are the patients diagnosed with low VWF. Difficulties in defining VWD subtypes occasionally occur because patients might have inherited genes for more than one hemostatic disorder; for example, they have sequence variations in the genes for both FVIII and VWF. Analyzing other members of the pedigree might provide clarity in this situation.

Differential Diagnosis

Qualitative and quantitative platelet disorders can simulate VWD. For example, a 13-year-old girl from the Middle East was admitted to the Emergency Department with massive vaginal bleeding in conjunction with her first menstrual period. She was assumed to have severe VWD and given four bags of cryoprecipitate while laboratory studies were underway. Her bleeding continued unabated; the tests showed a normal platelet count, no clot retraction, and absent platelet aggregation with collagen, ADP, and epinephrine. The diagnosis was revised to **Glanzmann Thrombasthenia** and a platelet transfusion was given. Bleeding dramatically decreased and subsequent studies showed a mutant $GP\alpha_{IIb}\beta_3$, consistent with thrombasthenia.

Another platelet disorder occasionally mistaken for VWD is the **Bernard-Soulier syndrome**. As in VWD, ristocetin-induced platelet aggregation is defective and platelet adhesion to subendothelium is impaired, but these defects are due to mutations in the platelet GpIb-IX-V complex [111,112]. The Bernard Soulier syndrome differs from VWD in being an autosomal recessive disorder associated with thrombocytopenia, giant platelets, and normal levels of VWF antigen and ristocetin cofactor.

VWD type 2B must be differentiated from **platelet-type VWD**, first described as "Pseudo-von Willebrand's disease" by Weiss et al. in 1982 [113]. It is an autosomal dominant, mild bleeding disorder characterized by intermittent thrombocytopenia, decreased FVIII/VWF, absence of HMWM, and increased ristocetin-induced platelet aggregation. A distinguishing feature is the aggregation of the patient's platelets by normal VWF even in the absence of ristocetin. These platelets have an increased affinity for VWF because of mutations in the $GP1b\alpha$ gene that enhance the binding to VWF [114].

Both VWD type 2B and platelet-type VWD can be misdiagnosed as **primary immune thrombocytopenia** (ITP). VWD might be suspected because of a lifelong history of bleeding, a positive family history of a bleeding disorder, and waxing and waning platelet counts unrelated to therapeutic interventions. VWD type 2B has also been misdiagnosed as **neonatal alloimmune thrombocytopenia**. The absence of alloantibodies and lack of response to intravenous immunoglobulin, as well as the marked reduction in ristocetin cofactor and normal level of VWF antigen will suggest the correct diagnosis, which can be confirmed by detecting mutations in exon 28 consistent with VWD type 2B [115]. As noted previously, VWD type 2N is often mistaken for mild-to-moderate **hemophilia**.

Other inherited bleeding disorders that could be confused with VWD are hereditary hemorrhagic telangiectasia and connective tissue diseases such as Ehlers-Danlos syndrome, but a careful examination of the bleeding sites will usually reveal the correct diagnosis. Lastly, a variety of disorders are associated with the development of **acquired VWD**; this condition is described in Chapter 12.

SUMMARY

The incidence of VWD ranges broadly from 1 to 100 per 10,000, depending on whether the disease is defined by a history of bleeding and a VWF level of $\leq 0.3\,IU/mL$, or by population screening and a VWF level of $\leq 0.5\,IU/mL$. The disorder is often overdiagnosed in African-Americans whose levels of ristocetin cofactor are often lower than in whites; many of these individuals have an exon 28 polymorphism that is not associated with an increase in bleeding risk. The classification of VWD includes three major types: those with a concordant decrease in VWF activity and antigen (type 1), discrepant activity and antigen (type 2), and complete or near-complete absence of the factor (type 3). Type 1 is the most frequently encountered and is associated with mutations that affect the synthesis, intracellular retention, or clearance of VWF. A subtype is low VWF $(0.3–0.5\,IU/mL)$. Type 2 consists of four subtypes: 2A has decreased synthesis or increased proteolysis of high-molecular-weight VWF multimers, type 2B has increased binding of the multimers to platelet GpIbα, type 2M has normal concentrations of multimers that are unable to bind to collagen or platelet GPIbα, and type 2N has mutations in the binding site for FVIII. Type 3 is uncommon and is usually due to large gene deletions that affect the propeptide, decreasing VWF synthesis and secretion.

One of the most valuable tests for the diagnosis of VWD is the bleeding assessment tool (BAT); it provides a quantitative description of bleeding severity. An elevated BAT in patients suspected of having a congenital bleeding disorder should be followed by laboratory tests that include the PFA-100 and assays of FVIII and VWF antigen. Depending on the results of these studies, more sophisticated tests can be ordered to determine the specific type of VWD, evaluate therapeutic options, and perform family studies. Diagnostic dilemmas are encountered in patients with mild-to-moderate bleeding disorders and VWF levels just below the normal range $(0.3–0.5\,IU/mL)$. It is often helpful to determine whether such individuals have blood group O, and to repeat the laboratory tests when they are not bleeding, taking aspirin, or ingesting other drugs that affect platelet function. Occasionally, older patients with a history of bleeding and prior low VWF levels are now found to have normal levels; such increases of VWF with age have been reported in patients with mild type 1 disease, but whether it is safe for them to undergo invasive procedures without prophylaxis is unclear. Lastly, genotyping is often required to distinguish the type 2 subtypes and for prenatal diagnosis in type 3 disease; however, when utilized as a screening test for type 1 disease, it is more often confusing than helpful.

Future research might:

- Identify the unique features of VWD that distinguish it from other hemostatic disorders. A clear definition would be helpful in estimating VWD prevalence.
- Determine whether "low VWF" is a congenital or acquired condition; if the latter, it should not be included in the current classification of VWD.
- Develop a functional assay for VWF that reflects its myriad biological activities.
- Conduct further studies of patients with typical phenotypic features of VWD but negative genotyping for VWF variants

References

[1] Nyman D, Eriksson AW, Blomback M, Frants RR, Wahlberg P. Recent investigations of the first bleeder family in Åland (Finland) described by von Willebrand. Thromb Haemost 1981;45:73–6.

[2] Weiss HJ, Ball AP, Mannucci PM. Incidence of severe von Willebrand's disease. N Engl J Med 1982;307:127.

[3] Bowman M, Hopman WM, Rapson D, Lillicrap D, James P. The prevalence of symptomatic von Willebrand disease in primary care practice. J Thromb Haemost 2010;8:213–6.

[4] Nichols WL, Hultin MB, James AH, Manco-Johnson MJ, Montgomery RR, Ortel TL, Rick ME, Sadler JE, Weinstein M, Yawn BP. von Willebrand disease (VWD): evidence-based diagnosis and management guidelines, the National Heart, Lung, and Blood Institute (NHLBI) Expert Panel report (USA). Haemophilia 2008;14(2):171–232.

[5] Flood VH, Gill JC, Morateck PA, Christopherson PA, Friedman KD, Haberichter SL, Branchford BR, Hoffmann RG, Abshire TC, Di Paola JA, Hoots WK, Leissinger C, Lusher JM, Ragni MV, Shapiro AD, Montgomery RR. Common VWF exon 28 polymorphisms in African Americans affecting the VWF activity assay by ristocetin cofactor. Blood 2010;116:280–6.

[6] Johnsen JM, Auer PL, Morrison AC, Jiao S, Wei P, Haessler J, Fox K, McGee SR, Smith JD, Carlson CS, Smith N, Boerwinkle E, Kooperberg C, Nickerson DA, Rich SS, Green D, Peters U, Cushman M, Reiner AP, Exome Sequencing Project NHLBI. Common and rare von Willebrand factor (VWF) coding variants, VWF levels, and factor VIII levels in African Americans: the NHLBI Exome Sequencing Project. Blood 2013;122:590–7.

[7] Miller CH, Haff E, Platt SJ, Rawlins P, Drews CD, Dilley AB, Evatt B. Measurement of von Willebrand factor activity: relative effects of ABO blood type and race. J Thromb Haemost 2003;1:2191–7.

[8] Flood VH, Friedman KD, Gill JC, Haberichter SL, Christopherson PA, Branchford BR, Hoffmann RG, Abshire TC, Dunn AL, Di Paola JA, Hoots WK, Brown DL, Leissinger C, Lusher JM, Ragni MV, Shapiro AD, Montgomery RR. No increase in bleeding identified in type 1 VWD subjects with D1472H sequence variation. Blood 2013;121:3742–4.

[9] Koutts J, Stott L, Sawers RJ, Firkin BG. Variant patterns in von Willebrand's disease. Thromb Res 1974;5:557–64.

[10] Ingram GIC. Classification of Von Willebrand's disease. Lancet 1978;ii:1364–5.

[11] Holmberg L, Nilsson IM. Two genetic variants of von Willebrand disease. N Engl J Med 1973;288:595–8.

[12] Peake IR, Bloom AL, Giddings JC. Inherited variants of factor-VIII-related protein in von Willebrand's disease. N Engl J Med 1974;291:113–7.

[13] Veltkamp JJ, van Tilburg NH. Autosomal haemophilia: a variant of von Willebrand's disease. Br J Haematol 1974;26:141–52.

[14] Gralnick HR, Coller BS, Sultan Y. Carbohydrate deficiency of the factor VIII/von Willebrand factor protein in von Willebrand's disease variants. Science 1976;192:56–9.

[15] Ewenstein BM, Inbal A, Pober JS, Handin RI. Molecular studies of von Willebrand disease: reduced von Willebrand factor biosynthesis, storage, and release in endothelial cells derived from patients with type I von Willebrand disease. Blood 1990;75:1466–72.

[16] Starke RD, Paschalaki KE, Dyer CEF, Harrison-Lavoie KJ, Cutler JA, McKinnon TAJ, Millar CM, Cutler DF, Laffan MA, Randi AM. Cellular and molecular basis of von Willebrand disease: studies on blood outgrowth endothelial cells. Blood 2013;121:2773–84.

[17] Wang J-W, Bouwens EAM, Pintao MC, Voorberg J, Safdar H, Valentijn KM, de Boer HC, Mertens K, Reitsma PH, Eikenboom J. Analysis of the storage and secretion of von Willebrand factor in blood outgrowth endothelial cells derived from patients with von Willebrand disease. Blood 2013;121:2762–72.

[18] Castaman G, Giacomelli SH, Jacobi PM, Obser T, Budde U, Rodeghiero F, Schneppenheim R, Haberichter SL. Reduced von Willebrand factor secretion is associated with loss of Weibel-Palade body formation. J Thromb Haemost 2012;10:951–8.

[19] Flood VH, Christopherson PA, Gill JC, Friedman KD, Haberichter SL, Bellissimo DB, Udani RA, Dasgupta M, Hoffmann RG, Ragni MV, Shapiro AD, Lusher JM, Lentz SR, Abshire TC, Leissinger C, Hoots WK, Manco-Johnson MJ, Gruppo RA, Boggio LN, Montgomery KT, Goodeve AC, James PD, Lillicrap D, Peake IR, Montgomery RR. Clinical and laboratory variability in a cohort of patients diagnosed with type 1 VWD in the United States. Blood 2016;127:2481–8.

[20] Bowen DJ, Collins PW. An amino acid polymorphism in von Willebrand factor correlates with increased susceptibility to proteolysis by ADAMTS13. Blood 2004;103:941–7.

[21] McKinnon TAJ, Chion ACK, Millington AJ, Lane DA, Laffan MA. N-linked glycosylation of VWF modulates its interaction with ADAMTS13. Blood 2008;111:3042–9.

[22] Gallinaro L, Cattini MG, Sztukowska M, Padrini R, Sartorello F, Pontara E, Bertomoro A, Daidone V, Pagnan A, Casonato A. A shorter von Willebrand factor survival in O blood group subjects explains how ABO determinants influence plasma von Willebrand factor. Blood 2008;111:3540–5.

[23] Mannucci PM, Lombardi R, Castaman G, Dent JA, Lattuada A, Rodeghiero F, Zimmerman TS. von Willebrand disease "Vicenza" with larger-than-normal (supranormal) von Willebrand factor multimers. Blood 1988;7:65–70.

[24] Rawley O, O'Sullivan JM, Chion A, Keyes S, Lavin M, van Rooijen N, Brophy TM, Fallon P, Preston RJS, O'Donnell JS. Von Willebrand factor arginine 1205 substitution results in accelerated macrophage-dependent clearance in vivo. J Thromb Haemost 2015;13:821–6.

[25] Rydz N, Swystun LL, Notley C, Paterson AD, Riches JJ, Sponagle K, Boonyawat B, Montgomery RR, James PD, Lilicrap D. The D-type lectin receptor CLEC4M binds, internalizes, and clears von Willebrand factor and contributes to the variation in plasma von Willebrand factor levels. Blood 2013;121:5228–37.

[26] Leebeek FWG, Eikenboom JCJ. Von Willebrand's disease. N Engl J Med 2016;375:2067–80.

[27] Lavin M, Aguila S, Schneppenheim S, Dalton N, Jones KL, O'Sullivan JM, O'Connell NM, Ryan K, White B, Byrne M, Rafferty M, Doyle MM, Nolan M, Preston RJS, Budde U, James P, Di Paola J, O'Donnell JS. Novel insights into the clinical phenotype and pathophysiology underlying low VWF levels. Blood 2017;130:2344–53.

[28] Rydz N, Grabell J, Lillicrap D, James PD. Changes in von Willebrand factor level and von Willebrand activity with age in type 1 von Willebrand disease. Haemophilia 2015;21:636–41.

[29] Abou-Ismail MY, Ogunbayo GO, Kouides PA. Outgrowing the laboratory diagnosis of type 1 von Willebrand's disease: a two decade study. Blood 2016;128:871 [abstract].

[30] Seaman CD, Ragni M. The effect of aging on von Willebrand factor levels and bleeding risk in type 1 von Willebrand disease. Blood 2016;128:2584.

[31] Sanders YV, Giezenaar MA, Laros-van Gorkom BA, Meijer K, van der Bom JG, Cnossen MH, Nijziel MR, Ypma PF, Fijnvandraat K, Eikenboom J, Mauser-Bunschoten EP, Leebeek FW, WiN study group. von Willebrand disease and aging: an evolving phenotype. J Thromb Haemost 2014;12:1066–75.

[32] Boender J, Eikenboom J, Fijnvandraat K, van Heerde W, Meijer K, Mauser-Bunschoten E, Cnossen M, de Meris J, Laros-van Gorkom B, van der Bom A, Leebeek F, for the WiN Study Group. No association between normalization of VWF levels and bleeding phenotype in patients with type 1 VWD-the WiN Study. Blood 2016;128:2577 [abstract].

[33] Stakiw J, Bowman M, Hegadorn C, Pruss C, Notley C, Groot E, Lenting PJ, Rapson D, Lillicrap D, James P. The effect of exercise on von Willebrand factor and ADAMTS13 in individuals with type 1 and type 2B von Willebrand disease. J Thromb Haemost 2008;6:90–6.

[34] Lillicrap D. Von Willebrand disease: advances in pathogenetic understanding, diagnosis, and therapy. Blood 2013;122:3735–40.

[35] Castaman G, Federici AB, Tosetto A, La Marca S, Stufano F, Mannucci PM, Rodeghiero F. Different bleeding risk in type 2A and 2M von Willebrand disease: a 2-year prospective study in 107 patients. J Thromb Haemost 2012;10:632–8.

[36] Hassenpflug WA, Budde U, Obser T, Angerhaus D, Drewke E, Schneppenheim S, Schneppenheim R. Impact of mutations in the von Willebrand factor A2 domain on ADAMTS13-dependent proteolysis. Blood 2006;107:2339–45.

[37] Jacobi PM, Gill J, Flood VH, Jakab DA, Friedman KD, Haberichter SL. Intersection of mechanisms of type 2A VWD through defects in VWF multimerization, secretion, ADAMTS-13 susceptibility, and regulated storage. Blood 2012;119:4543–53.

[38] Ruggeri ZM, Nilsson IM, Lombardi R, Holmberg L, Zimmerman TS. Aberrant multimeric structure of von Willebrand factor in a new variant of von Willebrand's disease (type IIC). J Clin Invest 1982;70:1124–7.

[39] Ledford MR, Rabinowitz I, Sadler JE, Kent JW, Civantos F. New variant of von Willebrand disease type II with markedly increased levels of von Willebrand factor antigen and dominant mode of inheritance: von Willebrand disease type IIC Miami. Blood 1993;82:169–75.

[40] Obser T, Ledford-Kraemer M, Oyen F, Brehm MA, Denis CV, Marschalek R, Montgomery RR, Sadler JE, Schneppenheim S, Budde U, Schneppenheim R. Identification and characterization of the elusive mutation causing the historical von Willebrand disease type IIC Miami. J Thromb Haemost 2016;14:1725–35.

[41] Rick ME, Williams SB, Sacher RA, McKeown LP. Thrombocytopenia associated with pregnancy in a patient with type IIB von Willebrand's disease. Blood 1987;69:786–9.

[42] Federici AB, Mannucci PM, Castaman G, Baronciani L, Bucciarelli P, Canciani MT, Pecci A, Lenting PJ, De Groot PG. Clinical and molecular predictors of thrombocytopenia and risk of bleeding in patients with von Willebrand disease type 2B: a cohort study of 67 patients. Blood 2009;113:526–34.

[43] Casari C, Du V, Wu YP, Kauskot A, de Groot PG, Christophe OD, Denis CV, de Laat B, Lenting PJ. Accelerated uptake of VWF/platelet complexes in macrophages contributes to VWD type 2B-associated thrombocytopenia. Blood 2013;122:2893–902.

[44] Flood VH, Schlauderaff AC, Haberichter SL, Slobodianuk TL, Jacobi PM, Belissimo DB, Christopherson PA, Friedman KD, Gill JC, Hoffmann RG, Montgomery RR. Crucial role for the VWF A1 domain in binding to type IV collagen. Blood 2015;125:2297–304.

[45] Howard MA, Perkin J, Salem HH, Firkin BG. The agglutination of human platelets by botrocetin: evidence that botrocetin and ristocetin act at different sites on the factor VIII molecule and platelet membrane. Br J Haematol 1984;57:25–35.

[46] Hillery CA, Mancuso DJ, Sadler JE, Ponder JW, Jozwiak MA, Christopherson PA, Gill JC, Scott JP, Montgomery RR. Type 2M von Willebrand disease: F606I and I662F mutations in the glycoprotein Ib binding domain selectively impair ristocetin- but not botrocetin-mediated binding of von Willebrand factor to platelets. Blood 1998;91:1572–81.

[47] Chen J, Hinckley JD, Haberichter S, Jacobi P, Montgomery RR, Flood VW, Wong R, Interlandi G, Chung DW, Lopez JA, Di Paola J. Variable content of von Willebrand factor mutant monomer drives the phenotypic variability in a family with von Willebrand disease. Blood 2015;126:262–9.

[48] Casonato A, Daidone V, Barbon G, Pontara E, Di Pasquale I, Galinaro L, Marullo L, Bertorelle G. A common ancestor more than 10,000 years old for patients with R854Q-related type 2N von Willebrand's disease in Italy. Haematologica 2013;98:147–52.

[49] Schneppenheim R, Budde U, Krey S, Drewke E, Bergmann F, Lechler E, Oldenburg J, Schwaab R. Results of a screening for von Willebrand disease type 2N in patients with suspected haemophilia A or von Willebrand disease type 1. Thromb Haemost 1996;76:598–602.

[50] Nishino M, Girma J-P, Rothschild C, Fressinaud E, Meyer D. New variant of von Willebrand disease with defective binding to factor VIII. Blood 1989;74:1591–9.

[51] Mazurier C, Dieval J, Jorieux J, Delobel J, Goudemand M. A new von Willebrand factor (vWF) defect in a patient with factor VIII (FVIII) deficiency but with normal levels and multimeric patterns of both plasma and platelet vWF. Characterization of abnormal VWF/FVIII interaction. Blood 1990;75:20–6.

[52] Mazurier C, Gaucher C, Jorieux S, Goudemand M. Biological effect of desmopressin in eight patients with type 2N ("Normandy") von Willebrand disease. Br J Haematol 1994;88:849–54.

[53] Sanders YV, Groeneveld D, Meijer K, Fijnvandraat K, Cnossen MH, van der Bom JG, Coppens M, de Meris J, Laros-van Gorkom BA, Mauser-Bunschoten EP, Leebeek FW, Eikenboom J, WiN study group. von Willebrand factor propeptide and the phenotypic classification of von Willebrand disease. Blood 2015;125:3006–13.

[54] Bowman M, Tuttle A, Notley C, Brown C, Tinlin S, Deforest M, Leggo J, Blanchette VS, Lillicrap D, James P, Association of Hemophilia Clinic Directors of Canada. The genetics of Canadian type 3 von Willebrand disease: further evidence for co-dominant inheritance of mutant alleles. J Thromb Haemost 2013;11:512–20.

[55] Zhang ZP, Blombäck M, Nyman D, Anvret M. Mutations of von Willebrand factor gene in families with von Willebrand disease in the Åland islands. Proc Natl Acad Sci U S A 1993;90:7937–40.

[56] Ramsay DM, Buist TAS, Macleod DAD, Heading RC. Persistent gastrointestinal bleeding due to angiodysplasia of the gut in von Willebrand's disease. Lancet 1976;ii:275–8.

[57] Ahr DJ, Rickles FR, Hoyer LW, O'Leary DS, Conrad ME. von Willebrand's disease and hemorrhagic telangiectasia: association of two complex disorders of hemostasis resulting in life-threatening hemorrhage. Am J Med 1977;62:452–8.

[58] Franchini M, Mannucci PM. Gastrointestinal angiodysplasia and bleeding in von Willebrand disease. Thromb Haemost 2014;112:427–31.

[59] Fressinaud E, Meyer D. International survey of patients with von Willebrand disease and angiodysplasia. Thromb Haemost 1993;70:546.

[60] Selvam SN, Casey LJ, Bowman ML, Hawke LG, Longmore AJ, Mewburn J, Ormiston ML, Archer SL, Maurice DH, James P. Abnormal angiogenesis in blood outgrowth endothelial cells derived from von Willebrand disease patients. Blood Coagul Fibrinolysis 2017;28:521–33.

[61] Starke RD, Ferraro F, Paschalaki KE, Dryden NH, McKinnon TAJ, Sutton RE, Payne EM, Haskard DO, Hughes AD, Cutler DF, Laffan MA, Randi AM. Endothelial von Willebrand factor regulates angiogenesis. Blood 2011;117:1071–80.

[62] Randi AM, Laffan MA. Von Willebrand factor and angiogenesis: basic and applied issues. J Thromb Haemost 2017;15:13–20.

[63] Tosetto A, Rodeghiero F, Castaman G, Goodeve A, Federici AB, Batlle J, Meyer D, Fressinaud E, Mazurier C, Goudemand J, Eikenboom J, Schneppenheim R, Budde U, Ingerslev J, Vorlova Z, Habart D, Holmberg L, Lethagen S, Pasi J, Peake I. A quantitative analysis of bleeding symptoms in type 1 von Willebrand disease: results from a multicenter European study (MCMDM-1 VWD). J Thromb Haemost 2006;4:766–73.

[64] Rodeghiero F, Tosetto A, Abshire T, Arnold DM, Coller B, James P, Neunert C, Lillicrap D, ISTH/SSC joint VWF and Perinatal/Pediatric Hemostasis Subcommittees Working Group. ISTH/SSC bleeding assessment tool: a standardized questionnaire and a proposal for a new bleeding score for inherited bleeding disorders. J Thromb Haemost 2010;8:2063–5.

[65] Bowman M, Mundell G, Grabell J, Hopman WM, Rapson D, Lillicrap D, James P. Generation and validation of the Condensed MCMDM-1VWD Bleeding Questionnaire for von Willebrand disease. J Thromb Haemost 2008;6:2062–6.

[66] Elbatarny M, Mollah S, Grabell J, Bae S, Deforest M, Tuttle A, Hopman W, Clark DS, Mauer AC, Bowman M, Riddel J, Christopherson PA, Montgomery RR, Zimmerman Program Investigators, Rand ML, Coller B, James PD. Normal range of bleeding scores for the ISTH-BAT: adult and pediatric data from the merging project. Haemophilia 2014;20:831–5.

[67] Federici AB, Bucciarelli P, Castaman G, Mazzucconi MG, Morfini M, Rocino A, Schiavoni M, Peyvandi F, Rodeghiero F, Mannucci PM. The bleeding score predicts clinical outcomes and replacement therapy in adults with von Willebrand disease. Blood 2014;123:4037–44.

[68] Cattaneo M, Federici AB, Lecchi A, Agati B, Lombardi R, Stabile F, Bucciarelli P. Evaluation of the PFA-100® system in the diagnosis and therapeutic monitoring of patients with von Willebrand disease. Thromb Haemost 1999;82:35–9.

[69] Lippi G, Franchini M, Brocco G, Manzato F. Influence of the ABO blood type on the platelet function analyzer PFA-100. Thromb Haemost 2001;85:369–70.

[70] Naik S, Teruya J, Dietrich JE, Jariwala P, Soundar E, Venkateswaran L. Utility of platelet function analyzer as a screening tool for the diagnosis of von Willebrand disease in adolescents with menorrhagia. Pediatr Blood Cancer 2013;60:1184–7.

[71] Chediak J, Maxey B, Telfer M. Determination of factor VIII-related antigen using commercial antisera. Am J Clin Pathol 1977;67:462–9.

[72] Counts RB. Solid-phase immunoradiometric assay of factor-VIII protein. Br J Haematol 1975;31:429–36.

[73] Ruggeri ZM, Mannucci PM, Jeffcoate SL, Ingram GIC. Immunoradiometric assay of factor VIII related antigen with observations in 32 patients with von Willebrand's disease. Br J Haematol 1976;33:221–32.

[74] Nilsson IM. Report of the working party on factor VIII-related antigens. Thromb Haemost 1978;39:511–20.

[75] Favaloro EJ, Koutts J. 2B or not 2B? Masquerading as von Willebrand disease? J Thromb Haemost 2012;10:317–9.

[76] Bodo I, Eikenboom J, Montgomery R, Patzke J, Schneppenheim R, Di Paola J. Platelet-dependent von Willebrand factor activity. Nomenclature and methodology: communication from the SSC of the ISTH. J Thromb Haemost 2015;13:1345–50.

[77] Favaloro EJ, Mohammed S, McDonald J. Validation of improved performance characteristics for the automated von Willebrand factor ristocetin cofactor activity assay. J Thromb Haemost 2010;8:2842–4.

[78] Flood VH, Friedman KD, Gill JC, Morateck PA, Wren JS, Scott JP, Montgomery RR. Limitations of the ristocetin cofactor assay in measurement of von Willebrand factor function. J Thromb Haemost 2009;7:1832–9.

[79] Stufano F, Lawrie AS, La Marca S, Berbenni C, Baronciani L, Peyvandi F. A two-centre comparative evaluation of new automated assays for von Willebrand factor ristocetin cofactor activity and antigen. Haemophilia 2014;20:147–53.

[80] Patzke J, Budde U, Huber A, Méndez A, Muth H, Obser T, Peerschke E, Wilkens M, Schneppenheim R. Performance evaluation and multicentre study of a von Willebrand factor activity assay based on GPIb binding in the absence of ristocetin. Blood Coagul Fibrinolysis 2014;25:860–70.

[81] Favaloro EJ, Mohammed S. Evaluation of a von Willebrand factor three test panel and chemiluminescent-based assay system for identification of, and therapy monitoring in, von Willebrand disease. Thromb Res 2016;141:202–11.

[82] Graf L, Moffat KA, Carlino SA, Chan AK, Iorio A, Giulivi A, Hayward CP. Evaluation of an automated method for measuring von Willebrand factor activity in clinical samples without ristocetin. Int J Lab Hematol 2014;36:341–51.

[83] Flood VH, Gill JC, Morateck PA, Christopherson PA, Friedman KD, Haberichter SL, Hoffmann RG, Montgomery RR. Gain-of-function GPIb ELISA assay for VWF activity in the Zimmerman Program for the Molecular and Clinical Biology of VWD. Blood 2011;117:e67–74.

[84] Read MS, Smith SV, Lamb MA, Brinkhous KM. Role of botrocetin in platelet agglutination: formation of an activated complex of botrocetin and von Willebrand factor. Blood 1989;74:1031–5.

[85] Fukuda K, Doggett T, Laurenzi IJ, Liddington RC, Diacovo TG. The snake venom protein botrocetin acts as a biological brace to promote dysfunctional platelet aggregation. Nat Struct Mol Biol 2005;12:152–9.

[86] Flood VH, Gill JC, Christopherson PA, Bellissimo DB, Friedman KD, Haberichter SL, Lentz SR, Montgomery RR. Critical von Willebrand factor AI domain residues influence type VI collagen binding. J Thromb Haemost 2012;10:1417–24.

[87] Flood VH, Gill JC, Christopherson PA, Wren JS, Friedman KD, Haberichter SL, Hoffmann RG, Montgomery RR. Comparison of type I, type III and type VI collagen binding assays in diagnosis of von Willebrand disease. J Thromb Haemost 2012;10:1425–32.

[88] Miller CH, Kelley L, Green D. Diagnosis of von Willebrand disease type 2N: a simplified method for measurement of factor VIII binding to von Willebrand factor. Am J Hematol 1998;58:311–8.

[89] Caron C, Mazurier C, Goudemand J. Large experience with a factor VIII binding assay of plasma von Willebrand factor using commercial reagents. Br J Hematol 2002;117:716–8.

[90] Vlot AJ, Koppelman SJ, Meijers JCM, Damas C, van den Berg HM, Bouma BN, Sixma JJ, Willems GM. Kinetics of factor VIII-von Willebrand factor association. Blood 1996;87:1809–16.

[91] Haberichter S. von Willebrand factor propeptide: biology and clinical utility. Blood 2015;126:1753–61.

[92] Eikenboom J, Federici AB, Dirven RJ, Castaman G, Rodeghiero F, Budde U, Schneppenheim R, Batlle J, Canciani MT, Goudeman J, Peake I, Goodeve A, MCMDM-1VWD Study Group. VWF propeptide and ratios between VWF, VWF propeptide, and FVIII in the characterization of type I von Willebrand disease. Blood 2013;121:2336–9.

[93] Haberichter SL, Balistreri M, Christopherson P, Morateck P, Gavazova S, Bellissimo DB, Manco-Johnson MJ, Gill JC, Montgomery RR. Assay of the von Willebrand factor (VWF) propeptide to identify patients with type 1 von Willebrand disease with decreased VWF survival. Blood 2006;108:3344–51.

[94] Archer NM, Samnaliev M, Grace R, Brugnara C. The utility of the desmopressin challenge test in children with low von Willebrand factor. Br J Haematol 2015;170:884–6.

[95] Castaman G, Lethagen S, Federici AB, Tosetto A, Goodeve A, Budde U, Batlle J, Meyer D, Mazurier C, Fressinaud E, Goudemand J, Eikenboom J, Schneppenheim R, Ingerslev J, Vorlova Z, Habart D, Holmberg L, Pasi J, Hill F, Peake I, Rodeghiero F. Response to desmopressin is influenced by the genotype and phenotype in type 1 von Willebrand disease (VWD): results from the European Study MCMDM-1VWD. Blood 2008;111:3531–9.

[96] Michiels JJ, Smejkal P, Penka M, Batorova A, Pricangova T, Budde U, Vangenechten I, Gadisseur A. Diagnostic differentiation of von Willebrand disease types 1 and 2 by von Willebrand factor multimer analysis and desmopressin challenge test. Clin Appl Thromb Hemost 2017;23:518–31.

[97] Holmberg L, Nilsson IM, Borge L, Gunnarsson M, Sjörin E. Platelet aggregation induced by 1-desamino-8-D-arginine vasopressin (desmopressin) in type IIB von Willebrand's disease. N Engl J Med 1983;309:816–21.

[98] Bellissimo DB, Christopherson PA, Flood VH, Gill JC, Friedman KD, Haberichter SL, Shapiro AD, Abshire TC, Leissinger C, Hoots WK, Lusher JM, Ragni MV, Montgomery RR. VWF mutations and new sequence variations identified in healthy controls are more frequent in the African-American population. Blood 2012;119:2135–40.

[99] Swystun LL, James PD. Genetic diagnosis in hemophilia and von Willebrand disease. Blood Rev 2017;31:47–56.

[100] Warner PE, Critchley HO, Lumsden MA, Campbell-Brown M, Douglas A, Murray GD. Menorrhagia I: measured blood loss, clinical features, and outcome in women with heavy periods: a survey with follow-up data. Am J Obstet Gynecol 2004;190:1216–23.

[101] Byams VR, Kouides PA, Kulkarni R, Baker JR, Brown DL, Gill JC, Grant AM, James AH, Konkle BA, Maahs J, Dumas MM, McAlister S, Nance D, Nugent D, Philipp CS, Soucie JM, Stang E, Haemophilia Treatment Centres Network Investigators. Surveillance of female patients with inherited bleeding disorders in United States Haemophilia Treatment Centres. Haemophilia 2011;17(Suppl. 1):6–13.

[102] James A, Matchar DB, Myers ER. Testing for von Willebrand disease in women with menorrhagia: a systematic review. Obstet Gynecol 2004;104:381–8.

[103] Mandalaki T, Louizou C, Dimitriadou C, Symeonidis P. Variations in factor VIII during the menstrual cycle in normal women. N Engl J Med 1980;302:1093–4.

[104] Gill JC, Endres-Brooks J, Bauer PJ, Marks Jr WJ, Montgomery RR. The effect of ABO blood group on the diagnosis of von Willebrand disease. Blood 1987;69:1691–5.

[105] Bucciarelli P, Siboni SM, Stufano F, Biguzzi E, Canciani MT, Baronciani L, Pagliari MT, La Marca S, Mistretta C, Rosendaal FR, Peyvandi F. Predictors of von Willebrand disease diagnosis in individuals with borderline von Willebrand factor plasma levels. J Thromb Haemost 2015;13:228–36.

[106] Weiss HJ. The bleeding tendency in patients with low von Willebrand factor and type 1 phenotype is greater in the presence of impaired collagen-induced platelet aggregation. J Thromb Haemost 2004;2:198–9.

[107] DiPaola J, Federici AB, Mannucci PM, Canciani MT, Kritzik M, Kunicki TJ, Nugent D. Low platelet $\alpha_2\beta_1$ levels in type I von Willebrand disease correlate with impaired platelet function in a high shear stress system. Blood 1999;93:3578–82.

[108] Mazurier C. von Willebrand disease masquerading as haemophilia A. Thromb Haemost 1992;67:391–6.

[109] James PD, Notley C, Hegadorn C, Leggo J, Tuttle A, Tinlin S, Brown C, Andrews C, Labelle A, Chirinian Y, O'Brien L, Othman M, Rivard G, Rapson D, Hough C, Lillicrap D. The mutational spectrum of type 1 von Willebrand disease: results from a Canadian cohort study. Blood 2007;109:145–54.

[110] Goodeve A, Eikenboom J, Castaman G, Rodeghiero F, Federici AB, Batlle J, Meyer D, Mazurier C, Goudemand J, Schneppenheim R, Budde U, Ingerslev J, Habart D,

Vorlova Z, Holmberg L, Lethagen S, Pasi J, Hill F, Hashemi Soteh M, Baronciani L, Hallden C, Guilliatt A, Lester W, Peake I. Phenotype and genotype of a cohort of families historically diagnosed with type 1 von Willebrand disease in the European study, Molecular and Clinical Markers for the Diagnosis and Management of Type 1 von Willebrand Disease (MCMDM-1VWD). Blood 2007;109:112–21.

[111] Caen JP, Nurden AT, Jeanneau C, Michel H, Tobelem G, Levy-Toledano S, Sultan Y, Valensi F, Bernard J. Bernard-Soulier syndrome: a new platelet glycoprotein abnormality. Its relationship with platelet adhesion to subendothelium and with the factor VIII von Willebrand protein. J Lab Clin Med 1976;87:586–96.

[112] Savoia A, Kunishima S, De Rocco D, Zieger B, Rand ML, Pujol-Moix N, Caliskan U, Tokgoz H, Pecci A, Noris P, Srivastava A, Ward C, Morel-Kopp MC, Alessi MC, Bellucci S, Beurrier P, de Maistre E, Favier R, Hézard N, Hurtaud-Roux MF, Latger-Cannard V, Lavenu-Bombled C, Proulle V, Meunier S, Négrier C, Nurden A, Randrianaivo H, Fabris F, Platokouki H, Rosenberg N, Hadj Kacem B, Heller PG, Karimi M, Balduini CL, Pastore A, Lanza F. Spectrum of the mutations in Bernard-Soulier syndrome. Hum Mutat 2014;35:1033–45.

[113] Weiss HJ, Meyer D, Rabinowitz R, Pietu G, Girma J-P, Vicic WJ, Rogers J. An intrinsic platelet defect with aggregation by unmodified human factor VIII/von Willebrand factor and enhanced adsorption of its high-molecular-weight multimers. N Engl J Med 1982;306:326–33.

[114] Russell SD, Roth GJ. Pseudo-von Willebrand disease: a mutation in the platelet glycoprotein Ibα gene associated with a hyperactive surface receptor. Blood 1993;81:1787–91.

[115] Penel-Page M, Meunier S, Fretigny M, Le Quellec S, Boisseau P, Vinciguerra C, Ternisien C, Rugeri L. Differential diagnosis of neonatal alloimmune thrombocytopenia: type 2B von Willebrand disease. Platelets 2017;24:1–4.

Recommended Reading

[1] Sharma R, Flood VH. Advances in the diagnosis and treatment of Von Willebrand disease. Blood 2017;29, https://doi.org/10.1182/blood-2017-05-782029.

[2] Bowman M, Hopman WM, Rapson D, et al. The prevalence of symptomatic von Willebrand disease in primary care practice. J Thromb Haemost 2010;8:213–6.

[3] Miller CH, Haff E, Platt SJ, et al. Measurement of von Willebrand factor activity: relative effects of ABO blood type and race. J Thromb Haemost 2003;1:2191–7.

[4] Lillicrap D. Von Willebrand disease: advances in pathogenetic understanding, diagnosis, and therapy. Blood 2013;122:3735–40.

[5] Elbatarny M, Mollah S, Grabell J, et al. Normal range of bleeding scores for the ISTH-BAT: adult and pediatric data from the merging project. Haemophilia 2014;20:831–5.

[6] Bodo I, Eikenboom J, Montgomery R, et al. Platelet-dependent von Willebrand factor activity. Nomenclature and methodology: communication from the SSC of the ISTH. J Thromb Haemost 2015;13:1345–50.

[7] Sanders YV, Groeneveld D, Meijer K, et al. von Willebrand factor propeptide and the phenotypic classification of von Willebrand disease. Blood 2015;125:3006–13.

[8] Flood VH, Christopherson PA, Gill JC, et al. Clinical and laboratory variability in a cohort of patients diagnosed with type 1 VWD in the United States. Blood 2016;127:2481–8.

[9] Leebeek FWG, Eikenboom JCJ. Von Willebrand's disease. N Engl J Med 2016;375:2067–80.

[10] Michiels JJ, Smejkal P, Penka M, et al. Diagnostic differentiation of von Willebrand disease types 1 and 2 by von Willebrand factor multimer analysis and desmopressin challenge test. Clin Appl Thromb Hemost 2017;23:518–31.

11

Treatment of Von Willebrand Disease

The treatment of Von Willebrand disease (VWD) is individualized because the intensity and pattern of bleeding, and the responsiveness to therapeutic agents, varies from patient to patient and between and within VWD types. However, there are a few principles that are broadly applicable. Aspirin and nonsteroidal antiinflammatory drugs that affect platelet function should be avoided; when considered essential, they are usually coadministered with proton pump inhibitors to prevent gastrointestinal bleeding. Intramuscular injections risk muscle hematomas, but vaccinations and subcutaneous injections are safe. Lastly, careful consideration should be given to the long-term consequences of interventions that are usually safe in most patients but could be deleterious in people with life-long bleeding disorders. For example, radiation therapy often induces colitis and cystitis, and can exacerbate bleeding in patients with inherited coagulopathies, as described in the following case vignette.

A premenopausal woman with type 3 VWD and intermittent vaginal bleeding was found to have carcinoma in situ of the cervix. A hysterectomy was performed, followed by routine postoperative radiation therapy. She developed radiation cystitis, proctitis, and enteritis, accompanied by persistent hematuria and melena. Transfusions of blood and megadoses of VWF concentrate were administered, but continued bleeding required performance of a colectomy and ileostomy. In addition, hemorrhage from the bladder was unremitting despite bladder irrigations with antifibrinolytic agents and other hemostatics, and slowed only after placement of bilateral nephrostomy tubes. Tests for the presence of alloantibodies were negative. Her condition finally stabilized, but VWF concentrate infusions were required twice daily to prevent recurrent bleeding from the gastrointestinal or genitourinary tract. There was no recurrence of the cervical cancer, but she died several years later from systemic infection arising from the nephrostomy tubes. In this patient, the adverse effects of radiation therapy greatly enhanced the mucosal bleeding characteristic of severe VWD and resulted in unacceptable morbidity and mortality.

The main therapeutic agents available for the treatment of VWD are desmopressin and VWF concentrates; the latter are prescribed for patients who do not have an adequate hemostatic response to desmopressin or in whom it is contraindicated.

DESMOPRESSIN

Desmopressin, 1-deamino-8-D-arginine vasopressin (DDAVP), can be administered by intravenous, subcutaneous, or intranasal routes, and stimulates the release of FVIII/VWF from endothelial cells. Intravenous doses of 0.3 μg/kg in 50 mL normal saline are infused over 20 min, or the same dose of concentrated desmopressin can be given subcutaneously. Peak plasma VWF concentrations are achieved within 30–60 min of intravenous infusion or subcutaneous injection, raising FVIII/VWF levels two- to fourfold. Measuring VWF level 4 h after the dose provides information about the clearance of the molecule. The intranasal spray is given as 150 μg in each nostril (or only one nostril if body weight is <50 kg) using a concentrated form of desmopressin (1.5 mg/mL, Stimate, CSL Behring); [Warning: the concentration of desmopressin supplied for diabetes insipidus or enuresis is only 0.15 mg/mL and should NOT be prescribed for VWD]. If bleeding recurs, additional doses can be given daily for up to 4 days, but the response generally becomes less vigorous with repeated dosing (tachyphylaxis). A good response is anticipated in most patients with the less severe forms of type 1 VWD; a test dose will establish the patient's responsiveness [1].

The adverse effects of desmopressin are flushing and headache; less commonly, there may be fluid retention sufficiently severe to cause hyponatremia and even seizures, especially if repeated doses are given to children or the elderly. Hyponatremia is also a concern in postoperative patients given desmopressin and is preventable by careful fluid management. There are rare reports of stroke or myocardial infarction in individuals with thrombotic risk factors receiving desmopressin infusions.

VWF CONCENTRATES

In the 1970s, studies of patients with severe VWD showed that infusions of either cryoprecipitate or a glycine-precipitated FVIII concentrate increased the levels of FVIII/VWF, but only cryoprecipitate controlled bleeding and shortened the skin bleeding time [2]. The increase in FVIII/VWF persisted for 48–72 h, but the bleeding time was shortened for only 6–8 h [3]. Although both the FVIII concentrate and cryoprecipitate contained VWF, only the latter had the relatively short-lived, but highly effective VWF high-molecular-weight multimers (HMWM); if HMWM are included in VWF concentrates, they are also effective for the treatment of VWD. Concentrates currently used in the United States will be described here; another product, Wilfactin (LFB S.A., Framingham, Massachusetts), a VWF concentrate with a very low FVIII content, is available outside the United States.

Alphanate (Fanhdi outside the United States; Grifols Biologicals Inc., Los Angeles, California) is prepared from pooled human plasma by cryoprecipitation of FVIII, fractional solubilization, and further purification using heparin-coupled, crosslinked agarose. It is stabilized with human albumin. Virus inactivation is achieved with solvent-detergent and heat treatment. The final product is labeled with VWF:RCo activity expressed in IU VWF:RCo/vial; the ratio of VWF to FVIII is 0.5:1. The product is indicated for the management of surgical and/or invasive procedures in adult and pediatric patients with VWD in whom desmopressin is either ineffective or contraindicated. The efficacy of the concentrate, using a standard dose of 40 IU/kg, was reported to be nearly 95% in a clinical trial of 150 bleeding episodes in 60 patients, and adequate hemostasis was secured in 98% of surgical procedures [4]. The preoperative dose for major surgery in adults is 60 IU/kg with subsequent doses of 40–60 IU/kg at 8–12 h intervals, and the pediatric dose is 75 IU/kg with subsequent doses of 50–75 IU/kg every 8–12 h. It is not indicated for patients with severe VWD (Type 3) undergoing major surgery. Alphanate was approved by the FDA in 1978.

Humate-P (Haemate-P outside the United States; CSL Behring LLC, Kankakee, Illinois) was approved for the treatment of VWD in 1986. It is prepared from the cold insoluble fraction of plasma, using glycine, aluminum hydroxide, and NaCl precipitation, followed by heating to inactivate viruses. The average ratio of ristocetin cofactor to FVIII is 2.4:1. It is indicated for the treatment of spontaneous and trauma-induced bleeding episodes, and the prevention of excessive bleeding during and after surgery in VWD patients in whom the use of desmopressin is known or suspected to be inadequate. The recommended dose to control bleeding is 40–80 IU/kg of ristocetin cofactor, which corresponds to 17–33 IU/kg of FVIII in Humate-P. The expected in vivo recovery is 1.5 IU/dL for each IU/kg of ristocetin cofactor infused and 2 IU/dL for each 1 IU/kg of FVIII infused. Doses are repeated every 8–12 h based on the extent and location of the hemorrhage. For major surgery, a loading dose sufficient to achieve a ristocetin cofactor level of 100 IU/dL is given 1–2 h preoperatively, followed by half the loading dose every 8–12 h. Approximately half these doses are needed for minor surgery, but trough ristocetin cofactor and FVIII levels should be measured daily in all patients and doses adjusted to avoid over- as well as underdosing. For major surgery, the target trough VWF/FVIII levels are >0.5 IU/mL for the first 3 postoperative days, decreasing to >0.3 IU/mL after day 3, and for minor surgery, ≥0.3 IU/mL for VWF up to day 3 and >0.3 IU/mL for FVIII after day 3. Humate-P is not indicated for the prophylaxis of spontaneous bleeding episodes in patients with VWD.

Wilate (Octapharma USA, Inc., Hoboken, New Jersey) is manufactured from cryoprecipitate. After aluminum hydroxide precipitation and two chromatography steps, the material is filtered and treated with solvent-detergent and terminal dry heat to remove viruses. It does not contain albumin or preservatives. The specific activity is ≥60 IU of ristocetin cofactor per mg of protein, and the ratio of ristocetin cofactor to FVIII is 1:1. Wilate is indicated for the on-demand treatment of hemorrhages and the perioperative management of bleeding. For major hemorrhages or surgery, an initial dose of ristocetin cofactor of 40–60 IU/kg is followed by a maintenance dose of 20–40 IU/kg at 12–24 h; for minor surgery, the loading dose is 30–60 IU/kg and the maintenance dose is 15–30 IU. Monitoring is performed 30 min after the loading dose and at least daily to assure hemostatic levels are achieved (trough VWF levels >0.5 IU/mL for major surgery and >0.3 IU/mL for minor surgery) and that excessive increases in FVIII do not occur.

The efficacy and safety of Wilate was evaluated in 28 patients undergoing 30 surgical procedures [5]. Most of the operations were major, and 21 patients had type 3 VWD. Treatment success was reported in 29 of the 30 procedures, as determined by measurements of blood loss, transfusion requirements, and postoperative bleeding. There was no accumulation of FVIII, no thromboembolic episodes, and inhibitors to VWF or FVIII did not develop. Wilate was approved by the FDA in 2009.

Vonvendi (Baxalta US, Inc., Westlake Village, California) is a recombinant Von Willebrand factor (rVWF) expressed in Chinese Hamster Ovary (CHO) cells. The final product contains trace amounts of mouse immunoglobulin derived from the immunoaffinity purification, CHO protein, and rFurin used for processing; in addition, glycine, mannitol, trehalose-dihydrate, and polysorbate 80 are added as stabilizers. Vonvendi contains HMWM because it is not exposed to proteolysis by ADAMTS13 during the manufacturing process, but has only a trace amount of rFVIII. It is approved for the on-demand treatment and control of bleeding episodes in adult patients with VWD.

The pharmacokinetics and safety of rVWF were examined in a phase 1 study of 32 patients with severe type 1 or type 3 VWD, and compared with a plasma-derived VWF concentrate [6]. Following infusion, the area under the plasma FVIII concentration curve was greater with rVWF than with plasma-derived VWF, and there was a larger secondary rise in FVIII at 72 h ($P < .01$). The two VWF products had similar multimer cleavage patterns attributable to ADAMTS13 proteolysis. Two patients with preexisting nonneutralizing anti-VWF-binding antibodies had a decrease in VWF multimers and activity, but no new inhibitors were detected.

Investigators conducted a phase 3 trial in 37 patients to examine the safety and efficacy of rVWF; each patient received up to four infusions for the treatment of bleeding episodes [7]. The first dose was combined with rFVIII and subsequent doses were given alone if FVIII levels were >0.4 IU/mL. The median doses of rVWF and rFVIII were 46.5 and 33.6 IU/kg, respectively. The HMWM of VWF increased 15 min after infusion and then declined over the next 12–24 h; the terminal half-life of the rVWF was 21.9 h. FVIII concentrations rose rapidly, achieving hemostatic levels within 6 h that were sustained for 72 h. Excellent control of bleeding was reported for nearly all hemorrhages and a single infusion was effective in 81.8%. One patient had two episodes of chest discomfort and tachycardia, which resolved during hospitalization, and no patient developed neutralizing antibodies. Vonvendi was approved by the FDA in 2015.

The recommended doses of VWF-containing concentrates for major and minor surgery and/or bleeding are displayed in Table 11.1.

TABLE 11.1   VWF Product Dosing, in IU/kg of Ristocetin Cofactor, for Major and Minor Surgery and/or Bleeding

Major surgery/bleeding			
Product	Loading dose (IU/kg) Goal: >1 IU/mL on day 1	Maintenance (IU/kg) Goal: >0.5 IU/mL	Duration (days)
Alphanate	60	40–60 q8–12 h	3–7
Humate-P	a	b	3[c]
Wilate	40–60	20–40 q12–24 h	6 or more
Vonvendi	50–80	40–60 q8–24 h	2–3
Minor surgery/bleeding			
Product	Loading dose (IU/kg) Goal >0.5–0.8 IU/mL on day 1	Maintenance (IU/kg) Goal: >0.3 IU/mL	Duration (days)
Alphanate	60	40–60 q8–12 h	1–3
Humate-P	a	b	>2
Wilate	30–60	15–30 q12–24 h	3
Vonvendi	40–50	40–50 q8–24 h	–

[a] *Manufacturer recommends that the dose be calculated as follows: change in ristocetin cofactor desired multiplied by body weight divided by in vivo recovery measured in the patient; dosing frequency varies from q6 to q12 h based on half-life.*
[b] *Maintenance dose is half the loading dose.*
[c] *Goal >0.3 IU/mL after day 3.*

It is recommended that laboratory monitoring be performed in all patients receiving VWF concentrates, but whether it is necessary to measure both VWF and FVIII, or just FVIII, is controversial. VWF activity in ristocetin cofactor units is displayed on the label of the concentrate vial used for dose preparation, and the treatment goals for major and minor surgery/bleeding are presented in ristocetin cofactor units (Table 11.1). This accounts for the recommendation that ristocetin cofactor activity be monitored. It is also necessary to monitor FVIII levels to detect accumulation of this procoagulant; levels >1.5 IU/mL (>150%) increase the risk of thrombosis. Mannucci and Franchini [8] argue that FVIII is the best predictor of hemostasis in surgical patients and measuring VWF is unnecessary. On the other hand, the content of FVIII varies among concentrates, and some patients might have normal basal levels of FVIII, albeit low ristocetin cofactor levels [9]. Therefore, current guidelines recommend that both activities be monitored [10].

All patients should have baseline assays of VWF (measured as ristocetin cofactor, GpIb-binding activity, or collagen-binding activity) and FVIII, and the measurements repeated after the loading dose of concentrate to assure that target levels are achieved. A trough level is obtained immediately preceding the next dose to determine whether hemostatic levels are being maintained; if inadequate, doses should be increased or the concentrate given more frequently. Peak and trough measurements should be obtained daily until wounds are healed and the risk of bleeding has subsided. To reduce the risk of thrombosis, trough levels of FVIII should not exceed 1.5 IU/mL; this is facilitated by the use of rVWF because it is possible to individually titrate the amount of rVWF and FVIII infused [6]. All patients having major surgery should receive anticoagulant prophylaxis postoperatively if they are not bleeding and their FVIII levels are >0.5 IU/mL.

Adverse reactions with plasma-derived concentrates are:

- Hypersensitivity reactions and anaphylaxis: the incidence is 5%–10% [11]
- Vasomotor reactions with rapid administration
- Alloantibodies: they develop in 5%–10% of type 3 VWD [12]
- Thromboembolic events: the incidence is 1.94 per 1000 [13]
- Hemolysis in patients with A, B, or AB blood groups receiving large doses
- Transmission of infectious agents not removed by the virucidal procedures

The adverse reactions with recombinant VWF include all of the aforementioned reactions except hemolysis and transmission of infectious agents. Alloantibodies inhibit the response to VWF concentrate and occur most often in type 3 patients with partial or complete VWF gene deletions and rarely in type 2B VWD [11,14]. Risk factors for alloantibody development are exposure to repeated doses of concentrate and a family history of anti-VWF antibodies. The alloantibodies are directed against epitopes on VWF and do not inhibit recombinant FVIII, which has been effective in controlling bleeding [15].

ADJUNCTIVE THERAPIES

Hormonal agents are often used to control heavy menstrual bleeding in women with congenital bleeding disorders. The levonorgestrel-releasing intrauterine system (Mirena, Bayer, New Jersey) provides effective nonsurgical management for heavy menstrual bleeding [16]. Scores derived from pictorial blood assessment charts showed a >70% reduction during the first 3 months of Mirena use that were sustained through 4 years. In other studies, the decrease in menstrual blood loss with the levonorgestrel system exceeded that of the combined oral contraceptive pill, although the latter is fairly effective and might be preferred by some patients. A pill containing estradiol valerate and dienogest was evaluated in 269 women through 7 cycles, and showed an 88% reduction in menstrual blood loss compared to 24% with placebo. Hormonal therapies are contraindicated in women with a history of venous thromboembolism, migraine with aura, and breast cancer.

Tranexamic acid and epsilon aminocaproic acid inhibit fibrinolysis by preventing plasminogen binding to fibrin. They retard the clot dissolution on mucosal surfaces that contributes to the hemorrhagic manifestations of VWD, and are often used in conjunction with desmopressin or VWF concentrate. Tranexamic acid is given in doses of 10–15 mg/kg every 8–12 h and epsilon aminocaproic acid is 50–60 mg/kg every 4–6 h, either orally or by intravenous infusion [17]. A 4.8% tranexamic acid solution can be used as a mouthwash to reduce bleeding in conjunction with oral surgery or trauma; a teaspoonful is swished every 6 h until a firm clot adheres to the site of injury.

Tranexamic acid is useful in the management of heavy menstrual bleeding. A metaanalysis of four clinical trials that compared antifibrinolytic drugs with other therapies such as hormonal agents noted that inhibiting fibrinolysis was associated with larger decreases in mean menstrual blood loss and greater improvements in measures of the quality of life as well as sex life [18]. Tranexamic acid (Lysteda, Ferring Pharmaceuticals Inc., Parsippany, New York) was evaluated in a randomized, placebo-controlled, phase 3 trial in 196 women with excessive menstrual bleeding [19]. Participants receiving tranexamic acid in a dose of 3.9 g/day had significantly greater reductions in blood loss than those on placebo (69.6 mL vs 12.6 mL, $P < .001$) and fewer limitations in work, social, and physical activities $(P < .01)$. There was no increase in hemoglobin or ferritin levels but the trial might have been too short to detect such changes. Adverse effects such as headache, back pain, and nausea were mild to moderate in severity. In particular, no participant developed thrombosis, the most serious complication of antifibrinolytic therapy. Tranexamic acid is FDA approved for the treatment of cyclic heavy menstrual bleeding.

Recombinant interleukin-11 (rIL-11, Neumega, Pfizer Laboratories, Philadelphia, Pennsylvania) is another agent with hemostatic efficacy in VWD. This cytokine is approved for the prevention of severe thrombocytopenia and the reduction in the need for platelet transfusions following myelosuppressive chemotherapy, but also appears to stimulate the synthesis of VWF. In patients with type 1 and type 2 VWD unresponsive to desmopressin, daily subcutaneous injections of rIL-11 for 1 week stimulated a 1.5- to 2-fold increase in VWF levels [20]. This was thought to represent new VWF synthesis as there was an increase in VWF mRNA and a further increase in VWF after desmopressin administration. A trial of rIL-11 in seven women with heavy menstrual bleeding refractory to hemostatic or hormonal agents reported a 50% decrease in pictorial blood assessment chart scores in 71% of patients and a shortening of bleeding duration by ≥ 2 days in 85% [21]. The agent was well tolerated with only minor fluid retention, flushing, and local bruising.

MANAGEMENT OF BLEEDING BY VWD TYPE

Low VWD and Type 1 VWD

The administration of desmopressin produces a two-fold increase in VWF/FVIII in 95% of patients with low VWF (0.3–0.5 IU/mL) and 86% of type 1 patients with levels of 0.2–0.3 IU/mL [22]. A retrospective study of 1234 patients reported that 69% of those with type 1 were responsive to desmopressin, but 30% also required VWF concentrate [23]. Concentrates are required for the remaining patients with type 1 disease, who generally have more severe bleeding events; some of these patients might benefit from prophylaxis (described in a subsequent paragraph).

Type 2 VWD

Nowhere is personalized medicine more important than for patients with type 2 VWD. Only a minority respond to desmopressin, and responses range from variable for type 2M to usually poor for type 2A, and the agent is contraindicated for type 2B [24]. Even in those individuals responsive to desmopressin, the response might be short lived because of a very short VWF/FVIII half-life; this is characteristic of patients with type 2N. Useful responses to desmopressin are levels >0.5 IU/mL at 1 h and

>0.3 IU/mL at 4 h; these levels might prevent bleeding after minor surgery or control nonmajor hemorrhages. Most patients with type 2 VWD are desmopressin unresponsive and require VWF concentrates; doses range from 30 IU/kg for a dental extraction to 50 IU/kg for major surgery, depending on the product (Table 11.1). An anti-Von Willebrand factor aptamer, ARC1779, was reported to increase VWF levels and platelet counts in three patients with type 2B VWD [25]. Its effectiveness in curbing bleeding awaits further study.

Type 3 VWD

These patients require concentrate therapy because they are almost always unresponsive to desmopressin. Procedures such as wisdom tooth extraction are considered major surgery for type 3 patients; the doses of concentrate displayed in Table 11.1 should be infused. Such patients are particularly predisposed to gastrointestinal bleeding and often need higher doses of concentrates to control hemorrhages. Unresponsiveness to concentrate therapy occasionally occurs because bleeding is arising from angiodysplastic vessels; angiodysplasia was observed by endoscopy in 38% of patients in one series [26]. Endoscopic laser therapy can provide temporary control of bleeding, but antiangiogenic agents such as thalidomide have not been effective.

Prophylaxis

The goals of prophylaxis are to improve the quality of life by reducing the frequency of bleeding episodes, hospitalizations, red cell transfusions, and consumption of VWF concentrates. Patients are selected based on the bleeding score, which predicts clinical outcomes and correlates with the intensity of on-demand treatment [27]. Doses have generally been 30–60 IU/kg given two to three times per week, using either plasma-derived or recombinant VWF concentrates. Long-term prophylactic therapy was evaluated in 105 patients receiving VWF concentrate one or more times per week, 45 weeks per year, or monthly for heavy menstrual bleeding [28]. All patients were unresponsive to desmopressin; 12% had type 1 (VWF $\leq$0.2 IU/mL), 36% type 2, and 51% type 3. Significant decreases in annual bleeding rates occurred in patients with gastrointestinal bleeding (from 9.3% to 6%), joint bleeding (from 11.9% to 0.8%), epistaxis in children (from 11.1% to 3.8%), and heavy menstrual bleeding (from 9.6% to 0). Individuals with gastrointestinal bleeding required prophylaxis three or more times per week and higher doses. In general, prophylaxis with VWF concentrates is appropriate for patients with a severe bleeding phenotype, and the intensity of treatment should be personalized.

OTHER MANAGEMENT ISSUES

Gastrointestinal Bleeding

Recurrent gastrointestinal bleeding occurs most often in patients with types 2A and 3 VWD and often arises from angiodysplastic lesions [29]. Infusions of VWF concentrates, administration of inhibitors of fibrinolysis, and local measures such as endoscopic ligation or embolization of bleeding vessels are appropriate. Other agents, such as recombinant FVIIa and thalidomide, might be effective for refractory disease [30].

Heavy Menstrual Bleeding

Heavy menstrual bleeding is reported in 60% of women with VWD and is usually managed with hormonal therapy, tranexamic acid, and desmopressin, as described earlier. When these measures fail, the administration of VWF concentrate, in doses ranging from 33 to 100 IU/kg on days 1–6 of the menstrual cycle, appears to be effective [31]. Prospective studies of other agents, including rIL-11, are anticipated.

Pregnancy and Postpartum

Many years ago, it was reported that VWF increases dramatically during pregnancy in healthy women, attaining levels as high as 3 IU/mL [32]. A recent study of 32 women (24 with type 1 VWD) reported that VWF levels reach 250% of baseline 4 h postpartum and subsequently approach baseline 1 week postpartum [33]. Although VWF levels are higher than the patients' baseline, they are considerably lower than the levels in healthy pregnant women at the same time points. The authors suggest that treatment is not required if VWF concentrations exceed 0.5 IU/mL in the third trimester; otherwise, VWF concentrate therapy should be infused during labor with a target level >1–2 IU/mL peripartum and maintained at >0.5 IU/mL for a minimum of 4–7 days postpartum [34]. Desmopressin might be substituted for VWF concentrate postpartum if responses reach target levels, but it is contraindicated in women with type 2B VWD. Patients with type 2B often have declining platelet counts during pregnancy and might need platelet transfusions if the platelet count falls below 50,000/μL.

VWD patients known to have a high bleeding score prior to pregnancy might benefit from tranexamic acid given postpartum. The first dose is administered intravenously, followed by oral therapy continuing after discharge until bleeding risks have abated. Postpartum hemorrhage is managed with fundal massage and other obstetrical measures, fluid and red cell resuscitation, oxytocin and misoprostol, continued VWF/FVIII replacement, intravenous tranexamic acid, fibrinogen replacement if levels are subphysiologic, and recombinant FVIIa for refractory bleeding [31]. Bleeding problems are infrequent in infants born of mothers with type 1 VWD. A study of nine such infants reported that the neonates had a mean VWF level of 96%, and none bled spontaneously or with procedures [35].

Future Approaches: Gene Therapy

Gene transfer has been considered for patients with severe VWD. Several years ago, two groups reported animal experiments testing the feasibility of this approach. A plasmid expressing the intact 8.4-kb murine VWF coding sequence, directed by the cytomegalovirus promoter/enhancer, was injected into VWF knockout mice [36]. Levels of VWF, including the HMWM, and FVIII rose to the normal range and persisted for 1 week, and the bleeding time was corrected. Similar results were obtained by the second group, using a hepatocyte-specific α1 antitrypsin promoter [37]. Maximum VWF levels were 10-fold higher than the mouse baseline levels, transiently corrected the bleeding time, and restored platelet adhesion and aggregation in a ferric-chloride-induced thrombosis model.

SUMMARY

The management of patients with VWD is individualized because the intensity of symptoms and the response to treatment vary considerably between and within each category of VWD. In general, most patients with type 1 disease have satisfactory responses to desmopressin given intranasally, subcutaneously, or intravenously. However, those with severe type 1 deficiency, and all patients with type 3 disease, require VWF concentrates, administered either on-demand or prophylactically. Four concentrates are approved in the United States: three are plasma derived and contain FVIII as well as VWF, and all have undergone viral inactivation steps. The fourth is a recombinant product that contains only trace amounts of FVIII. Following concentrate administration, the levels of ristocetin cofactor and FVIII should be monitored to ensure that therapeutic levels have been achieved; when giving concentrates containing FVIII, care must be taken to avoid accumulation of the procoagulant and inducing thrombosis. Current adjunctive therapies include hormonal agents and inhibitors of fibrinolysis. Combined oral contraceptive agents and the levonorgestrel-releasing intrauterine system are effective in controlling heavy menstrual bleeding; alternatively, uterine hemorrhage can be managed by giving tranexamic acid in daily doses of 3.9 g. Heavy menstrual bleeding refractory to the above agents has responded to subcutaneous injections of recombinant interleukin 11, but is still under investigation. Tranexamic acid is also effective as a mouth rinse for limiting bleeding with oral surgical procedures. Pregnant VWD patients usually have an increase in VWF levels as they approach term, but those with persistently low levels and a history of bleeding are given VWF concentrate during labor, and treatment is continued postpartum for at least 4–7 days. In some patients, postpartum blood loss can be limited by the administration of desmopressin or tranexamic acid. In general, intramuscular injections and analgesics containing aspirin or nonsteroidal antiinflammatory drugs should be avoided in patients with VWD, and if nasal spray desmopressin is prescribed, the clinician should ensure that the concentration dispensed is no less than 1.5 mg/mL.

Future research might include the following:

- Developing delivery system to enable desmopressin to be effective orally; for example, fillable microparticles have recently been described [38] that might deliver sufficient drug for oral absorption.
- Conducting prospective studies to identify doses and dose intervals for prophylactic therapy.
- Modifying recombinant VWF concentrates to prolong their half-life, possibly by inhibiting binding to LRP1 or CLEM4M receptors on macrophages and endothelial cells.
- Synthesizing analogs of interleukin-11 that stimulate VWF release but are orally available and have fewer adverse reactions.
- Initiating human trials of VWF gene therapy for patients with severe type 1 and type 3 disease.

References

[1] Federici AB. The use of desmopressin in von Willebrand disease: the experience of the first 30 years (1977–2007). Haemophilia 2008;14(Suppl. 1):5–14.
[2] Blatt PM, Brinkhous KM, Culp HR, Krauss JS, Roberts HR. Antihemophilic factor concentrate therapy in von Willebrand disease. JAMA 1976;236:2770–2.
[3] Chediak JR, Telfer MC, Green D. Platelet function and immunologic parameters in von Willebrand's disease following cryoprecipitate and factor VIII concentrate infusion. Am J Med 1977;62:369–76.
[4] Hernandez-Navarro F, Quintana M, Jimenez-Yuste V, Alvarez MT, Fernandez-Morata R. Clinical efficacy in bleeding and surgery in von Willebrand patients treated with Fanhdi® a highly purified, doubly inactivated FVIII/VWF concentrate. Haemophilia 2008;14:963–7.
[5] Srivastava A, Serban M, Werner S, Schwartz BA, Kessler CM, Wonders Study Investigators. Efficacy and safety of a VWF/FVIII concentrate (Wilate®) in inherited von Willebrand disease patients undergoing surgical procedures. Haemophilia 2017;23:264–72.

[6] Mannucci PM, Kempton C, Millar C, Romond E, Shapiro A, Birschmann I, Ragni MV, Gill JC, Yee TT, Klamroth R, Wong WY, Chapman M, Engl W, Turecek PL, Suiter TM, Ewenstein BM, rVWF Ad Hoc Study Group. Pharmacokinetics and safety of a novel recombinant human von Willebrand factor manufactured with a plasma-free method: a prospective clinical trial. Blood 2013;122:648–57.

[7] Gill JC, Castaman G, Windyga J, Kouides P, Ragni M, Leebeek FW, Obermann-Slupetzky O, Chapman M, Fritsch S, Pavlova BG, Presch I, Ewenstein B. Hemostatic efficacy, safety, and pharmacokinetics of a recombinant von Willebrand factor in severe von Willebrand disease. Blood 2015;126:2038–46.

[8] Mannucci PM, Franchini M. Von Willebrand's disease. N Engl J Med 2017;376:701.

[9] Leebeek FWG, Eikenboom JCJ. Von Willebrand's disease. N Engl J Med 2017;376:701–2.

[10] Laffan MA, Lester W, O'Donnell JS, Will A, Tait RC, Goodeve A, Millar CM, Keeling DM. The diagnosis and management of von Willebrand disease: a United Kingdom Haemophilia Centre Doctors Organization guideline approved by the British Committee for Standards in Haematology. Br J Haematol 2014;167:453–65.

[11] Franchini M, Makris M, Santagostino E, Coppola A, Mannucci PM. Non-thrombotic, non-inhibitor-associated adverse reactions to coagulation factor concentrates for treatment of patients with hemophilia and von Willebrand's disease: a systematic review of prospective studies. Haemophilia 2012;18:e164–72.

[12] James PD, Lillicrap D, Mannucci PM. Alloantibodies in von Willebrand disease. Blood 2013;122:636–40.

[13] Coppola A, Franchini M, Makris M, Santagostino E, DiMinno G, Mannucci PM. Thrombotic adverse events to coagulation factor concentrates for treatment of patients with haemophilia and von Willebrand disease: a systematic review of prospective studies. Haemophilia 2012;18:e173–87.

[14] Baaij M, van Galen KPM, Urbanus RT, Nigten J, Eikenboom JHC, Schutgens REG. First report of inhibitor VWF alloantibodies in type 2B VWD. Br J Haematol 2015;171:424–7.

[15] Franchini M, Gandini G, Giuffrida A, de Gironcoli M, Federici AB. Treatment for patients with type 3 von Willebrand disease and alloantibodies: a case report. Haemophilia 2008;14:645–6.

[16] Davies J, Kadir RA. Heavy menstrual bleeding: an update on management. Thromb Res 2017;151(Suppl. 1):S70–7.

[17] Mannucci PM. Treatment of von Willebrand's disease. N Engl J Med 2004;351:683–94.

[18] Lethaby A, Farquhar C, Cooke I. Antifibrinolytics for heavy menstrual bleeding. Cochrane Database Syst Rev 2000;4.

[19] Lukes AS, Moore KA, Muse KN, Gersten JK, Hecht BR, Edlund M, Richter HE, Eder SE, Attia GR, Patrick DL, Rubin A, Shangold GA. Tranexamic acid treatment for heavy menstrual bleeding. Obstet Gynecol 2010;116:865–75.

[20] Ragni MV, Novelli EM, Murshed A, Merricks EP, Kloos MT, Nichols TC. Phase II prospective open-label trial of recombinant interleukin-11 in desmopressin-unresponsive von Willebrand disease and mild or moderate haemophilia A. Thromb Haemost 2013;109:248–54.

[21] Ragni MV, Jankowitz RC, Jaworski K, Merricks EP, Kloos MT, Nichols TC. Phase II prospective open-label trial of recombinant interleukin-11 in women with mild von Willebrand disease and refractory menorrhagia. Thromb Haemost 2011;106:641–5.

[22] Archer NM, Samnaliev M, Grace R, Brugnara C. The utility of the DESMOPRESSIN challenge test in children with low von Willebrand factor. Br J Haematol 2015;170:884–6.

[23] Federici AB, Bucciarelli P, Castaman G, Baronciani L, Canciani MT, Mazzucconi MG, Morfini M, Rocino A, Schiavoni M, Oliovecchio E, Iorio A, Mannucci PM. Management of inherited von Willebrand disease in Italy: results from the retrospective study on 1234 patients. Semin Thromb Hemost 2011;37:511–21.

[24] Tosetto A, Castaman G. How I treat type 2 variant forms of von Willebrand disease. Blood 2015;125:907–14.

[25] Jilma-Stohlawetz P, Knöbl P, Gilbert JC, Jilma B. The anti-von Willebrand factor aptamer ARC1779 increases von Willebrand factor levels and platelet counts in patients with type 2B von Willebrand disease. Thromb Haemost 2012;108:284–90.

[26] Makris M, Federici AB, Mannucci PM, Bolton-Maggs PH, Yee TT, Abshire T, Berntorp E. The natural history of occult or angiodysplastic gastrointestinal bleeding in von Willebrand disease. Haemophilia 2015;21:338–42.

[27] Federici AB. Prophylaxis in patients with von Willebrand disease: who, when, how? J Thromb Haemost 2015;13:1581–4.

[28] Holm E, Abshire TC, Bowen J, Álvarez MT, Bolton-Maggs P, Carcao M, Federici AB, Gill JC, Halimeh S, Kempton C, Key NS, Kouides P, Lail A, Landorph A, Leebeek F, Makris M, Mannucci P, Mauser-Bunschoten EP, Nugent D, Valentino LA, Winikoff R, Berntorp E. Changes in bleeding patterns in von Willebrand disease after institution of long-term replacement therapy: results from the von Willebrand Disease Prophylaxis Network. Blood Coagul Fibrinolysis 2015;26:383–8.

[29] Randi AM. Endothelial dysfunction in von Willebrand disease: angiogenesis and angiodysplasia. Thromb Res 2016;141(Suppl. 2):S55–8.

[30] Franchini M, Mannucci PM. Gastrointestinal angiodysplasia and bleeding in von Willebrand disease. Thromb Haemost 2014;112:427–31.

[31] Ragni MV, Machin N, Malec LM, James AH, Kessler CM, Konkle BA, Kouides PA, Neff AT, Philipp CS, Brambilla DJ. Von Willebrand factor for menorrhagia: a survey and literature review. Haemophilia 2016;22:397–402.

[32] Van Royen EA, Flier OTN, ten Cate JW. Von-Willebrand-factor activity in pregnancy. Lancet 1974;ii:657.

[33] James AH, Konkle BA, Kouides P, Ragni MV, Thames B, Gupta S, Sood S, Fletcher SK, Philipp CS. Postpartum von Willebrand factor levels in women with and without von Willebrand disease and implications for prophylaxis. Haemophilia 2015;21:81–7.

[34] Kouides PA. An update on the management of bleeding disorders during pregnancy. Curr Opin Hematol 2015;22:397–405.

[35] Sood SL, James AH, Ragni MV, Shapiro AD, Witmer C, Vega R, Bolgiano D, Konkle BA. A prospective study of von Willebrand factor levels and bleeding in pregnant women with type 1 von Willebrand disease. Haemophilia 2016;22:e562–64.

[36] Pergolizzi RB, Jin G, Chan D, Pierre L, Bussel J, Ferris B, Leopold PL, Crystal RG. Correction of a murine model of von Willebrand diseas by gene transfer. Blood 2006;108:862–9.

[37] De Meyer SF, Vandeputte N, Pareyn I, Petrus I, Lenting PJ, Chuah MKL, Vanden Driessche T, Deckmyn H, Vanhoorelbeke K. Restoration of plasma von Willebrand factor deficiency is sufficient to correct thrombus formation after gene therapy for severe von Willebrand disease. Arterioscler Thromb Vasc Biol 2008;28:1621–6.

[38] McHugh KJ, Nguyen TD, Linehan AR, Yang D, Behrens AM, Rose S, Tochka ZL, Tzeng SY, Norman JJ, Anselmo AC, Xu X, Tomasic S, Taylor MA, Lu J, Guarecuco R, Langer R, Jaklenec A. Fabrication of fillable microparticles and other complex 3D microstructures. Science 2017;357:1138–42.

Recommended Reading

[1] Laffan MA, Lester W, O'Donnell JS, et al. The diagnosis and management of von Willebrand disease: a United Kingdom Haemophilia Centre Doctors Organization guideline approved by the British Committee for Standards in Haematology. Br J Haematol 2014;167:453–65.

[2] Gill JC, Castaman G, Windyga J, et al. Hemostatic efficacy, safety, and pharmacokinetics of a recombinant von Willebrand factor in severe von Willebrand disease. Blood 2015;126:2038–46.

[3] Tosetto A, Castaman G. How I treat type 2 variant forms of von Willebrand disease. Blood 2015;125:907–14.

[4] Holm E, Abshire TC, Bowen J, et al. Changes in bleeding patterns in von Willebrand disease after institution of long-term replacement therapy: results from the von Willebrand Disease Prophylaxis Network. Blood Coagul Fibrinolysis 2015;26:383–8.

[5] James AH, Konkle BA, Kouides P, et al. Postpartum von Willebrand factor levels in women with and without von Willebrand disease and implications for prophylaxis. Haemophilia 2015;21:81–7.

[6] Ragni MV, Machin N, Malec LM, et al. Von Willebrand factor for menorrhagia: a survey and literature review. Haemophilia 2016;22:397–402.

[7] Sood SL, James AH, Ragni MV, et al. A prospective study of von Willebrand factor levels and bleeding in pregnant women with type 1 von Willebrand disease. Haemophilia 2016;22:e562–64.

12

Acquired Von Willebrand Syndrome

Von Willebrand disease (VWD) is generally an inherited condition, but the disorder can be acquired by people with a diverse group of diseases; the major categories are listed in Table 12.1.

These disorders might seem quite disparate, but the unifying theme is that they all are associated with a decrease in the high-molecular-weight multimers (HMWM) of Von Willebrand factor (VWF), although the pathophysiology is specific for each disease. Immunologically mediated acquired Von Willebrand Syndrome (aVWS) is due to allo- or autoantibodies that bind to VWF multimers and hasten their removal from the circulation; some antibodies also impair the function of the molecule. Accelerated proteolysis occurs when the VWF is subject to shear stresses that increase its vulnerability to cleavage by ADAMTS13. Proteolysis of HMWM occurs in myeloproliferative disorders such as essential thrombocythemia, polycythemia vera, and chronic myelogenous leukemia, but there is also evidence that the HMWM become bound to the excessive numbers of platelets. Decreased protein synthesis probably is responsible for the reduced levels of VWF in patients with hypothyroidism, and various mechanisms cause the reduced levels of HMWM in patients exposed to certain drugs. As noted in previous chapters, the HMWM of VWF are the most hemostatically effective, and their loss partly accounts for the bleeding manifestations of aVWS. In addition, the loss of VWF from endothelial cells is associated with vascular malformations; gastrointestinal bleeding from angiodysplasia is often the presenting feature of acquired VWD. In this chapter, the various disorders reported to affect VWF production, circulation, and clearance will be described, and the diagnostic steps and therapeutic options for patients with aVWS discussed.

TABLE 12.1   Disorders Associated With Acquired Von Willebrand Syndrome

Immunologically mediated

  Autoimmune disease

  Systemic lupus erythematosus, other connective tissue diseases

  Monoclonal gammopathies of undetermined significance (MGUS)

  Multiple myeloma, Waldenstrom's macroglobulinemia

  Chronic lymphocytic leukemia, lymphoma

  Nonhematologic neoplasms (Wilms tumor)

Accelerated proteolysis of Von Willebrand factor

  Aortic stenosis (Heyde's syndrome)

  Malfunctioning prosthetic valves

  Left ventricular assist devices

  Hypertrophic cardiomyopathy

  Congenital heart disease

Myeloproliferative disorders

  Thrombocythemia

  Polycythemia vera

  Chronic myeloid leukemia

Genetic defects

  Glycogen storage disease

  Mesenchymal dysplasia

Miscellaneous

  Hypothyroidism

  Drug induced (ciprofloxacin, valproic acid, hydroxyethyl starch)

IMMUNOLOGICALLY MEDIATED LOSS OF VWF

Alloantibodies develop in up to 10% of patients with type 3 VWD, mainly in those with partial or complete *VWF* gene deletions, and/or a family history of VWF antibodies [1]. Most patients have a history of treatment with therapeutic products containing VWF, and become refractory to further therapy. The alloantibodies can be assessed using either an ELISA method or the Bethesda assay, although the latter might not recognize nonfunctional antibodies that are still capable of enhancing the clearance of the molecule. Because the antibodies specifically target VWF, patients have normal levels of FVIII; but similar to FVIII inhibitors, the antibodies most often are of the IgG4 subclass and do not activate complement. Occasionally, high titer antibodies have precipitated infused VWF; in some patients, re-exposure to VWF has resulted in immune complex formation, complement fixation, and anaphylaxis. Therefore, treatment with products containing VWF should be avoided. On the other hand, a recombinant FVIII free of VWF, given by continuous infusion, might provide effective hemostasis [1]. Patients failing FVIII can be treated with recombinant factor VIIa, but would be at increased risk of thrombotic complications [2]. Elimination of the alloantibody, by inducing immune tolerance or administering immunosuppressive agents, is also an option [3].

Autoantibodies to VWF occasionally complicate diseases such as systemic lupus erythematosus (SLE), monoclonal gammopathies, and lymphoproliferative syndromes. In 1968, Simone et al. [4] reported repeated episodes of bleeding in a 12-year-old boy with no previous personal or family history of bleeding. The laboratory studies were consistent with VWD. Within a year, he presented with typical signs and symptoms of SLE involving the skin, joints, and kidneys. The bleeding and manifestations of SLE remitted after therapy with corticosteroids. A similar sequence of events was described in a 17-year-old girl with bleeding due to an IgG anti-VWF antibody followed by the appearance of typical SLE [5]. Treatment with prednisone led to a cessation of bleeding and restoration of normal levels of VWF and HMWM. Patients failing steroids have remitted with intravenous immunoglobulin [6].

IgG autoantibodies in patients with monoclonal gammopathies of undetermined significance (MGUS) are often associated with increased VWF clearance and decreased VWF antigen, ristocetin cofactor, and HMWM levels [7–9]. Decreased levels of FVIII, VWF antigen, and ristocetin-induced platelet aggregation, accompanied by a loss of VWF HMWM, have also been described in patients with lymphoma. In some of these patients, VWF on the malignant cells was demonstrated by specific immunohistochemical staining, and successful elimination of these cells led to remission of the aVWS [10]. In the patient reported by Joist et al. [11], bleeding subsided and all VWF-related activities, including the HMWM, returned to normal following radiation therapy. Another report describes a patient with non-Hodgkin lymphoma and monoclonal gammopathy [12]. This patient had an IgM autoantibody that inhibited VWF collagen binding activity by reacting with epitopes present on the glycoprotein Ib and A3 domains of VWF. Impaired collagen binding has also been reported in patients with Waldenstrom macroglobulinemia and IgM antibodies.

aVWS that occurs in patients with multiple myeloma has been attributed to a variety of mechanisms. Some patients have anti-VWF antibodies [13]; such antibodies can accelerate the plasma clearance of VWF or interfere with the binding of VWF to platelet GP1b [14]. Another mechanism is the absorption of VWF by the malignant plasma cells [15], and in another patient with amyloidosis, VWF degradation was due to excessive plasmin-mediated fibrinolysis [13]. Chemotherapy that induces a remission in the myeloma ameliorates the aVWS. A patient with IgG-lambda mycloma and bleeding due to aVWS was given bortezomib and dexamethasone; after 3 months, the IgG levels declined, the serum free light-chain ratio normalized, bleeding stopped and FVIII, VWF antigen, and ristocetin cofactor returned to normal [16]. The remission of the aVWS was sustained for 3 years. Another patient, described in a clinical problem-solving exercise, received dexamethasone and the immunomodulatory agent, thalidomide [17]. Within 4 months of initiating therapy, both the myeloma and the aVWS remitted.

Laboratory studies are essential for confirming the diagnosis and should include VWF multimer analysis if available; a loss of HMWM is almost always observed. In patients with monoclonal gammopathies, prolonged PFA-100 closure times might provide early evidence of the hemostatic abnormality. In a study of 36 patients, 21 of whom had Waldenstrom macroglobulinemia, PFA-100 closure times were prolonged in all, but levels of FVIII $\leq$20 IU/dL or ristocetin cofactor $\leq$10 IU/dL were recorded in only 31% and 44%, respectively [18]. Unfortunately, the PFA-100 is often uninterpretable because of concomitant anemia or thrombocytopenia. Prolongation of the activated partial thromboplastin time (aPTT) is infrequent because many patients have levels of FVIII $\geq$30 IU/dL. Tiede et al. [19] suggest that a VWF antigen <50 IU/dL, ristocetin cofactor to antigen ratio of <0.7, and VWF collagen binding to antigen ratio of <0.8 is the most sensitive combination of laboratory measurements for diagnosing the disorder. Measurement of the VWF propeptide is often useful because propeptide concentrations that are higher than VWF antigen levels suggest increased clearance of VWF.

The diagnostic approach to patients with lymphoproliferative disorders and new-onset bleeding begins with a strong clinical suspicion for aVWS. Laboratory studies of FVIII/VWF in a typical patient might reveal a FVIII of 11 IU/dL, VWF antigen of 19 IU/dL, and ristocetin cofactor of <5 U/dL. Next, the VWF propeptide should be measured; a value within the normal range, and a ratio of propeptide to VWF antigen $\geq$2, indicates increased clearance or inactivation of VWF, and would be consistent with a VWF autoantibody associated with the lymphoid neoplasm.

The treatment of patients with immunologically mediated aVWS is twofold: control of bleeding and elimination of the pathologic autoantibodies. The options for controlling bleeding include desmopressin,

FVIII/VWF concentrates, and intravenous immunoglobulin. Desmopressin is the simplest and least costly, transiently increases FVIII/VWF, and can decrease bleeding [20]. However, it is only effective in 50% of patients with demonstrable inhibitors of VWF [21]. If a test dose produces therapeutic FVIII/VWF levels, desmopressin is given prophylactically for patients requiring minimally invasive procedures such as catheter insertion. Otherwise, FVIII/VWF concentrates are often effective in controlling bleeding, but larger and more frequent doses are infused than when they are given for congenital VWD, and levels should be checked often and doses adjusted accordingly [22].

Many reports describe the benefit of intravenous immunoglobulin (IVIg) in patients with immunologically mediated aVWS, usually administered in doses of 1 g/kg daily [17,19,23]. However, doses as low as 0.3 g/kg for 3 days have also been effective [24]. The response rate in 89 patients with lymphoproliferative disorders enrolled in an international registry was 37% [25]. Patients with IgM paraproteins were less likely to respond to IVIg. The responses to IVIg often occur within 12–72 h, with cessation of bleeding and improvement in VWF levels persisting for up to 20 days [26,27]. Doses are repeated every 21 days until there is complete remission of the disease. The rapid responses to IVIg are consistent with several possible modes of action. The large amounts of immunoglobulin might saturate Fcγ receptors on immune cells, preventing the uptake of paraprotein-VWF complexes, or inhibit the clearance of VWF by upregulating the expression of the inhibitory receptor, FCγRIIB. Additionally, there might be modulation of dendritic cells, expansion of regulatory T cells, reduction of proinflammatory molecules, inhibition of the complement cascade, and direct neutralization of autoantibodies by antiidiotypic antibodies contained within the IVIg preparation [28]. Enthusiasm for treatment with IVIg is tempered by its cost and adverse effects. Most patients complain of headache during the infusion, and the high concentration of immunoglobulin increases blood viscosity, which is especially problematic for patients whose paraproteins are associated with hyperviscosity. Prior to infusing IVIg, the concentration of the paraprotein should be decreased by plasmapheresis or chemotherapy, and all patients should be adequately hydrated to avoid injury to the kidneys. IVIg infusions have occasionally been associated with thrombosis, especially in patients with cardiovascular disease. Other adverse effects are hemolysis, noncardiogenic pulmonary edema, transmission of infections, and hypersensitivity reactions, including anaphylaxis.

There are a few other options for patients refractory to IVIg. Platelet transfusions might be helpful because VWF in the platelets is protected from antibodies in the circulation and could provide hemostatic benefit when released at the site of bleeding [1]. Extracorporeal immunoabsorption successfully removed a VWF antibody in a patient with chronic lymphocytic leukemia, and resolved bleeding that had been refractory to a variety of clotting factor concentrates [29]. More recently, successful treatment with immunomodulatory agents has been reported; thalidomide induced remission in gastrointestinal bleeding, and lenalidomide decreased the accelerated clearance of VWF and produced sustained increases in VWF levels in two patients with monoclonal gammopathies [30,31].

Wilms tumor is one of the most common primary malignant tumors of childhood, with an incidence of 7–10 cases per million; aVWS occurs in 4%–8% of patients with Wilms tumor [32]. The association has been described in an infant only 4 months old [33]. Patients do not always present with excessive bleeding, but major bleeding can occur with invasive procedures. The principal laboratory abnormalities are a prolonged aPTT, decreased FVIII, and low levels of VWF antigen, ristocetin cofactor, and collagen-binding activity [31]. VWF might bind to the tumor tissue, because the aVWS remits when the tumor is removed. However, studies to confirm this possibility have not been reported. To prevent surgical bleeding, desmopressin or FVIII/VWF concentrates should be evaluated preoperatively and selected for infusion if they achieve therapeutic plasma levels [34]. Alternatively, there are anecdotal reports that IVIg is effective [26].

CARDIOVASCULAR DISEASE

On July 24, 1958, the *New England Journal of Medicine* published a letter from Doctor E.C. Heyde describing 10 patients with calcific aortic stenosis and massive gastrointestinal bleeding [35]. Subsequently, other physicians confirmed the relationship between aortic valve disease and gastrointestinal bleeding, which has been designated "Heyde's Syndrome" [36]. In 1992, Warkentin et al. [37] suggested that the aortic valve disease and the gastrointestinal bleeding were linked by aVWS; the valvular heart disease was responsible for the loss of the HMWM of VWF, and angiodysplasia of the bowel developed because of the decrease in VWF multimers.

A variety of cardiovascular diseases have been associated with aVWS. Among the congenital cardiac disorders are ventricular septal defect, atrial septal defect, and aortic stenosis [38]. Acquired conditions associated with aVWS are aortic stenosis [39], mitral valve regurgitation [40] and prolapse [41], hypertrophic obstructive cardiomyopathy [42], mechanical valve prostheses [43], and left ventricular assist devices [44,45]. Turbulent blood flow and high shear stresses in these entities unravel the tightly packed, globular VWF. Stretching of the VWF molecule exposes sites vulnerable to cleavage by the protease, ADAMTS13, and the resultant loss of HMWM produces the bleeding phenotype [35,46].

Support for this view comes from clinical studies: VWF abnormalities are related to the magnitude of the outflow obstruction in patients with hypertrophic cardiomyopathy [41] and occur almost immediately after implantation of ventricular assist devices associated with high shear stress [47]. In addition, blocking the VWF-ADAMTS13 interaction in a prosthetic-valve-based whole blood perfusion system prevents excessive degradation of HMWM [48].

Gastrointestinal hemorrhage arising from submucosal angiodysplasia is an integral component of Heyde's syndrome and other disorders associated with aVWS. Intestinal angiodysplasia or telangiectasias are reported in congenital VWD [49,50], so it is not surprising to find them in aVWS as well [51,52]. The pathophysiology of the vascular malformations is still not well understood. Early on, investigators discovered that 60% of patients with aVWS had mitral valve prolapse, and suggested that both conditions were expressions of a heritable disorder of connective tissue [40]. Recent work has focused on blood vessels, observing that VWF-deficient mice have increased vascularization [53]. In vitro, inhibiting endothelial VWF expression enhances angiogenesis, decreases integrin $\alpha v \beta 3$, increases vascular endothelial growth factor receptor-2 (VEGFR2)-dependent proliferation and migration of endothelial cells, and augments the release of angiopoietin-2 (Ang-2). Integrin $\alpha v \beta 3$ is required for the binding of VWF to endothelial cells and regulates the sensitivity of these cells to VEGFR2. In the absence of VWF, $\alpha v \beta 3$ is internalized; its loss from the cell surface likely enhances VEGFR2 signaling, contributing to the formation of the immature, fragile blood vessels characteristic of angiodysplasia [54]. Ang-2 is normally present in Weibel-Palade (W-P) bodies; absence of VWF results in the disassembly of W-P bodies and the release of Ang-2. Ang-2 can then bind to the Tie-2 receptor, antagonizing angiopoietin-1 and destabilizing cell-cell junctions, thereby contributing to vessel malformation. The specific loss of HMWM in aVWS suggests that VWF monomers probably play an important role in the development of the vascular disease [55,56], but the specific pathophysiologic link between VWF deficiency and the development of angiodysplasia remains a subject of continuing interest and investigation.

The diagnosis of cardiovascular-disease-associated aVWS is based on a history of a congenital or acquired cardiac disorder and bleeding, often from the gastrointestinal tract. Routine laboratory studies are often unremarkable, but a recent study of patients with aortic regurgitation after transcatheter aortic valve replacement found prolonged closure times with adenosine diphosphate using the PFA-100 device [57]. Closure times >180 s were 92% sensitive for predicting postimplantation aortic regurgitation. Also available is an enzyme-linked immunosorbent assay (ELISA) that has been developed to detect VWF fragmentation; patients with aortic stenosis and left ventricular assist devices had significantly increased

VWF proteolysis as compared to controls [58]. When aVWS is suspected, an analysis of VWF multimers should be requested; a decrease in HMWM confirms the diagnosis.

Therapy that corrects the cardiovascular disease is often effective in terminating the aVWS. Shortly after Warkentin et al. [37] published their hypothesis that aVWS was the link between aortic stenosis and gastrointestinal bleeding, Anderson et al. [59] reported two patients whose aVWS and gastrointestinal bleeding remitted after aortic valve replacement, and Warkentin et al. [60] described two additional patients whose aVWS was reversed by aortic valve replacement. Septal reduction therapy for hypertrophic obstructive cardiomyopathy has led to remissions in gastrointestinal bleeding in other patients with aVWS [61]. Removal and replacement of defective prosthetic devices also terminate the syndrome. The treatment of gastrointestinal bleeding with infusions of VWF concentrates is of limited value, and therapies specifically targeting the gastrointestinal angiodysplasia are of only temporary benefit because new bleeding lesions soon appear. Desmopressin is reported to have reduced postoperative blood loss by 42% in patients undergoing aortic valve replacement [62], but a test dose should be given preoperatively to confirm that it is effective. If the underlying cause of the aVWS cannot be directly addressed, there are a few drugs that appear to have antiangiogenic properties (reviewed in reference [53]). These include tamoxifen, an antiestrogen; the immunomodulatory drugs, thalidomide and lenalidomide; and the HMG-CoA reductase inhibitor, atorvastatin; however, none have yet been FDA-approved for the treatment of angiodysplasia in aVWS.

MYELOPROLIFERATIVE DISORDERS

In 1986, Budde et al. [63] reported a relative decrease in the HMWM of VWF in three patients with essential thrombocythemia (ET), and observed that the intravenous infusion of desmopressin in one patient only transiently increased HMWM. They interpreted these findings as consistent with enhanced proteolysis. In a subsequent study, they reported the absence of HMWM in 12 of 19 patients with myeloproliferative disorders, as well as in 12 patients with postsplenectomy reactive thrombocytosis [64]. They concluded that an increase in the number of platelets favored adsorption of HMWM onto the platelet membrane, a conclusion supported by the observation that platelet cytoreduction restored plasma HMWM levels [65]. It remains unclear whether the predominant mechanism contributing to the aVWS is fragmentation of VWF by platelet proteolytic enzymes or adsorption of HMWM to the platelet membrane.

A recent survey reported that 64 of 116 (55%) patients with ET and 28 of 57 (49%) of those with polycythemia vera had evidence of aVWS, and more than half of those with aVWS had platelet counts of less than 1 million/μL [66]. The Syndrome was more likely to occur in patients with the *JAK2 V617F* mutation than in those with mutations in the calreticulin (*CALR*) gene (70% vs 46%, $P = .02$). Factors predicting the development of aVWS in patients with ET were younger age, higher platelet count, hemoglobin level, and the *JAK2* mutation; only a higher platelet count predicted aVWS in polycythemic patients. Aspirin is recommended for patients with *JAK2*-positive ET, age $\geq$60, a history of cardiovascular disease or thrombosis, or platelet count $\geq$1.5 million/μL [67], although it can provoke bleeding complications if patients have decreased HMWM [68]. Overall, the relative risk of bleeding is raised with antiplatelet therapy, but the increases are not statistically significant (any bleeding, 1.95 [95% confidence interval, 0.48–11.04]; major bleeding, 1.3 [0.48–5.17]) [69]. Treatment with desmopressin transiently increases HMWM and might provide temporary control of bleeding [64]. Cytoreduction with hydroxyurea or interferon α is recommended for the longer-term management of aVWS in patients with these myeloproliferative disorders.

GLYCOGEN STORAGE DISEASE

As many as 60% of children with deficiency of glucose-6-phosphatase and glycogen storage disease type Ia have abnormal hemostasis [70]. The PFA-100 closure time is prolonged, VWF antigen levels are decreased, and ristocetin-induced platelet aggregation is impaired. Abnormal VWF multimer patterns that were described in four of five patients corrected after infusion of desmopressin [71]. One group of investigators suggested that the impairment in ristocetin-induced platelet aggregation occurred because patient fibrinogen inhibited the binding of VWF to platelets [72]. Mühlhausen et al. [70] speculated that the glycosylation of VWF is altered by the metabolic disorder, and recent studies have shown that terminal sialic acid residues on VWF glycans protect the protein from cleavage by serine and cysteine proteases [73]. Whether the terminal sialylation of VWF is altered in glycogen storage disease has not been reported.

MISCELLANEOUS DISORDERS

The bleeding tendency associated with hypothyroidism has been recognized for many years, but first was attributed to aVWS in 1987 when clinicians described features of VWD in three patients with severe primary hypothyroidism [74]. They noted that the bleeding and coagulation abnormalities remitted after treatment with thyroid hormone. A confirmatory study found laboratory evidence of VWD in 5 of 11 patients with hypothyroidism, which resolved with thyroid replacement in all but one [75]. The investigators also measured the levels of plasminogen activator and its inhibitor and concluded that the reduced levels of VWF were due to decreased protein synthesis. However, a report that triiodothyronine increases VWF synthesis by cultured umbilical vein endothelial cells has not been confirmed [76,77].

A recent prospective study observed that the frequency of aVWS in patients with overt hypothyroidism was 33% (29 of 90) [78]. Significant increases in VWF, FVIII, and endogenous thrombin potential occurred after restoration of the euthyroid condition. The investigators commented that the bleeding tendency in most patients was mild and might not stimulate an evaluation of hemostasis, but such studies are indicated because hemorrhaging after invasive procedures or trauma is common in these patients [79]. Treatment options for acute bleeding episodes include desmopressin and VWF concentrates while awaiting the salutary effects of thyroid hormone replacement.

aVWS has been reported as an adverse effect of several drugs, including ciprofloxacin, valproic acid, and hydroxyethyl starch. Ciprofloxacin-associated aVWS is especially well documented; investigators described two patients having the acute onset of bleeding after exposure to the drug [80,81]. Both patients had markedly diminished levels of VWF activity and antigen, loss of HMWM, and increased VWF subunit fragments; these observations are consistent with accelerated VWF proteolysis. Inhibitory antibodies were not detected. Restudy of the patients after they recovered revealed that all the laboratory abnormalities had completely resolved.

Valproic acid can cause dose-dependent suppression of the bone marrow, resulting in leukopenia, thrombocytopenia, and red cell aplasia, as well as platelet dysfunction, hypofibrinogenemia, and FXIII deficiency [82,83]. This antiepileptic agent has also been implicated as a cause of aVWS. Bleeding and decreased levels of VWF were reported in 67% of those receiving the drug in one study [84], but in only 21% in another report [85], and were not dose dependent. Other studies suggest an even lower incidence [86]. Nevertheless, in view of the broad range of hematologic abnormalities, patients receiving valproic acid should be closely monitored and have hemostatic studies if unusual bleeding occurs or invasive procedures are contemplated.

Hydroxyethyl starch (hetastarch, HES) is a synthetic colloid composed of amylopectin; some of the polymers have molecular weights of up to 2.6 million [87]. It is a plasma expander that is used for the management of hypovolemia. Case reports of aVWS began appearing in the 1980s and 1990s, and a study of six patients receiving repeated infusions of HES showed that all molecular weight multimers of VWF were uniformly decreased [88–90]. An analysis of nine patients given a pentastarch preparation with a long half-life noted that the cumulative dose was higher in patients with bleeding; four patients had cerebral hemorrhages and three died [91]. FVIII levels ranged from 14% to 43% and VWF antigen was 13%–42%. The investigators suggested that the aVWS was due to elimination of FVIII-VWF complexes that had become bound to the starch polymers. Newer preparations with a lower molecular weight (130/0.4 compared with 200/0.5) appear to have less effect on VWF [92], but patients receiving repeated doses of HES should be closely monitored for the development of a coagulopathy.

SUMMARY

Acquired Von Willebrand syndrome is suspected in patients with new-onset bleeding and no antecedent history of a hemorrhagic disorder, especially with these clinical scenarios:

- Monoclonal gammopathies of undetermined significance (MGUS)
- Wilms Tumor
- Gastrointestinal bleeding from angiodysplasia in patients with aortic valve disease
- Bleeding after heart valve replacement or with left ventricular assist devices (LVAD)
- Hematomas and hemorrhages in patients with myeloproliferative disorders
- Hypothyroid patients with easy bruising and bleeding
- Bleeding in patients receiving plasma volume expanders

The diagnosis is suspected if the PFA-100 closing time using the ADP reagent is prolonged, the levels of FVIII, VWF antigen, and ristocetin-induced platelet aggregation are decreased; and VWF fragments are detected using an ELISA method. The definitive study is VWF multimer analysis, which usually shows the loss of HMWM. Treatments with proven efficacy are desmopressin and intravenous immunoglobulin (IVIg) for patients with MGUS, resection of Wilms tumors, repair of incompetent or stenotic valves, removal or replacement of LVADs, reduction of excessive platelet numbers, and replacement of thyroid hormone.

Future research activities might include the following:

- Identify the component in IVIg that prevents the accelerated clearance of HMWM in people with MGUS. The use of a specific agent might be more effective and avoid some of the adverse effects of the current preparations of IVIg.

- Demonstrate that the neoplastic cells of Wilms tumor actually bind HMWM; do they have receptors for VWF or is there another mechanism?
- Clarify which mechanism accounts for the loss of HMWM in ET binding to platelets or enhanced proteolysis. If the latter, what platelet component is proteolytic; alternatively, do the abnormal platelets fail to produce an inhibitor of proteolysis?
- Determine whether sialylation of the VWF is reduced in children with glycogen storage disease type Ia, and if the endothelial cell processing of VWF is impaired in patients with hypothyroidism.

References

[1] James PD, Lillicrap D, Mannucci PM. Alloantibodies in von Willebrand disease. Blood 2013;122:636–40.

[2] Boyer-Neumann C, Dreyfus M, Wolf M, Veyradier A, Meyer D. Multi-therapeutic approach to manage delivery in an alloimmunized patient with type 3 von Willebrand disease. J Thromb Haemost 2003;1:190–2.

[3] Pergantou H, Xafaki P, Adamtziki E, Koletsi P, Komitopoulou A, Platokouki H. The challenging management of a child with type 3 von Willebrand disease and antibodies to von Willebrand factor. Haemophilia 2012;18:e66–7.

[4] Simone JV, Cornet JA, Abildgaard CF. Acquired von Willebrand's syndrome in systemic lupus erythematosus. Blood 1968;31:806–12.

[5] Soff GA, Green D. Autoantibody to von Willebrand factor in systemic lupus erythematosus. J Lab Clin Med 1993;121:424–30.

[6] Hanley D, Arkel YS, Lynch J, Kamiyama M. Acquired von Willebrand's syndrome in association with a lupus-like anticoagulant corrected by intravenous immunoglobulin. Am J Hematol 1994;46:141–6.

[7] Guerin V, Ryman A, Velez F. Acquired von Willebrand disease: potential contribution of the von Willebrand factor collagen-binding to the identification of functionally inhibiting auto-antibodies to von Willebrand factor: a rebuttal. J Thromb Haemost 2008;6:1051–2.

[8] Siaka C, Rugeri L, Caron C, Goudemand J. A new ELISA assay for diagnosis of acquired von Willebrand syndrome. Haemophilia 2003;9:303–8.

[9] Luboshitz J, Lubetsky A, Schliamser L, Kotler A, Tamarin I, Inbal A. Pharmacokinetic studies with FVIII/von Willebrand factor concentrate can be a diagnostic tool to distinguish between subgroups of patients with acquired von Willebrand syndrome. Thromb Haemost 2001;85:806–9.

[10] Michiels JJ, Budde U, van der Planken M, van Vliet HHDM, Schroyens W, Berneman Z. Acquired von Willebrand syndromes: clinical features, aetiology, pathophysiology, classification and management. Best Pract Res Clin Haematol 2001;14:401–36.

[11] Joist JH, Cowan JF, Zimmerman TS. Evidence for a quantitative and qualitative factor VIII disorder. N Engl J Med 1978;298:988–91.

[12] Van Genderen PJJ, Vink T, Michiels JJ, van't Veer MB, Sixma JJ, van Vliet HHDM. Acquired von Willebrand disease caused by an autoantibody selectively inhibiting the binding of von Willebrand factor to collagen. Blood 1994;84:3378–84.

[13] Mohri H, Tanabe J, Ohtsuka M, Yoshida M, Motomura S, Nishida S, Fujimura Y, Okubo T. Acquired von Willebrand disease associated with multiple myeloma; characterization of an inhibitor to von Willebrand factor. Blood Coagul Fibrinolysis 1995;6 (6):561.

[14] Dicke C, Schneppenheim S, Holstein K, Spath B, Bokemeyer C, Dittmer R, Budde U, Langer F. Distinct mechanisms account for acquired von Willebrand syndrome in plasma cell dyscrasias. Ann Hematol 2016;95:945–57.

[15] Richard C, Cuadrado MA, Prieto M, Batlle J, López Fernández MF, Rodriguez Salazar ML, Bello C, Recio M, Santoro T, Gomez Casares MT, et al. Acquired von Willebrand disease in multiple myeloma secondary to absorption of von Willebrand factor by plasma cells. Am J Hematol 1990;35:114–7.

[16] Katagiri S, Akahane D, Amano K, Ohyashiki K. Long-term remission of acquired von Willebrand syndrome associated with multiple myeloma using bortezomib and dexamethasone therapy. Haemophilia 2016;22:e545–75.

[17] Cuker A, Connors JM, Katz JT, Levy BD, Loscalzo J. A bloody mystery. N Engl J Med 2009;361:1887–94.

[18] Voisin S, Hamidou M, Lefrançois A, Sigaud M, Mahé B, Trossaërt M. Acquired von Willebrand syndrome associated with monoclonal gammopathy: a single-center study of 36 patients. Medicine (Baltimore) 2011;90:404–11.

[19] Tiede A, Priesack J, Werwitzke S, Bohlmann K, Oortwijn B, Lenting P, Eisert R, Ganser A, Budde U. Diagnostic workup of patients with acquired von Willebrand syndrome: a retrospective single-centre cohort study. J Thromb Haemost 2008;6:569–76.

[20] Federici AB, Stabile F, Castaman G, Canciani MT, Mannucci PM. Treatment of acquired von Willebrand syndrome in patients with monoclonal gammopathy of uncertain significance: comparison of three different therapeutic approaches. Blood 1998;92:2707–11.

[21] Mohri H, Motomura S, Kanamori H, Matsuzaki M, Watanabe S, Maruta A, Kodama F, Okubo T. Clinical significance of inhibitors in acquired von Willebrand syndrome. Blood 1998;91:3623–9.

[22] Alvarez MT, Jimenez-Yuste V, Gracia J, Quintana M, Hernandez-Navarro F. Acquired von Willebrand syndrome. Haemophilia 2008;14:856–8.

[23] Agarwal N, Klix MM, Burns CP. Successful management with intravenous immunoglobulins of acquired von Willebrand disease associated with monoclonal gammopathy of undetermined significance. Ann Intern Med 2004;141:83–4.

[24] Hayashi T, Yagi H, Suzuki H, Nonaka Y, Nomura T, Sakurai Y, Shibata M, Matsumoto M, Yamamoto Y, Fujimura Y. Low-dosage intravenous immunoglobulin in the management of a patient with acquired von Willebrand syndrome associated with monoclonal gammopathy of undetermined significance. Pathophysiol Haemost Thromb 2002;32:33–9.

[25] Federici AB, Rand JH, Bucciarelli P, Budde U, van Genderen PJJ, Mohri H, Meyer D, Rodeghiero F, Sadler JE. Acquired von Willebrand syndrome: data from an international registry. Thromb Haemost 2000;84:345–9.

[26] Tiede A, Rand JH, Budde U, Ganser A, Federici AB. How I treat the acquired von Willebrand syndrome. Blood 2011;17:6777–85.

[27] Federici AB. Use of intravenous immunoglobulin in patients with acquired von Willebrand syndrome. Hum Immunol 2005;66:422–30.

[28] Chaigne B, Mouthon L. Mechanisms of action of intravenous immunoglobulin. Transfus Apher Sci 2017;56:45–9.

[29] Uehlinger J, Rose E, Aledort LM, Lerner R. Successful treatment of an acquired von Willebrand factor antibody by extracorporeal immunoadsorption. N Engl J Med 1989;320:254–5.

[30] Engelen ET, van Galen KP, Schutgens RE. Thalidomide for treatment of gastrointestinal bleedings due to angiodysplasia: a case report in acquired von Willebrand syndrome and review of the literature. Haemophilia 2015;21:419–29.

[31] Lavin M, Brophy TM, Rawley O, O'Sullivan JM, Hayden PJ, Browne PV, Ryan K, O'Connell N, O'Donnell JS. Lenalidomide as a novel treatment for refractory acquired von Willebrand syndrome associated with monoclonal gammopathy. J Thromb Haemost 2016;14:1200–5.

[32] Fosbury E, Szychot E, Slater O, Mathias M, Sibson K. An 11-year experience of acquired von Willebrand syndrome in children diagnosed with Wilms tumour in a tertiary referral Centre. Pediatr Blood Cancer 2017;64.

[33] Han P, Lou J, Wong HB. Wilms' tumour with acquired von Willebrand's disease. Aust Paediatr J 1987;23:253–5.

[34] Callaghan MU, Wong TE, Federici AB. Treatment of acquired von Willebrand syndrome in childhood. Blood 2013;122:2019–22.

[35] Heyde EC. Gastrointestinal bleeding in aortic stenosis. N Engl J Med 1958;259:196.

[36] Loscalzo J. From clinical observation to mechanism-Heyde's syndrome. N Engl J Med 2012;367:1954–6.

[37] Warkentin TE, Moore JC, Morgan DG. Aortic stenosis and bleeding gastrointestinal angiodysplasia: is acquired von Willebrand's disease the link? Lancet 1992;340:35–7.

[38] Gill JC, Wilson AD, Endres-Brooks J, Montgomery RR. Loss of the largest von Willebrand factor multimers from the plasma of patients with congenital cardiac defects. Blood 1986;67:758–61.

[39] Vincentelli A, Susen S, Le Tourneau T, Six I, Fabre O, Juthier F, Bauters A, Decoene C, Goudemand J, Prat A, Jude B. Acquired von Willebrand syndrome in aortic stenosis. N Engl J Med 2003;349:343–9.

[40] Blackshear JL, Wysokinska EM, Safford RE, Thomas CS, Shapiro BP, Ung S, Stark ME, Parikh P, Johns GS, Chen D. Shear stress-associated acquired von Willebrand syndrome in patients with mitral regurgitation. J Thromb Haemost 2014;12:1966–74.

[41] Pickering NJ, Brody JI, Barrett MJ. Von Willebrand syndromes and mitral-valve prolapse. N Engl J Med 1981;305.131–4.

[42] Le Tourneau T, Susen S, Caron C, Millaire A, Maréchaux S, Polge AS, Vincentelli A, Mouquet F, Ennezat PV, Lamblin N, de Groote P, Van Belle E, Deklunder G, Goudemand J, Bauters C, Jude B. Functional impairment of von Willebrand factor in hypertrophic cardiomyopathy: relation to rest and exercise obstruction. Circulation 2008;118:1550–7.

[43] Smith RE, Santamaria J. Unexplained bleeding in a patient with mechanical valve prostheses. Am J Med 1988;85:748.

[44] Heilmann C, Geisen U, Beyersdorf F, Nakamura L, Benk C, Berchtold-Herz M, Trummer G, Schlensak C, Zieger B. Acquired von Willebrand syndrome in patients with ventricular assist device or total artificial heart. Thromb Haemost 2010;103:962–7.

[45] Uriel N, Pak SW, Jorde UP, Jude B, Susen S, Vincentelli A, Ennezat PV, Cappleman S, Naka Y, Mancini D. Acquired von Willebrand syndrome after continuous-flow mechanical device support contributes to a high prevalence of bleeding during long-term support and at the time of transplantation. J Am Coll Cardiol 2010;56:1207–13.

[46] Crawley JT, de Groot R, Xiang Y, Luken BM, Lane DA. Unraveling the scissile bond: how ADAMTS13 recognizes and cleaves von Willebrand factor. Blood 2011;118:3212–21.

[47] Heilmann C, Geisen U, Beyersdorf F, Nakamura L, Trummer G, Berchtold-Herz M, Schlensak C, Zieger B. Acquired Von Willebrand syndrome is an early-onset problem in ventricular assist device patients. Eur J Cardiothorac Surg 2011;40:1328–33.

[48] Rauch A, Legendre P, Christophe OD, Goudemand J, van Belle E, Vincentelli A, Denis CV, Susen S, Lenting PJ. Antibody-based prevention of von Willebrand factor degradation mediated by circulatory assist devices. Thromb Haemost 2014;112:1014–23.

[49] Ramsay DM, Buist TAS, Macleod DAD, Heading RC. Persistent gastrointestinal bleeding due to angiodysplasia of the gut in von Willebrand's disease. Lancet 1976;ii:275–8.

[50] Ahr DJ, Rickles FR, Hoyer LW, O'Leary DS, Conrad ME. von Willebrand's disease and hemorrhagic telangiectasia: association of two complex disorders of hemostasis resulting in life-threatening hemorrhage. Am J Med 1977;62:452–8.

[51] Wautier J-L, Caen JP, Rymer R. Angiodysplasia in acquired von Willebrand disease. Lancet 1976;ii:973.

[52] McGrath KM, Johnson CA, Stuart JJ. Acquired von Willebrand disease associated with an inhibitor of factor VIII antigen and gastrointestinal telangiectasia. Am J Med 1979;67:693–6.

[53] Starke RD, Ferraro F, Paschalaki KE, Dryden NH, McKinnon TAJ, Sutton RE, Payne EM, Haskard DO, Hughes AD, Cutler DF, Laffan MA, Randi AM. Endothelial von Willebrand factor regulates angiogenesis. Blood 2011;117:1071–80.

[54] Randi AM, Laffan MA. Von Willebrand factor and angiogenesis: basic and applied issues. J Thromb Haemost 2017;15:13–20.

[55] Veyradier A, Balian A, Wolf M, Giraud V, Montembault S, Obert B, Dagher I, Chaput JC, Meyer D, Naveau S. Abnormal Von Willebrand factor in bleeding angiodysplasias of the digestive tract. Gastroenterology 2001;120:346–53.

[56] Franchini M, Mannucci PM. Von Willebrand disease-associated angiodysplasia: a few answers, still many questions. Br J Hematol 2013;161:177–82.

[57] Van Belle E, Rauch A, Vincent F, Robin E, Kibler M, Labreuche J, Jeanpierre E, Levade M, Hurt C, Rousse N, Dally JB, Debry N, Dallongeville J, Vincentelli A, Delhaye C, Auffray JL, Juthier F, Schurtz G, Lemesle G, Caspar T, Morel O, Dumonteil N, Duhamel A, Paris C, Dupont-Prado A, Legendre P, Mouquet F, Marchant B, Hermoire S, Corseaux D, Moussa K, Manchuelle A, Bauchart JJ, Loobuyck V, Caron C, Zawadzki C, Leroy F, Bodart JC, Staels B, Goudemand J, Lenting PJ, Susen S. Von Willebrand factor multimers during transcatheter aortic-valve replacement. N Engl J Med 2016;375:335–44.

[58] Rauch A, Caron C, Vincent F, Jeanpierre E, Ternisien C, Boisseau P, Zawadzki C, Fressinaud E, Borel-Derlon A, Hermoire S, Paris C, Lavenu-Bombled C, Veyradier A, Ung A, Vincentelli A, van Belle E, Lenting PJ, Goudemand J, Susen S. A novel ELISA-based diagnosis of acquired von Willebrand disease with increased VWF proteolysis. Thromb Haemost 2016;115:950–9.

[59] Anderson RP, McGrath K, Street A. Reversal of aortic stenosis, bleeding gastrointestinal angiodysplasia, and von Willebrand syndrome by aortic valve replacement. Lancet 1996;347:689–90.

[60] Warkentin TE, Moore JC, Morgan DG. Gastrointestinal angiodysplasia and aortic stenosis. N Engl J Med 2002;347:858–9.

[61] Blackshear JL, Stark ME, Agnew RC, Moussa ID, Safford RE, Shapiro BP, Waldo OA, Chen D. Remission of recurrent gastrointestinal bleeding after septal reduction therapy in patients with hypertrophic obstructive cardiomyopathy-associated acquired von Willebrand syndrome. J Thromb Haemost 2015;13:191–6.

[62] Steinlechner B, Zeidler P, Base E, Birkenberg B, Ankersmit HJ, Spannagl M, Quehenberger P, Hiesmayr M, Jilma B. Patients with severe aortic valve stenosis and impaired platelet function benefit from preoperative desmopressin infusion. Ann Thorac Surg 2011;91:1420–6.

[63] Budde U, Dent JA, Berkowitz SD, Ruggieri ZM, Zimmerman TS. Subunit composition of plasma von Willebrand factor in patients with the myeloproliferative syndrome. Blood 1986;68:1213–7.

[64] Budde U, Scharf RE, Franke P, Hartmann-Budde K, Dent J, Ruggeri ZM. Elevated platelet count as a cause of abnormal von Willebrand factor multimer distribution in plasma. Blood 1993;82:1749–57.

[65] van Genderen PJ, Prins FJ, Lucas IS, van de Moesdijk D, van Vliet HH, van Strik R, Michiels JJ. Decreased half-life time of plasma von Willebrand factor collagen binding activity in essential thrombocythaemia: normalization after cytoreduction of the increased platelet count. Br J Haematol 1997;99:832–6.

[66] Rottenstreich A, Kleinstern G, Krichevsky S, Varon D, Lavie D, Kalish Y. Factors related to the development of acquired von Willebrand syndrome in patients with essential thrombocythemia and polycythemia vera. Eur J Intern Med 2017;41:49–54.

[67] Rumi E, Cazzola M. How I treat essential thrombocythemia. Blood 2016;128:2403–14.

[68] van Genderen PJ, van Vliet HH, Prins FJ, van de Moesdijk D, van Strik R, Zijlstra FJ, Budde U, Michiels JJ. Excessive prolongation of the bleeding time by aspirin in essential thrombocythemia is related to a decrease of large von Willebrand factor multimers in plasma. Ann Hematol 1997;75:215–20.

[69] Chu DK, Hillis CM, Leong DP, Anand SS, Siegal DM. Benefits and risks of antithrombotic therapy in essential thrombocythemia—a systematic review. Ann Intern Med 2017;167:170–80.

[70] Mühlhausen C, Schneppenheim R, Budde U, Merkel M, Muschol N, Ullrich K, Santer R. Decreased plasma concentration of von Willebrand factor antigen (VWF:Ag) in patients with glycogen storage disease type Ia. J Inherit Metab Dis 2005;28:945–50.

[71] Marti GE, Rick ME, Sidbury J, Gralnick HR. Desmopressin infusion in five patients with type Ia glycogen storage disease and associated correction of prolonged bleeding times. Blood 1986;68:180–4.

[72] Kao K-J, Coleman RA, Pizzo SV. The bleeding diathesis in human glycogen storage disease type I: in vitro identification of a naturally occurring inhibitor of ristocetin-induced platelet aggregation. Thromb Res 1980;18:683–92.

[73] McGrath RT, McKinnon TAJ, Byrne B, O'Kennedy R, Terraube V, McRae E, Preston RJS, Laffan MA, O'Donnell JS. Expression of terminal α2-6-linked sialic acid on von Willebrand factor specifically enhances proteolysis by ADAMTS13. Blood 2010;115:2666–73.

[74] Dalton RG, Savidge GF, Matthews KB, Dewar MS, Kernoff PBA, Greaves M, Preston FE. Hypothyroidism as a cause of acquired von Willebrand's disease. Lancet 1987;i:1007–9.

[75] Levesque H, Borg JY, Cailleux N, Vasse M, Daliphard S, Gancel A, Monconduit M, Courtois H. Acquired von Willebrand's syndrome associated with decrease of plasminogen activator and its inhibitor during hypothyroidism. Eur J Med 1993;2:287–8.

[76] Baumgartner-Parzer SM, Wagner L, Reining G, Sexl V, Nowotny P, Müller M, Brunner M, Waldhäusl W. Increase by tri-iodothyronine of endothelin-1, fibronectin and von Willebrand factor in cultured endothelial cells. J Endocrinol 1997;154:231–9.

[77] Diekman MJ, Zandieh Doulabi B, Platvoet-Ter Schiphorst M, Fliers E, Bakker O, Wiersinga WM. The biological relevance of thyroid hormone receptors in immortalized human umbilical vein endothelial cells. J Endocrinol 2001;168:427–33.

[78] Stuijver DJ, Piantanida E, van Zaane B, Galli L, Romualdi E, Tanda ML, Meijers JC, Büller HR, Gerdes VE, Squizzato A. Acquired von Willebrand syndrome in patients with overt hypothyroidism: a prospective cohort study. Haemophilia 2014;20:326–32.

[79] Federici AB. Acquired von Willebrand syndrome associated with hypothyroidism: a mild bleeding disorder to be further investigated. Semin Thromb Hemost 2011;37:35–40.

[80] Castaman G, Rodeghiero F. Acquired transitory von Willebrand syndrome with ciprofloxacin. Lancet 1994;343:492.

[81] Castaman G, Lattuada A, Mannucci PM, Rodeghiero F. Characterization of two cases of acquired transitory von Willebrand syndrome with ciprofloxacin: evidence for heightened proteolysis of von Willebrand factor. Am J Hematol 1995;49:83–6.

[82] Acharya S, Bussel JB. Hematologic toxicity of sodium valproate. J Pediatr Hematol Oncol 2000;22:62–5.

[83] Koenig S, Gerstner T, Keller A, Teich M, Longin E, Dempfle CE. High incidence of vaproate-induced coagulation disorders in children receiving valproic acid: a prospective study. Blood Coagul Fibrinolysis 2008;19:375–82.

[84] Kreuz W, Linde R, Funk M, Meyer-Schrod R, Föll E, Nowak-Göttl U, Jacobi G, Vigh Z, Scharrer I. Valproate therapy induces von Willebrand disease type I. Epilepsia 1992;33:178–84.

[85] Serdaroglu G, Tütüncüoglu S, Kavakli K, Tekgül H. Coagulation abnormalities and acquired von Willebrand's disease type 1 in children receiving valproic acid. J Child Neurol 2002;17:41–3.

[86] Eberl W, Budde U, Bentele K, Christen HJ, Knapp R, Mey A, Schneppenheim R. Acquired von Willebrand syndrome as side effect of valproic acid therapy in children is rare. Hamostaseologie 2009;29:137–42.
[87] Accessed from www.bbraunusa.com [July 2017].
[88] Sanfelippo MJ, Suberviola PD, Geimer NF. Development of a von Willebrand-like syndrome after prolonged use of hydroxethyl starch. Am J Clin Pathol 1987;88:653–85.
[89] Dalrymple-Hay M, Aitchison R, Collins P, Sekhar M, Colvin B. Hydroxyethyl starch induced acquired von Willebrand's disease. Clin Lab Haematol 1992;14:209–11.
[90] Treib J, Haass A, Pindur G, Miyachita C, Grauer MT, Jung F, Wenzel E, Schimrigk K. Highly substituted hydroxyethyl starch (HES200/0.62) leads to type-1 von Willebrand syndrome after repeated administration. Haemostasis 1996;26:210–3.
[91] Jonville-Bera A-P, Autret-Leca E, Gruel Y. Acquired type 1 von Willebrand's disease associated with highly substituted hydroxyethyl starch. N Engl J Med 2001;345:622–3.
[92] Ertmer C, Wulf H, Van Aken H, Friederich P, Mahl C, Bepperling F, Westphal M, Gogarten W. Efficacy and safety of 10% HES 130/0.4 versus 10% HES 200/0.5 for plasma volume expansion in cardiac surgery patients. Minerva Med 2012;103:111–22.

Recommended Reading

[1] Federici AB, Budde U, Castaman G, Rand JH, Tiede A. Current diagnostic and therapeutic approaches to patients with acquired von Willebrand syndrome: a 2013 update. Semin Thromb Hemost 2013;39:191–201.
[2] Loscalzo J. From clinical observation to mechanism-Heyde's syndrome. N Engl J Med 2012;367:1954–6. Letter to the Editor, N Engl J Med 2013;368:579–80.
[3] Randi AM, Laffan MA. Von Willebrand factor and angiogenesis: basic and applied issues. J Thromb Haemost 2017;15:13–20.
[4] Fosbury E, Szychot E, Slater O, Mathias M, Sibson K. An 11-year experience of acquired von Willebrand syndrome in children diagnosed with Wilms tumor in a tertiary referral centre. Pediatr Blood Cancer 2017;64.

Factor VIII and Thrombosis

In previous chapters, the theme was low FVIII and bleeding; in this chapter, the focus is on high FVIII and thrombosis. Thrombosis is a pathological condition characterized by clot formation within blood vessels that obstructs blood flow and produces tissue ischemia. Although FVIII is absolutely necessary for normal hemostasis, it is unclear whether elevated FVIII levels cause or augment thrombus formation. FVIII concentrations in healthy persons vary over a broad range, from 0.5 to 1.5 U/mL; values as low as 0.3 U/mL prevent excessive bleeding with trauma. It might be argued that healthy people have up to a fivefold increase in FVIII above the levels required for normal hemostasis, suggesting that such concentrations are not a risk factor for thrombosis. Furthermore, although FVIII is increased in a variety of clinical disorders complicated by intravascular clotting, concomitant increases in fibrinolysis or other antithrombotic activities might occur in these conditions as well, counterbalancing the effect of FVIII on thrombogenesis [1]. However, accumulating epidemiological and other evidence favors the view that FVIII is a major contributor to thrombus formation and will be presented in this chapter.

Increases in FVIII are reported with specific polymorphisms of *F8*, *VWF*, and other genes; the extent of glycation of FVIII and VWF (ABO group); and factors that promote the release of VWF from endothelial cells [2]. The affected genes are involved in the synthesis, release, and clearance of the protein. In addition, enhanced release of VWF from endothelial cells or decreased clearance because of changes in its sialylation can account for elevated FVIII levels. For example, the Atherosclerosis Risk in Communities study identified three single-nucleotide polymorphisms in the *ST3GAL4* sialyltransferase gene that were associated with FVIII and VWF levels [3]. Finally, high FVIII concentrations could be a consequence of factors reducing the activity of protein C, which normally degrades activated FVIII. Other mutations increasing FVIII/VWF levels are described in the following paragraphs.

Polymorphisms in the promoter or polyadenylation signal region of the *F8* gene might enhance FVIII transcription, but searches of these sites

Hemophilia and Von Willebrand Disease
https://doi.org/10.1016/B978-0-12-812954-8.00013-8

failed to discover candidate mutations that associate with elevated levels of FVIII or VTE [4]. However, a single-nucleotide polymorphism (SNP) encoding the B-domain substitution D1241E, was significantly associated with the FVIII level [5]. Subsequently, four intronic single-nucleotide polymorphisms (SNPs) and two haplotypes were found to associate with FVIII activity, but only in men of European ancestry [6]. Another study discovered additional SNPs affecting FVIII activity in both European and African-Americans [7].

FVIII circulates in a high-affinity complex with VWF; examination of the *VWF* gene identified three sets of independent SNPs, spanning 24 SNPs, that affected FVIII levels [7]. In addition, there are reports of a rare variant in the *VWF* gene, rs7962217, associated with higher FVIII levels [8]. Polymorphisms in genes other than those for FVIII, VWF, and ABO have also been found to influence FVIII levels; these include the genes for syntaxin-binding protein 5, scavenger receptor class A, sta-bilin 2, high-molecular-weight kininogen, trimethyllysine dioxygenase, and in African-Americans, hepatic methionine adenosyltransferase I/III [9]. Significant associations with venous thrombosis were observed for SNP rs1039084 in the gene for syntaxin-binding protein 5 ($P = .005$), but not with the scavenger receptor class A or stabilin 2 variants [10]. Further studies are needed to determine the relative contributions of the various genetic polymorphisms to FVIII levels in health and disease.

The low-density lipoprotein-receptor-related protein-1 (LRP-1) receptor on macrophages removes FVIII/VWF from the plasma [11,12]; mutations that impair the function of this receptor increase FVIII levels by reducing its clearance. A genetic polymorphism of the LRP-1 gene, 633C > T, has been independently associated with elevated FVIII activity in a multivariate analysis that included blood group, age, and VWF antigen [13]. Glycans and the extent of sialylation affect the susceptibility of FVIII and VWF to proteolysis and account for the higher levels of FVIII/VWF in people with non-O blood groups [14]. High levels of VWF propeptide, consistent with increased release of VWF from endothelial cells, are reported in patients with high levels of FVIII [15].

Elevated FVIII levels are found in a variety of disorders [16]. Table 13.1 lists these conditions and the suspected drivers of the increased FVIII; in most, VWF levels are also increased.

Increases in FVIII after epinephrine infusion or strenuous exercise were noted in Chapter 1 and are attributed to beta-adrenergic stimulation. Inhalation of the short-acting selective β2-adrenoreceptor agonist, salbuta-mol, increased FVIII by $11 \pm 3\%$ and VWF by $7 \pm 1\%$ in healthy volunteers; a concomitant rise in d-dimer and prothrombin fragments indicated activation of the clotting system [17]. FVIII concentrations correlate with free T4 levels in patients with hyperthyroidism ($r = .35$, $P < .05$) [18].

TABLE 13.1 Clinical Conditions Associated With Elevated FVIII/VWF, and Their Likely Drivers

Clinical condition	Mechanism
Epinephrine infusion, strenuous exercise, fever induction	Beta-adrenergic stimulation
Hyperthyroidism	Enhanced protein synthesis
Cushing syndrome, therapy with corticosteroids	Cortisol-induced upregulation of promoter polymorphisms in *VWF* gene
Late pregnancy, postpartum, use of oral contraceptives	Estradiol upregulates VWF mRNA expression by endothelial cells
Following major surgery or trauma (i.e., spinal cord injury)	Cytokines and release of adrenal hormones
Pancreatitis and diabetes mellitus	Increased release from activated endothelial cells
Hyperlipidemia and coronary artery disease	Inflammatory cytokines associated with atherosclerosis
Renal failure, nephrotic syndrome	Probably enhanced release from activated endothelial cells
Neoplasms (solid tumors, hematologic malignancies)	Cytokines, tumor products, and synthesis by cancer cells
Hemoglobinopathies, hereditary spherocytosis, acquired hemolytic anemia	Release from activated endothelial cells, decreased clearance by macrophages

High FVIII levels in some patients with Cushing syndrome are associated with *VWF* gene promoter polymorphisms that are sensitive to upregulation by cortisol [19,20]. A mechanism for the increased levels in pregnancy and hormonal users might be upregulation of endothelial VWF mRNA expression by estradiol [21]. Cytokines and release of adrenal hormones probably account for the elevated levels after surgery or trauma. Increasing levels of VWF and its propeptide occur as patients with type 2 diabetes age, and more of the VWF is in the glycoprotein Ib-binding conformation [22]. Elevated levels of VWF are reported in diabetics with, but not without, albuminuria, consistent with endothelial cell activation and release of the factor [23].

Interleukin-6 (IL-6) and other inflammatory biomarkers play major roles in atherogenesis. Several fold increases in FVIII mRNA occur when cultured cells are exposed IL-6, but not IL-1 or IL-2 [24]. Renal failure is associated with a sustained increase in FVIII/VWF, but the mechanism is uncertain; it could be due to enhanced synthesis or decreased clearance of the complex [25]. Raised levels of FVIII/VWF associated with neoplasms are attributed to cytokines and tumor products; in addition, direct expression of VWF by glial and osteosarcoma cells has been demonstrated [26]. Finally, the complex might be released from the hypoxic endothelium associated with microvascular infarcts in sickle cell disease or be elevated because of decreased macrophage clearance in other hemolytic anemias.

VENOUS THROMBOSIS

Thrombi most often occur in the superficial or deep veins of the extremities, and less frequently in the cerebral, retinal, axillary, mesenteric, and other veins. Thrombi developing in the lower extremities commonly embolize to the lungs (venous thromboembolism; VTE). There appear to be specific factors that contribute to thrombus formation at a particular site. For example, infection triggers thrombi in superficial veins, trauma or surgery provokes deep-vein thrombosis in the legs, and cancer is associated with thrombosis at less common sites. Quantitative or qualitative alterations in FVIII could augment the development of thrombi at any of these sites, or increase the risk of recurrent thrombosis.

In 1983, Stead et al. [27] reported venous thrombosis, pulmonary emboli, and mesenteric vein thrombosis in 5 male members in two generations of a family; their ages ranged from 14 to 38 at the time of thrombosis. FVIII/VWF was measured in 4 of the 5 members and values were 1.8–3.4 U/mL and 1.3–3.5 U/mL, respectively; in addition, they had decreased plasminogen activator. The authors suggested that endothelial cell dysfunction accounted for the coagulation abnormalities that were responsible for the venous thrombosis in this family.

Evidence that specifically associates increased FVIII levels with venous thrombosis has come from the Leiden Thrombophilia Study [28]. FVIII concentrations were measured in 301 consecutive individuals with a first diagnosed venous thrombosis and 301 age- and sex-matched healthy controls. In multivariate analyses that included blood group and VWF levels, only FVIII was found to be a significant risk factor. The adjusted odds ratio was 4.8 for FVIII concentrations above 1.5 U/mL, and the risk increased with increasing FVIII levels. C-reactive protein (CRP) levels were higher in thrombosis patients than in controls, but high levels of FVIII continued to predict thrombosis even after adjustment for CRP [29]. Another group of investigators confirmed these findings, reporting that an elevated FVIII level was the single most common abnormality detected in 25% of a group of 260 patients with unexplained VTE [30]. FVIII concentrations did not correlate with CRP or the erythrocyte sedimentation rate, indicating that the increased levels were not simply an acute-phase response to the venous thrombosis. Follow-up over an extended period (median, 8 months; range, 3–39 months) showed that 94% of patients continued to have an elevated FVIII, leading to the conclusion that raised FVIII levels represented a constitutional risk factor for VTE [31]. On the other hand, a more recent study found that IL-6 levels were also increased in patients with deep-vein thrombosis, remained high for more than 2 years, and were independently associated with FVIII [32].

Still other investigators found that the risk for a single episode of VTE increased by 10% for each 0.1 U/mL increment in FVIII; for recurrent episodes, the risk increased by 24% [33]. Adjustments for fibrinogen, C-reactive protein, and other known thrombophilic factors did not change the association of elevated FVIII with thrombosis. Other studies have confirmed that high levels of FVIII increase the risk for repeated episodes of VTE. Kyrle et al. [34] observed that the mean levels of FVIII were significantly higher in patients having recurrent VTE (1.82 U/mL vs 1.57 U/mL, $P = .009$). Furthermore, the likelihood of recurrence at 2 years was higher in patients with FVIII >90th percentile (37% vs 5%, $P < .001$). The MEGA study measured levels of FVIII and VWF at least 3 months after a VTE in 2242 patients [35]. Over a median of 6.9 years, 343 individuals developed recurrent thromboses; recurrence rates (events per 100 patient years) increased from 1.4 to 5.1 as FVIII levels rose from <1 to >2 U/mL. Patients with the highest FVIII levels had a threefold higher VTE recurrence rate than those with the lowest levels, and all-cause mortality increased in a dose-response fashion with increasing percentiles of FVIII levels [36]. Finally, many of the disorders with increased FVIII listed in Table 13.1, such as Cushing syndrome [37], have been clearly associated with venous thrombosis.

Hereditability of high FVIII levels has been demonstrated in families having one or more members with elevated FVIII and a history of VTE; genetic analysis found linkage to imprinted loci on chromosomes 5 and 11, consistent with a complex trait caused by several genetic factors [38,39]. Another study showed that the heritability estimate was highly significant for FVIII coagulant activity and antigen, but only the antigen remained significant when household effects were removed [40]. The investigators suggested that basal FVIII antigen levels are under greater genetic control while FVIII activity is more influenced by environmental factors. A prospective family cohort study recruited 532 first-degree relatives of consecutive patients with elevated FVIII and VTE or arterial vascular events prior to age 50 [41]. Participants were followed for a mean of 31 months. The 190 relatives with FVIII >1.5 U/mL had a 5.5-fold increased risk of developing a VTE as compared to relatives without elevated FVIII levels. As noted earlier, a polymorphism in the syntaxin-binding protein 5 gene is strongly associated with elevated FVIII levels and venous thrombosis [10].

People with classical hemophilia rarely experience deep-vein thrombosis, but VTE is an occasional postoperative complication if levels of FVIII become supertherapeutic during the prolonged administration of clotting factor concentrate [42]. In people with cancer, elevated FVIII levels are an additional risk factor for thrombosis. A prospective study of patients with solid tumors or hematologic malignancies reported that the cumulative probability of VTE during 6 months follow-up was significantly higher in those with elevated FVIII levels (14% vs 4%, $P = .001$) [43]. High FVIII levels add to thrombosis susceptibility in cancer patients even when other risk factors are present, such as low levels of protein C or the factor V Leiden mutation [44,45]. Even in individuals without cancer but with other risk factors for VTE, high levels of FVIII add to the risk; for example, carriers of factor V Leiden with high levels of FVIII have an adjusted odds ratio of 2.7 for VTE [46].

Lastly, a pediatric study showed that a FVIII >1.5 U/mL at the time of VTE diagnosis was strongly predictive of a lack of thrombus resolution, recurrent thrombosis, and the postthrombotic syndrome [47]. A firm conclusion from all these investigations is that high levels of FVIII are not just a consequence of VTE, but actually predict recurrent disease; whether FVIII is the spark that ignites the fire, or the fodder that feeds the flames, remains unclear.

ARTERIAL THROMBOSIS

Arterial thrombi differ from venous thrombi in several important respects. In arteries, thrombi form in a pulsatile, rapid blood flow, high shear environment, while venous thrombi develop in relatively static, but turbid blood within the cusps of vein valves. Platelets are essential for the formation and growth of arterial thrombi, but are less involved in venous thrombosis. FVIII is found in proximity to macrophages and smooth muscle cells in atheromatous regions, especially those associated with deposits of oxidized low-density lipoprotein cholesterol [48]. However, FVIII appears to play a smaller role in arterial than venous thrombosis; studies of patients with FVIII deficiency have shown that venous thrombosis is rare in patients with severe hemophilia, but fibrin-platelet thrombi in the coronary arteries and aorta have been reported even in the presence of a potent FVIII autoantibody and no exogenous source of FVIII [49].

Animal studies show that mice doubly-deficient in apolipoprotein E and FVIII develop fewer atheromatous lesions compared to knockouts for apo-E alone, suggesting that FVIII might be a necessary participant in the early stages of atherogenesis [50]. However, FVIII deficiency does not retard the progression of atherosclerosis in mice lacking the LDL receptor, demonstrating that atherosclerosis can develop in the complete absence of

FVIII [51]. Hemophilic patients with risk factors for atherosclerosis, such as obesity, have coronary artery calcification, carotid intima media thickness, and brachial flow-mediated dilatation measurements no different from those of age-matched nonhemophiliacs [52,53].

Most arterial thrombi occur in association with ulcerated atheromatous plaques; tissue factor is present in these plaques and contributes to thrombus formation following plaque disruption [54]. FVIII is not required for tissue-factor-induced clot initiation, as implied by the normal prothrombin time test in patients with severe hemophilia. However, FVIII does enhance the spatial growth of clots initiated by tissue factor [55], and FVIII deficiency might retard thrombus formation on atheromatous plaques and resultant vascular occlusion [56]. In support of this hypothesis, a study from the Netherlands reported that carriers of hemophilia have a reduction in overall mortality and deaths from ischemic heart disease as compared to the general Dutch female population adjusted for age and calendar period [57]. In general, individuals with FVIII deficiency hemophilia have a nonsignificant decrease in mortality due to arterial thrombosis [58].

FVIII is increased by inflammation and there is considerable evidence that atherogenesis is an inflammatory process. The concentrations of several inflammatory cytokines and other acute-phase reactants are increased in patients with atherosclerotic vascular disease and are thought to reflect the local inflammatory process in the artery [59]. Associations between inflammatory markers and coagulation proteins abound; for example, CRP and the levels of FVIII, FX, and plasminogen activator inhibitor-1 are significantly correlated, tissue factor expression by monocytes is stimulated by IL-6 and CRP, and thrombin stimulates monocyte IL-6 expression [60]. The observation that FVIII is increased by inflammation does not exclude the possibility that it might also contribute to thrombus formation in patients with atheromatous disease.

Epidemiological evidence supports a role for FVIII in atherothrombosis. The Northwick Park Heart Study reported that FVIII levels at the time of recruitment were higher in middle-aged white men subsequently dying of cardiovascular disease [61]. The Cardiovascular Health Study enrolled more than 5000 subjects over age 65; cross-sectional study showed that a variety of cardiac risk factors were associated with elevated levels of factors VII, FVIII, and fibrinogen [62]. Furthermore, fibrinogen and to a lesser extent FVIII showed positive associations with a variety of subclinical cardiovascular disease measures, and FVIII was significantly associated with coronary heart disease events and mortality in men, and with stroke and transient ischemic attacks (TIA) in women [63,64]. And in a prospective analysis, coronary heart disease events and mortality in men, and stroke/TIA in women, were associated with FVIII [65]. These associations have been confirmed by a number of other epidemiologic studies [66,67]. Also, it has been reported that FVIII levels are significantly higher in patients dying of myocardial infarction (MI) than in survivors, independent of infarct size and severity [68]. FVIII is strongly associated with sudden and nonsudden cardiac death as well as nonfatal MI [69]. On the other hand, the Multiethnic Study of Atherosclerosis could not confirm that FVIII was an independent risk factor for cardiovascular disease [70]. Nevertheless, high levels of FVIII might augment thrombus formation in patients with atheromatous disease and might serve as an indicator of thrombosis risk. Kamphuisen et al. [71] have calculated that an elevated level of FVIII accounts for 4% of arterial thrombotic events.

There are several explanations for the elevated levels of FVIII in atherosclerosis. The disease is typically associated with older age, diabetes, and hypercholesterolemia, conditions that also feature elevated levels of FVIII. The changes in FVIII with aging are described in Chapter 3; the increases are modest, and, in healthy aging, levels generally do not exceed 1.85 U/mL even in those >100 years of age [72]. Increased FVIII is reported in diabetics with peripheral vascular disease [73], and a prospective case-control study found high levels of FVIII and enhanced thrombin generation in diabetic patients prior to the development of macrovascular complications [74]. The Atherosclerosis Risk in Communities (ARIC) Study reported a statistically significant association of FVIII with the development of diabetes, which they attributed to endothelial dysfunction [75]. Other investigators confirmed the association of elevated FVIII with type 2 diabetes, but did not find evidence of endothelial dysfunction or inflammation [76].

Low-density lipoprotein (LDL) cholesterol, a major player in atheromatous plaque formation, is cleared from the circulation by binding to its receptor on endothelial cells (LDLR). LDLR cooperates with LRP-1 in regulating plasma FVIII levels in vivo [77]. Patients with familial hypercholesterolemia due to loss-of-function mutations in the LDLR gene have increased levels of FVIII compared with their unaffected relatives [78]. In addition, carriers of the LDLR-rs688 allele have higher levels of FVIII as well as a significantly greater risk for coronary artery disease (odds ratio, 1.48), and the risk is independent of lipid levels [79]. The genetic influence on FVIII levels and arterial thrombosis is supported by a study reporting that two haplotypes in the *F8* gene are significantly associated with ischemic events [80]. Family studies offer further evidence that elevated FVIII contributes to atherogenesis.

FVIII levels are increased in individuals with a family history of coronary heart disease [81], and first thrombotic events are more frequent in the relatives of people with high FVIII concentrations. Bank et al. [82] performed a retrospective study of 584 first-degree relatives of 177 patients with elevated factor VIII and a first MI or peripheral arterial thrombosis. Those with high FVIII, as compared to relatives with normal levels, had an annual incidence of a first arterial thrombotic event that was threefold increased. In particular, the risks for a first MI (odds ratio: 4.3) and first peripheral arterial thrombosis (odds ratio: 8.6) were significantly increased ($P=.046$). While this study suggests that high levels of factor VIII predispose to arterial disease, it is possible that in addition to inheriting a propensity for an elevated factor VIII level, these relatives may have inherited other risk factors for atherosclerosis. These observations are consistent with the concept that atherosclerosis is a polygenic disorder, and its development and progression are likely influenced by a variety of factors.

TREATMENT OF THROMBOSIS IN PATIENTS WITH FVIII DEFICIENCY

Venous thrombosis in people with hemophilia most often occurs in association with other risk factors, such as indwelling catheters, intensive concentrate therapy, or postoperatively. Catheters that become nonfunctional are removed and replaced; if the vein is thrombosed, therapeutic anticoagulation can be initiated while infusing concentrate to provide hemostatic levels of FVIII. Thrombosis that occurs as a complication of concentrate therapy or postoperatively is treated with therapeutic doses of anticoagulants but without a loading dose, along with a FVIII concentrate to maintain trough activity of $\geq 0.3\,U/mL$ [83]. One month of anticoagulation might be sufficient; the status of the thrombus can be monitored by ultrasound and treatment resumed if progression is observed. Temporary inferior vena cava filters are used for patients in whom anticoagulation is not an option, such as those with FVIII inhibitors. Patients with arterial thrombosis usually require antiplatelet therapy, and most will need concomitant factor replacement to prevent bleeding. A target trough level of $0.1–0.15\,U/mL$ is suggested for patients receiving dual antiplatelet therapy. Stents for stenotic coronary arteries have differing requirements for long-term antiplatelet therapy and should be selected in consultation with cardiovascular experts (described in Chapter 7).

Another thrombotic disorder that might require anticoagulant therapy is atrial fibrillation; the goal of treatment is the prevention of an embolic stroke. If the baseline FVIII is $\geq 0.2\,U/mL$, and stroke risk is moderate, an oral anticoagulant is suggested [84]. Catheter ablation should be considered for patients with FVIII $<0.2\,U/mL$ and a high risk for stroke; if the patient is not a candidate for ablation, low-dose aspirin is recommended. No treatment is advised for patients at low risk, including those with FVIII $<0.01\,U/mL$.

SUMMARY

Factor VIII levels are elevated in a variety of disorders complicated by thrombosis. In some of these diseases, inflammatory cytokines such as interleukin-6 are active and stimulate increases in FVIII and other acute-phase reactants. In addition, the plasma concentrations of FVIII are under genetic control and are strongly dependent on the levels of VWF. People with non-O blood group have higher levels of VWF and FVIII, and an increased risk of thrombosis. Venous thromboembolism (VTE) in particular is strongly associated with increased FVIII levels, and the hereditable component of FVIII for venous thrombosis is highly significant $(P < .0001)$. It is estimated that FVIII concentrations $\geq 1.5\,U/mL$ account for 16% of VTE. Elevated FVIII levels often persist for years after thrombosis and are associated with recurrent VTE. Even hemophilic patients are at risk for VTE if their levels exceed $1.5\,U/mL$ while receiving FVIII concentrates.

A role for FVIII in atherothrombosis is supported by epidemiologic and family studies. Elevated levels augment the risk of arterial thrombosis in patients with other risk factors such as older age, diabetes, and hyperlipidemia. Family studies show that first-degree relatives of patients with high FVIII and thrombotic disease have a threefold increased risk of first arterial events if their own FVIII levels are increased. In addition, the risk for coronary artery disease is increased in individuals with elevated FVIII due to loss-of-function mutations in the *LDLR* gene. In summary, a concordance of studies suggests that high concentrations of FVIII augment thrombotic disease in both veins and arteries.

Future research might include the following:

- Determine the relative contribution of the various genetic polymorphisms to FVIII levels
- Identify the basis for elevated FVIII levels in diabetes, nephrotic syndrome, and cancer
- Search for additional *F8* polymorphisms in families with a high incidence of atherothrombotic disease
- Examine the safety and effectiveness of antithrombotic therapies in patients with hemophilia

References

[1] Stewart GJ. The role of hypercoagulability in thrombosis. Br J Haematol 1978;40:359–62.
[2] Jenkins PV, Rawley O, Smith OP, O'Donnell JS. Elevated factor VIII levels and risk of venous thrombosis. Br J Haematol 2012;157:653–63.
[3] Song J, Xue C, Preisser JS, Cramer DW, Houck KL, Liu G, Folsom AR, Couper D, Yu F, Dong JF. Association of single nucleotide polymorphisms in the ST3GAL4 gene with VWF antigen and factor VIII activity. PLoS ONE 2016;11.
[4] Mansvelt EPG, Laffan M, McVey JH, Tuddenham EGD. Analysis of the F8 gene in individuals with high plasma factor VIII:C levels and associated venous thrombosis. Thromb Haemost 1998;80:56165.
[5] Viel KR, Machiah DK, Warren DM, Khachidze M, Buil A, Fernstrom K, Souto JC, Peralta JM, Smith T, Blangero J, Porter S, Warren ST, Fontcuberta J, Soria JM, Flanders WD, Almasy L, Howard TE. A sequence variation scan of the coagulation

factor VIII (FVIII) structural gene and associations with plasma FVIII activity levels. Blood 2007;109:3713–24.

[6] Campos M, Buchanan A, Yu F, Barbalic M, Xiao Y, Chambless LE, Wu KK, Folsom AR, Boerwinkle E, Dong J-F. Influence of single nucleotide polymorphisms in factor VIII and von Willebrand factor genes on plasma factor VIII activity: the ARIC Study. Blood 2012;119:1929–34.

[7] Tang W, Cushman M, Green D, Rich SS, Lange LA, Yang Q, Tracy RP, Tofler GH, Basu S, Wilson JG, Keating BJ, Weng LC, Taylor HA, Jacobs Jr DR, Delaney JA, Palmer CD, Young T, Pankow JS, O'Donnell CJ, Smith NL, Reiner AP, Folsom AR. Gene-centric approach identifies new and known loci for FVIII activity and VWF antigen levels in European Americans and African Americans. Am J Hematol 2015;90:534–40.

[8] Huffman JE, de Vries PS, Morrison AC, Sabater-Lleal M, Kacprowski T, Auer PL, Brody JA, Chasman DI, Chen MH, Guo X, Lin LA, Marioni RE, Müller-Nurasyid M, Yanek LR, Pankratz N, Grove ML, de Maat MP, Cushman M, Wiggins KL, Qi L, Sennblad B, Harris SE, Polasek O, Riess H, Rivadeneira F, Rose LM, Goel A, Taylor KD, Teumer A, Uitterlinden AG, Vaidya D, Yao J, Tang W, Levy D, Waldenberger M, Becker DM, Folsom AR, Giulianini F, Greinacher A, Hofman A, Huang CC, Kooperberg C, Silveira A, Starr JM, Strauch K, Strawbridge RJ, Wright AF, McKnight B, Franco OH, Zakai N, Mathias RA, Psaty BM, Ridker PM, Tofler GH, Völker U, Watkins H, Fornage M, Hamsten A, Deary IJ, Boerwinkle E, Koenig W, Rotter JI, Hayward C, Dehghan A, Reiner AP, O'Donnell CJ, Smith NL. Rare and low-frequency variants and their association with plasma levels of fibrinogen, FVII, FVIII, and vWF. Blood 2015;126:e19–29.

[9] Smith NL, Chen MH, Dehghan A, Strachan DP, Basu S, Soranzo N, Hayward C, Rudan I, Sabater-Lleal M, Bis JC, de Maat MP, Rumley A, Kong X, Yang Q, Williams FM, Vitart V, Campbell H, Mälarstig A, Wiggins KL, Van Duijn CM, WL MA, Pankow JS, Johnson AD, Silveira A, McKnight B, Uitterlinden AG, Wellcome Trust Case Control Consortium, Aleksic N, Meigs JB, Peters A, Koenig W, Cushman M, Kathiresan S, Rotter JI, Bovill EG, Hofman A, Boerwinkle E, Tofler GH, Peden JF, Psaty BM, Leebeek F, Folsom AR, Larson MG, Spector TD, Wright AF, Wilson JF, Hamsten A, Lumley T, Witteman JC, Tang W, O'Donnell CJ. Novel associations of multiple genetic loci with plasma levels of factor VII, factor VIII, and von Willebrand factor: the CHARGE (Cohorts for Heart and Aging Research in Genome Epidemiology) Consortium. Circulation 2010;121:1382–92.

[10] Smith NL, Rice KM, Bovill EG, Cushman M, Bis JC, McKnight B, Lumley T, Glazer NL, van Hylckama Vlieg A, Tang W, Dehghan A, Strachan DP, O'Donnell CJ, Rotter JI, Heckbert SR, Psaty BM, Rosendaal FR. Genetic variation associated with plasma von Willebrand factor levels and the risk of incident venous thrombosis. Blood 2011;117:6007–11.

[11] Saenko EL, Yakhyaev AV, Mikhailenko I, Strickland DK, Sarafanov AG. Role of the low density lipoprotein-related protein receptor in mediation of factor VIII catabolism. J Biol Chem 1999;274:37685–92.

[12] Lenting PJ, Neels JG, van den Berg BM, Clijsters sPP, Meijerman DW, Pannekoek H, van Mourik JA, Mertens K, van Zonneveld AJ. The light chain of factor VIII comprises a binding site for low density lipoprotein receptor-related protein. J Biol Chem 1999;274:23734–9.

[13] Vormittag R, Bencur P, Ay C, Tengler T, Vukovich T, Quehenberger P, Mannhalter C, Pabinger I. Low-density lipoprotein receptor-related protein 1 polymorphism 663 C > T affects clotting factor VIII activity and increases the risk of venous thromboembolism. J Thromb Haemost 2007;5:497–502.

[14] McGrath RT, McKinnon TAJ, Byrne B, O'Kennedy R, Terraube V, McRae E, Preston RJS, Laffan MA, O'Donnell JS. Expression of terminal α2-6-linked sialic acid on von Willebrand factor specifically enhances proteolysis by ADAMTS13. Blood 2010;115:2666–73.

[15] Nossent AY, Van Marion V, VAN Tilburg NH, Rosendaal FR, Bertina RM, Van Mourik JA, Eikenboom HC. Von Willebrand factor and its propeptide: the influence of secretion and clearance on protein levels and the risk of venous thrombosis. J Thromb Haemost 2006;4:2556–62.

[16] Green D. Factor VIII (anti-hemophilic factor). J Chronic Dis 1970;23:213–25.

[17] Ali-Saleh M, Sarig G, Ablin JN, Brenner B, Jacob G. Inhalation of a short-acting β2-adrenoreceptor agonist induces a hypercoagulable state in healthy subjects. PLoS ONE 2016;11.

[18] Erem C, Ersoz HO, Karti SS, Ukinc K, Hacihasanoglu A, Deger O, Telatar M. Blood coagulation and fibrinolysis in patients with hyperthyroidism. J Endocrinol Invest 2002;25:345–50.

[19] Dal Bo Zanon R, Fornasiero L, Boscaro M, Cappellato G, Fabris F, Girolami A. Increased factor VIII associated activities in Cushing's syndrome: a probable hypercoagulable state. Thromb Haemost 1982;47:116–7.

[20] Daidone V, Boscaro M, Pontara E, Cattini MG, Occhi G, Scaroni C, Mantero F, Casonato A. New insight into the hypercoagulability of Cushing's syndrome. Neuroendocrinology 2011;93:121–5.

[21] Powazniak Y, Kempfer AC, Pereyra JC, Palomino JP, Lazzari MA. VWF and ADAMTS13 behavior in estradiol-treated HUVEC. Eur J Haematol 2011;86:140–7.

[22] Chen SF, Xia ZL, Han JJ, Wang YT, Wang JY, Pan SD, Wu YP, Zhang B, Li GY, Du JW, Gao HQ, de Groot PG, de Laat B, Hollestelle MJ. Increased active von Willebrand factor during disease development in the aging diabetic patient population. Age (Dordr) 2013;35:171–7.

[23] Vischer UM, Emeis JJ, Bilo HJ, Stehouwer CD, Thomsen C, Rasmussen O, Hermansen K, Wollheim CB, Ingerslev J. von Willebrand factor (vWf) as a plasma marker of endothelial activation in diabetes: improved reliability with parallel determination of the vWf propeptide (vWf:AgII). Thromb Haemost 1998;80:1002–7.

[24] Stirling D, Hannant WA, Ludlam CA. Transcriptional activation of the factor VIII gene in liver cell lines by interleukin-6. Thromb Haemost 1998;79:74–8.

[25] Mannucci PM. Von Willebrand factor: a marker of endothelial damage? Arterioscler Thromb Vasc Biol 1998;18:1359–62.

[26] Mojiri A, Stoletov K, Carrillo MA, Willetts L, Jain S, Godbout R, Jurasz P, Sergi CM, Eisenstat DD, Lewis JD, Jahroudi N. Functional assessment of von Willebrand factor expression by cancer cells of non-endothelial origin. Oncotarget 2017;8:13015–29.

[27] Stead N, Bauer KA, Kinney TR, Lewis JG, Campbell EE, Shifman MA, Rosenberg RD, Pizzo SV. Venous thrombosis in a family with defective release of vascular plasminogen activator and elevated plasma factor VIII/von Willebrand's factor. Am J Med 1983;74:33–9.

[28] Koster T, Blann AD, Briet E, Vandenbroucke JP, Rosendaal FR. Role of clotting factor VIII in effect of von Willebrand factor on occurrence of deep-vein thrombosis. Lancet 1995;345:152–5.

[29] Kamphuisen PW, Eikenboom JCJ, Vos HL, Pablo R, Sturk A, Bertina RM, Rosendaal FR. Increased levels of factor VIII and fibrinogen in patients with venous thrombosis are not caused by acute phase reactions. Thromb Haemost 1999;81:680–3.

[30] O'Donnell J, Tuddenham EGD, Manning R, Kemball-Cook G, Johnson D, Laffan M. High prevalence of elevated factor VIII levels in patients referred for thrombophilia screening: role of increased synthesis and relationship to the acute phase reaction. Thromb Haemost 1997;77:825–8.

[31] O'Donnell J, Mumford AD, Manning R, Laffan M. Elevation of FVIII:C in venous thromboembolism is persistent and independent of the acute phase response. Thromb Haemost 2000;83:10–3.

[32] Bittar LF, Bde M, Orsi FL, Collela MP, De Paula EV, Annichino-Bizzacchi JM. Long-term increased factor VIII levels are associated to interleukin-6 levels but not to post-

thrombotic syndrome in patients with deep venous thrombosis. Thromb Res 2015;135:497–501.

[33] Kraaijenhagen RA, in't Anker PS, MMW K, Reitsma PH, Prins MH, van den Ende A, Buller HR. High plasma concentration of factor VIIIc is a major risk factor for venous thromboembolism. Thromb Haemost 2000;83:5–9.

[34] Kyrle PA, Minar E, Hirschl M, Bialonczyk C, Stain M, Schneider B, Weltermann A, Speiser W, Lechner K, Eichinger S. High plasma levels of factor VIII and the risk of recurrent venous thromboembolism. N Engl J Med 2000;343:457–62.

[35] Timp JF, Lijfering WM, Flinterman LE, van Hylckama Vlieg A, Le Cessie S, Rosendaal FR, Cannegieter SC. Predictive value of factor VIII levels for recurrent venous thrombosis: results from the MEGA follow-up study. J Thromb Haemost 2015;13:1823–32.

[36] Yap ES, Timp JF, Flinterman LE, van Hylckama Vlieg A, Rosendaal FR, Cannegieter SC, Lijfering WM. Elevated levels of factor VIII and subsequent risk of all-cause mortality: results from the MEGA follow-up study. J Thromb Haemost 2015;13:1833–42.

[37] Koutroumpi S, Daidone V, Sartori MT, Cattini MG, Albiger NM, Occhi G, Ferasin S, Frigo A, Mantero F, Casonato A, Scaroni C. Venous thromboembolism in patients with Cushing's syndrome: need of a careful investigation of the prothrombotic risk profile. Pituitary 2013;16:175–81.

[38] Schambeck CM, Hinney K, Haubitz I, Taleghani BM, Wahler D, Keller F. Familial clustering of high factor VIII levels in patients with venous thromboembolism. Arterioscler Thromb Vasc Biol 2001;21:289–92.

[39] Berger M, Mattheisen M, Kulle B, Schmidt H, Oldenburg J, Bickeboller H, Walter U, Lindner TH, Strauch K, Schambeck CM. High factor VIII levels in venous thromboembolism show linkage to imprinted loci on chromosomes 5 and 11. Blood 2005;105:638–44.

[40] Kreuz W, Stoll M, Junker R, Heinecke A, Schobess R, Kurnik K, Kelsch R, Nowak-Göttl U. Familial elevated factor VIII in children with symptomatic venous thrombosis and post-thrombotic syndrome: results of a multicenter study. Arterioscler Thromb Vasc Biol 2006;26:1901–6.

[41] Bank I, van de Poel MH, Coppens M, Hamulyak K, Prins MH, van der Meer J, Veeger NJ, Buller HR, Middeldorp S. Absolute annual incidences of first events of venous thromboembolism and arterial vascular events in individuals with elevated FVIII:c. A prospective family cohort study. Thromb Haemost 2007;98:1040–4.

[42] Hermans C. Venous thromboembolic disease in patients with haemophilia. Thromb Res 2012;130(Suppl. 1):S50–52.

[43] Vormittag R, Simanek R, Ay C, Dunkler D, Quehenberger P, Marosi C, Zielinski C, Pabinger I. High factor VIII levels independently predict venous thromboembolism in cancer patients: the cancer and thrombosis study. Arterioscler Thromb Vasc Biol 2009;29:2176–81.

[44] Tafur AJ, Dale G, Cherry M, Wren JD, Mansfield AS, Comp P, Rathbun S, Stoner JA. Prospective evaluation of protein C and factor VIII in prediction of cancer-associated thrombosis. Thromb Res 2015;136:1120–5.

[45] Kovac M, Kovac Z, Tomasevic Z, Vucicevic S, Djordjevic V, Pruner I, Radojkovic D. Factor V Leiden mutation and high FVIII are associated with an increased risk of VTE in women with breast cancer during adjuvant tamoxifen - results from a prospective, single center, case control study. Eur J Intern Med 2015;26:63–7.

[46] Libourel EJ, Bank I, Meinardi JR, Baljé-Volkers CP, Hamulyak K, Middeldorp S, Koopman MM, van Pampus EC, Prins MH, Büller HR, van der Meer J. Co-segregation of thrombophilic disorders in factor V Leiden carriers; the contributions of factor VIII, factor XI, thrombin activatable fibrinolysis inhibitor and lipoprotein(a) to the absolute risk of venous thromboembolism. Haematologica 2002;87:1068–73.

[47] Goldenberg NA, Knapp-Clevenger R, Manco-Johnson MJ, for the Mountain States Regional Thrombophilia Group. Elevated plasma factor VIII and d-dimer levels as predictors of poor outcomes of thrombosis in children. N Engl J Med 2004;351:1081–8.

[48] Ananyeva NM, Kouiavskaia DV, Shima M, Saenko EL. Intrinsic pathway of blood coagulation contributes to thrombogenicity of atherosclerotic plaque. Blood 2002;99:4475–85.

[49] Green D, Rizza CR. Myocardial infarction in a patient with a circulating anticoagulant. Lancet 1967;ii:434–6.

[50] Khallou-Laschet J, Caligiuri G, Tupin E, Gaston A-T, Poirier B, Groyer E, Urbain D, Maisnier-Patin S, Sarkar R, Kaveri SV, Lacroix-Desmazes S, Nicoletti A. Role of the intrinsic coagulation pathway in atherogenesis assessed in hemophilic apolipoprotein E knockout mice. Arterioscler Thromb Vasc Biol 2005;25:e123–6.

[51] Fabri DR, De Paula EV, Cost DSP, Annichino-Bizzacchi JM, Arruda VR. Novel insights into the development of atherosclerosis in hemophilia A mouse models. J Thromb Haemost 2011;9:1556–61.

[52] Biere-Rafi S, Tuinenburg A, Haak BW, Peters M, Huijgen R, De Groot E, Verhamme P, Peerlinck K, Visseren FL, Kruip MJ, Laros-Van Gorkom BA, Gerdes VE, Buller HR, Schutgens RE, Kamphuisen PW. Factor VIII deficiency does not protect against atherosclerosis. J Thromb Haemost 2012;10:30–7.

[53] Zwiers M, Lefrandt JD, Mulder DJ, Smit AJ, Gans RO, Vliegenthart R, Brands-Nijenhuis AV, Kluin-Nelemans JC, Meijer K. Coronary artery calcification score and carotid intima–media thickness in patients with hemophilia. J Thromb Haemost 2012;10:23–9.

[54] Toschi V, Gallo R, Lettino M, Fallon JT, Gertz SD, Fernández-Ortiz A, Chesebro JH, Badimon L, Nemerson Y, Fuster V, Badimon JJ. Tissue factor modulates the thrombogenicity of human atherosclerotic plaques. Circulation 1997;95:594–9.

[55] Ovanesov MV, Lopatina EG, Saenko EL, Ananyeva NM, Ul'yanova LI, Plyushch OP, Butilin AA, Ataullakhanov FI. Effect of factor VIII on tissue factor-initiated spatial clot growth. Thromb Haemost 2003;89:235–42.

[56] Kamphuisen PW, ten Cate H. Cardiovascular risk in patients with hemophilia. Blood 2014;123:1297–301.

[57] Sramek A, Kriek M, Rosendaal FR. Decreased mortality of ischaemic heart disease among carriers of haemophilia. Lancet 2003;362:351–4.

[58] Biere-Rafi S, Zwiers M, Peters M, van der Meer J, Rosendaal FR, Büller HR, Kamphuisen PW. The effect of haemophilia and von Willebrand disease on arterial thrombosis: a systematic review. Neth J Med 2010;68:207–14.

[59] Hansson GK. Inflammation, atherosclerosis, and coronary artery disease. N Engl J Med 2005;352:1685–95.

[60] Tracy RP. Thrombin, inflammation, and cardiovascular disease: an epidemiologic perspective. Chest 2003;124:49S–57S.

[61] Meade TW, Chakrabarti R, Haines SP, North WRS, Stirling Y, Thompson SG. Haemostatic function and cardiovascular death: early results of a prospective study. Lancet 1980;i:1050–4.

[62] Cushman M, Yanez D, Psaty BM, Fried LP, Heiss G, Lee M, Polak JF, Savage PJ, Tracy RP. Association of fibrinogen and coagulation factor VII and VIII with cardiovascular risk factors in the elderly: the Cardiovascular Health Study. Am J Epidemiol 1996;143:665–76.

[63] Tracy RP, Bovill EG, Yanez D, Psaty BM, Fried LP, Heiss G, Lee M, Polak JF, Savage PJ. Fibrinogen and factor VIII, but not Factor VII, are associated with measures of subclinical cardiovascular disease in the elderly. Arterioscl Thromb Vasc Biol 1995;15:1269–79.

[64] Tracy RP, Arnold AM, Ettinger W, Fried L, Meilahn E, Savage P. The relationship of fibrinogen and Factor VII and VIII to incident cardiovascular disease and death in the elderly: results from the Cardiovascular Health Study. Arterioscler Thromb Vasc Biol 1999;19:1776–83.

[65] Zakai NA, Katz R, Jenny NS, Psaty BM, Reiner AP, Schwartz SM, Cushman M. Inflammation and hemostasis biomarkers and cardiovascular risk in the elderly. J Thromb Haemost 2007;5:1128–35.

[66] Cortellaro M, Boschetti C, Cofrancesco E, Zanussi C, Catalano M, de Gaetano G, Gabrielli L, Lombardi B, Specchia G, Tavazzi L, Tremoli E, della Volpe A, Polli E, the PLAT Study Group. The PLAT Study: hemostatic function in relation to atherothrombotic ischemic events in vascular disease patients. Principal results. PLAT Study Group. Progetto Lombardo Atero-Trombosi (PLAT) Study Group. Arterioscler Thromb 1992;12:1063–70.

[67] Martinelli I. von Willebrand factor and factor VIII as risk factors for arterial and venous thrombosis. Semin Hematol 2005;42:49–55.

[68] Haines AP, Howarth D, North WRS, Goldenberg E, Stirling Y, Meade TW, Rafery EB, Millar Craig MW. Haemostatic variables and the outcome of myocardial infarction. Thromb Haemost 1983;50:800–3.

[69] Kucharska-Newton AM, Couper DJ, Pankow JS, Prineas RJ, Rea TD, Sotoodehnia N, Chakravarti A, Folsom AR, Siscovick DS, Rosamond WD. Hemostasis, inflammation, and fatal and nonfatal coronary heart disease: long-term follow-up of the atherosclerosis risk in communities (ARIC) cohort. Arterioscler Thromb Vasc Biol 2009;29:2182–90.

[70] Folsom AR, Delaney JA, Lutsey PL, Zakai NA, Jenny NS, Polak JF, Cushman M, Multiethnic Study of Atherosclerosis Investigators. Associations of factor VIIIc, D-dimer, and plasmin-antiplasmin with incident cardiovascular disease and all-cause mortality. Am J Hematol 2009;84:349–53.

[71] Kamphuisen PW, Eikenboom CJ, Bertina RM. Elevated factor VIII levels and the risk of thrombosis. Arterioscler Thromb Vasc Biol 2001;21:731–8.

[72] Mari D, Mannucci PM, Coppola R, Bottasso B, Bauer KA, Rosenberg RD. Hypercoagulability in centenarians: the paradox of successful aging. Blood 1995;85:3144–9.

[73] Hermanns MI, Grossmann V, Spronk HM, Schulz A, Jünger C, Laubert-Reh D, Mazur J, Gori T, Zeller T, Pfeiffer N, Beutel M, Blankenberg S, Münzel T, Lackner KJ, Ten Cate-Hoek AJ, Ten Cate H, Wild PS. Distribution, genetic and cardiovascular determinants of FVIII:c—data from the population-based Gutenberg Health Study. Int J Cardiol 2015;187:166–74.

[74] Kim HK, Kim JE, Park SH, Kim YI, Nam-Goong IS, Kim ES. High coagulation factor levels and low protein C levels contribute to enhanced thrombin generation in patients with diabetes who do not have macrovascular complications. J Diabetes Complications 2014;28:365–9.

[75] Duncan BB, Schmidt MI, Offenbacher S, Wu KK, Savage PJ, Heiss G, for the ARIC Investigators. Factor VIII and other hemostasis variables are related to incident diabetes in adults. Diabetes Care 1999;22:767–72.

[76] Soares AL, Kazmi RS, Borges MA, Rosario PW, Fernandes AP, Sousa MO, Lwaleed BA, Carvalho MG. Elevated plasma factor VIII and von Willebrand factor in women with type 2 diabetes: inflammatory reaction, endothelial perturbation or else? Blood Coagul Fibrinolysis 2011;22:600–5.

[77] Bovenschen N. LDL receptor polymorphisms revisited. Blood 2010;116:5439–40.

[78] Huijgen R, Kastelein JJP, Meijers JCM. Increased coagulation factor VIII activity in patients with familial hypercholesterolemia. Blood 2011;118:6990–1.

[79] Martinelli N, Girelli D, Lunghi B, Pinotti M, Marchetti G, Malerba G, Pignatti PF, Corrocher R, Olivieri O, Bernardi F. Polymorphisms at LDLR locus may be associated with coronary artery disease through modulation of coagulation factor VIII activity and independently from lipid profile. Blood 2010;116:5688–97.

[80] Smith NL, Bis JC, Biagiotti S, Rice K, Lumley T, Kooperberg C, Wiggins KL, Heckbert SR, Psaty BM. Variation in 24 hemostatic genes and associations with non-fatal myocardial infarction and ischemic stroke. J Thromb Haemost 2008;6:45–53.

[81] Pankow JS, Folsom AR, Province MA, Rao DC, Eckfeldt J, Heiss G, Shahar E, Wu KK, on behalf of the Atherosclerosis Risk in Communities Investigators and Family Heart Study Research Group. Family history of coronary heart disease and hemostatic variables in middle-aged adults. Thromb Haemost 1997;77:87–93.

[82] Bank I, Libourel EJ, Middeldorp S, Hamulyak K, van Pampus ECM, Koopman MMW, Prins MH, van der Meer J, Buller HR. Elevated levels of FVIII:C within families are associated with an increased risk for venous and arterial thrombosis. J Thromb Haemost 2005;3:79–84.

[83] Martin K, Key NS. How I treat patients with inherited bleeding disorders who need anticoagulant therapy. Blood 2016;128:178–84.

[84] Schutgens REG, van der Heijden JF, Mauser-Bunshoten EP, Mannucci PM. New concepts for anticoagulant therapy in persons with hemophilia. Blood 2016;128:2471–4.

Recommended Reading

[1] Campos M, et al. Influence of single nucleotide polymorphisms in factor VIII and von Willebrand factor genes on plasma factor VIII activity: the ARIC Study. Blood 2012;119:1929–34.

[2] Bank I, et al. Absolute annual incidences of first events of venous thrombo-embolism and arterial vascular events in individuals with elevated FVIII:c. A prospective family cohort study. Thromb Haemost 2007;98:1040–4.

[3] Jenkins PV, Rawley O, Smith OP, O'Donnell JS. Elevated factor VIII levels and risk of venous thrombosis. Br J Haematol 2012;157:653–63.

[4] Yap ES, Timp JF, et al. Predictive value of factor VIII levels for recurrent venous thrombosis; elevated levels of factor VIII and subsequent risk of all-cause mortality: results from the MEGA follow-up study. J Thromb Haemost 2015;13:1823–32. 1833–42.

[5] Martin K, Key NS. How I treat patients with inherited bleeding disorders who need anticoagulant therapy. Blood 2016;128:178–84.

[6] Smith NL, Chen MH, Dehghan A, et al. Novel associations of multiple genetic loci with plasma levels of factor VII, factor VIII, and von Willebrand factor: the CHARGE (Cohorts for Heart and Aging Research in Genome Epidemiology) Consortium. Circulation 2010;121:1382–92.

14

Von Willebrand Factor and Thrombosis

Von Willebrand factor (VWF) is a major participant in normal hemostasis and contributes to the pathophysiology of thrombosis. As described in Chapter 9, VWF monomers released from endothelial cells polymerize to form long threads. These threads adhere to subendothelial connective tissue, where they provide binding sites for platelets. Under shear stress, the binding of VWF to platelet GpIbα receptors promotes an influx of Ca^{2+} and induces platelet aggregation [1]. Traces of thrombin that form on the platelet surface dissociate FVIII from VWF, providing FVIII for the tenase complex and the generation of sufficient thrombin to make a hemostatic plug. Thrombosis is usually initiated by processes that activate or strip endothelial cells from the vessel wall, thereby releasing VWF and inducing platelet adhesion, aggregation, and thrombus formation.

Elevated levels of VWF are regularly observed in patients with thrombotic disorders, raising the question of whether VWF is a cause or consequence of the disease. In support of the latter, it is reported that within 48 h of an acute coronary event, there is a significant increase in VWF and circulating endothelial cells [2]. The VWF levels likely remain elevated for prolonged periods, as has been reported after other inflammatory stimuli [3], and the high levels of VWF might augment thrombus formation. In this chapter, studies will be described that have examined the role of VWF in atherosclerosis, venous thrombosis, and thrombus formation in atrial fibrillation.

ATHEROTHROMBOSIS

While FVIII plays an essential role in venous thrombosis, VWF participates in arterial thrombosis by mediating platelet adherence to the injured vessel wall. Denudation of the endothelium and exposure of the

Hemophilia and Von Willebrand Disease
https://doi.org/10.1016/B978-0-12-812954-8.00014-X

subendothelium are key events in the pathogenesis of atherosclerosis [4]. Atheromatous plaques typically form at bends and branches of the arterial tree, sites where the endothelium is subject to the stress of turbulent blood flow and becomes vulnerable to injury. Hyperlipidemia, cytokines, immune complexes, and components of tobacco smoke contribute to the endothelial damage. Macrophages and platelets are drawn to the site of injury, foam cells appear, and smooth muscle cells proliferate. The role of VWF in the genesis of the atheromatous plaque has recently been reviewed by Wu et al. [5] (Fig. 14.1).

VWF released from the Weibel-Palade bodies of activated endothelial cells dramatically increases the concentration of platelets along the vessel wall through an interaction with GpIbα [6], and mediates platelet adhesion in areas of high shear stress—the branching points where atheroma typically form [7]. Furthermore, fluid shear increases the effective size of VWF bound to GpIbα by self-association, contributing to platelet aggregation and thrombus growth [8]. Microfluidic chambers that mimic the flow conditions in stenotic arteries display enhanced platelet aggregation that is critically dependent on blood-borne VWF [9]. Other in vitro studies show that platelets subject to high shear forces develop membrane tethers that adhere to VWF and generate procoagulant microparticles [10].

Studies in animal models further delineate the role of VWF in atherogenesis. Mice with a knockout of the *low-density lipoprotein receptor (LDLR)* and the *apo-lipoprotein B editing enzyme genes* develop fatty streaks and early plaques displaying platelet-endothelial cell interactions that are mediated by VWF [11]. *LDLR*-knockout mice fed a diet rich in saturated fat and cholesterol develop severe atheromatous plaques, but if the *VWF* gene is knocked out as well, the plaques are much less developed [12]. In addition, VWF-knockout mice sustain smaller brain infarcts and less severe neurologic deficits after transient occlusion of the middle cerebral artery [13].

Furthermore, mice deficient in ADAMTS13, the VWF protease, have enlarged infarcts and signs of acute brain inflammation [14]. Rabbits on a hypercholesterolemic diet had more VWF and recruited fivefold more platelets to vascular fatty streaks than control animals [15]. On the other hand, pigs with very severe Von Willebrand Disease (VWD) remain free of atherosclerosis despite the feeding of a high-cholesterol diet [16], and VWF-deficient pigs and healthy pigs on a normal diet develop plaques of similar size in partially occluded, shear-stressed arteries [17], suggesting that VWF needs other risk factors, such as hyperlipidemia, to enhance plaque growth. The conclusion from these studies is that VWF is necessary but not sufficient for atherogenesis.

An examination of coronary thrombi removed from patients with myocardial infarction (MI) shows the prominent presence of VWF at sites of platelet accumulation, along with tissue factor and fibrin, linking VWF

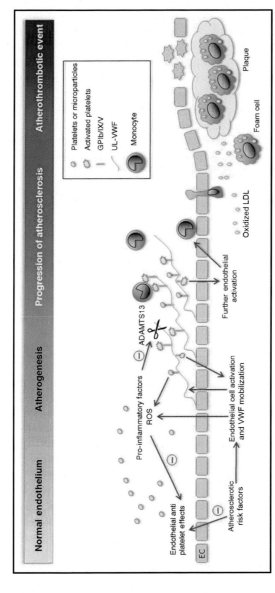

FIG. 14.1 Von Willebrand factor, ADAMTS13, and atherogenesis. *LDL*, low-density lipoprotein; *ROS*, reactive oxygen species. *Reproduced with permission from Wu MD, Atkinson TM, Lindner JR. Platelets and von Willebrand factor in atherogenesis. Blood 2017;129:1415–19.*

with thrombus formation in human arteries [18]. In addition, women during the acute stage of myocardial infarction have conformational changes in VWF making it more reactive with platelets [19]. Indirect evidence that VWF participates in arterial thrombus formation comes from studies of ADAMTS13. Recall that ADAMTS13 cleaves the secreted, and very thrombogenic, high-molecular-weight strands of VWF that remain attached to endothelial cells. ADAMTS13-knockout mice fed a high-fat diet have increased atherosclerotic plaque in the aorta and aortic sinus, as well as more macrophages in the plaques, than control mice [20], and exhibited larger myocardial infarct size after ischemia/reperfusion injury [21]. Treatment of the knockout mice with recombinant ADAMTS13 resulted in smaller infarcts and a reduction in infiltrating neutrophils [22]. This animal research is consistent with epidemiologic studies reporting a 4.4-fold increase in arterial thrombosis in middle-aged men and women with low levels of ADAMTS13, and a 7.7-fold increase if low ADAMTS13 was combined with high VWF [23]. Aging is also associated with an increase in the VWF to ADAMTS13 ratio, due to an age-related decline in ADAMTS13 levels [24].

Other evidence supporting an association of VWF with arterial thrombotic events is the observation that the diurnal and seasonal concentrations of VWF are highest when coronary deaths are most prevalent (noon and winter) [25–27]. Also, populations known to have increased VWF levels, such as the elderly and people with non-O blood group, have higher incidences of cardiovascular events than younger persons and those with blood group O [28].

Epidemiologic Studies

Clinical studies also support a role for VWF in atherosclerotic cardiovascular disease. In the Caerphilly Heart Study, multivariate analyses showed that VWF as well as FVIII was significantly associated with incident major ischemic heart disease [29]. Other studies found that elevated levels of VWF, fibrinogen, and tissue plasminogen activator were independent predictors of acute coronary syndromes in patients with angina [30], and VWF and fibrinogen concentrations were significantly higher in women with than without coronary atheroma [31]. High coronary plaque burden was reported to be associated with elevated VWF levels in patients with angina and predicted 1-year all-cause death and hospitalization for acute coronary syndromes [32]. In addition, elevated VWF concentrations were associated with increased carotid intimal-medial thickness (IMT) in patients with rheumatoid arthritis [33], and IMT and microalbuminuria in middle-aged people free of overt cardiovascular disease [34]. VWF was also positively associated with subclinical brain infarcts as detected by magnetic resonance imaging [35]. Other research has described an association of VWF with ischemic stroke [36,37], and with the occurrence of coronary events and vascular death after a stroke [38]. Ischemic stroke susceptibility in children is associated with single-nucleotide polymorphisms in the *ADAMTS13* gene that affect ADAMTS13 levels [39]. Young women with high VWF, low ADAMTS13, and oral contraceptive use have an increase in the risk of ischemic stroke and myocardial infarction [40]. Furthermore, high levels of VWF predict the appearance of cardiovascular disease in people with type 2 diabetes and insulin resistance, but not in those who are insulin sensitive [41].

On the other hand, several studies report weaker associations of VWF with ischemic cardiovascular disease. A prospective investigation reported that the increased levels of VWF recorded in an at-risk patient population did not predict coronary events after adjustment for established risk factors [42]. Other studies found that the association of high levels of VWF with an increased risk of cardiovascular death was attenuated after adjustment for conventional risk factors and C-reactive protein [43–45]. A prospective study of patients with peripheral vascular disease reported a nonsignificant trend toward an increased VWF to ADAMTS13 ratio prior to major cardiovascular events [46]. Lastly, a Dutch study sequenced the *VWF* gene and reported that three single-nucleotide polymorphisms were associated with VWF antigen concentrations but not with coronary heart disease (CHD), leading the authors to suggest that VWF is not causal but rather a marker for CHD [47].

Studies of people with VWD could provide more convincing evidence of the association of VWF with atherothrombotic disease, but the data are conflicting. Although carotid IMT and coronary artery calcification scores are similar in patients with VWD and controls [48,49], the prevalence of cardiovascular disease in those with VWD is significantly decreased [50,51]. An explanation for these observations is that atherosclerosis in humans is a multifactorial disease, of which VWF is an expendable contributor. But atherothrombotic events, such as myocardial infarction and stroke, probably require the active participation of VWF.

VENOUS THROMBOSIS

Venous thrombi consist mainly of erythrocytes trapped in a fibrin scaffold. Recently, it was shown that red cells adhere to VWF, suggesting that VWF might mediate the binding of erythrocytes to fibrin and play a role in the pathogenesis of venous thrombosis [52]. Indirect evidence that VWF is important in venous thrombosis comes from studies of patients with congenital VWD. Venous thrombosis is rare in such patients and usually occurs as a complication of VWF replacement therapy [53]. Epidemiologic studies show that the levels of VWF are higher in patients with venous thrombosis than in controls, but the association is attenuated when adjusted for other risk factors.

The Leiden Thrombophilia Study reported that both FVIII and VWF were increased in patients with venous thromboembolism (VTE), but after adjustment for blood group, only FVIII was a risk factor for thrombosis [54]. Further study revealed that patients with high levels of FVIII and VWF had high levels of VWF propeptide, and the propeptide was more elevated in thrombosis patients than in controls [55]. High propeptide levels reflect increased secretory rates of FVIII/VWF. Because the release of VWF is stimulated by the binding of arginine vasopressin (AVP) to its endothelial cell receptor (AVR2), investigators considered whether increased binding accounted for the raised VWF levels, and searched for polymorphisms in the AVR2 gene. They discovered a variant AVR2 gene with a gain-of-function mutation [56], but the population frequency of this mutation and its relationship to incident venous thrombosis were not reported.

The genes for VWF and stabilin-2 are located on chromosome 12, and the stabilin-2 scavenger receptor participates in VWF clearance. STAB2 mutations were reported to be more frequent in VTE cases than controls, and individuals with loss-of-function variants have significantly higher levels of VWF than those without variants, identifying STAB2 as a risk factor for VTE [57]. Other workers identified two other gene variants associated with elevated VWF and venous thrombosis: syntaxin-binding protein-5 (STXBP5) and an SNP in the VWF gene [58]. STXBP5 is expressed by endothelial cells and the unmutated gene product inhibits the exocytosis of VWF [59]. In addition, patients with deep-vein thrombosis have been found to have an excess of rare coding SNPs of the ADAMTS13 gene [60]. A reduced level of ADAMTS13 was associated with venous thrombosis in women [61].

Although VWF is not considered an independent risk factor for venous thrombosis, it does act in concert with other factors to promote VTE. High levels of VWF, along with increased FVIII, are associated with a heightened risk of VTE in African-Americans [62]. The two factors also increase the risk of venous thrombosis in people with impaired kidney function as well as in those with cancer [63,64]. In addition, raised concentrations of free thyroxine (FT4) increase the levels of FVIII and VWF, explaining the association of venous thrombosis with hyperthyroidism [65].

Falciparum malaria is associated with erythrocyte-platelet occlusion of the brain microvasculature as well as cerebral venous thrombosis. Binding of infected erythrocytes to brain endothelium appears to be the key pathologic event. Malaria-infected children have elevated levels of VWF [66], and in vitro studies show that infected, but not uninfected, erythrocytes bind to platelet-decorated VWF strings; the erythrocytes bind to platelet

CD36, and platelet glycoprotein Ibα attaches to the A1 domain of VWF [67]. Intact VWF is essential because cleavage of VWF by ADAMTS13 abolishes the proadhesive state of infected erythrocytes.

ATRIAL FIBRILLATION/EMBOLIC STROKE

Atrial fibrillation (AF) affects more than two million Americans [68–70] with projections that it will affect 8–12 million people in the United States by 2050 [71]. The chief complication of AF is thromboembolic stroke; the thrombi mostly arise from the left atrium, and especially the left atrial appendage. The major factor contributing to thrombus formation is stasis due to diminished left atrial appendage blood flow [72–75], which is affected by the presence of atrial myopathy, left atrial geometry, and the frequency and duration of AF episodes [76]. How these atrial characteristics promote stasis is under active investigation.

Coagulation factors also contribute to the development of atrial thrombi. Platelet activation is significantly greater, and platelet-monocyte aggregates more frequent, in left atrial than right atrial blood in patients with spontaneous echocardiographic contrast [77]. Patients with chronic AF have increased thrombin generation in left atrial blood compared to peripheral blood [78]. Activated protein C is a potent natural anticoagulant that is generated by the thrombin-thrombomodulin complex; a decrease in activated protein C increases thrombin generation and fibrin formation. Cervero et al. [79] report significant underexpression of thrombomodulin by the left as compared to the right atrial endocardium ($P = .028$), with less than half the ability to activate protein C.

D-dimer is formed when fibrin undergoes dissolution and is a marker for thrombotic risk. Elevated levels of d-dimer are significantly associated with the risk of stroke and cardiovascular death in patients with AF [80,81]. In those with new-onset AF, d-dimer levels increase above the normal range within 12 h and reach a plateau at the 18th hour [82]. Furthermore, the range of d-dimer increase is characteristic for each patient with chronic AF and is stable over time [83]. Additional evidence for fibrin formation and lysis is provided by the observation that plasmin-antiplasmin complexes are increased in AF patients at high risk for thromboembolism [84]. Finally, a pig model of AF found increased plasminogen activator inhibitor-1 concentrations in the left atrium and decreased nitric oxide synthase and nitric oxide bioavailability [85].

Within minutes of the onset of AF, atrial blood shows evidence of platelet and endothelial cell activation [86]. The concentrations of VWF in left atrial blood are increased in patients with persistent AF and higher than in the left atria of those with paroxysmal AF or controls [87]. They are also associated with spontaneous echocardiographic contrast, and are higher in AF patients with than without left atrial thrombi (200 vs 155, $P = .0006$) [88]. The level of VWF is also associated with the extent of periatrial epicardial fat, but not body mass index or epicardial adipose tissue, suggesting local adiposity affects VWF concentrations [89]. In healthy persons, the ratio of VWF antigen to ADAMTS13 is unity; elevated ratios indicate an increase in the more thrombogenic multimers of VWF. The VWF: ADAMTS13 ratio is significantly higher in patients with chronic AF than in those with paroxysmal AF ($P < .01$) or controls ($P < .0001$) [90], and a high ratio independently predicts major adverse cardiovascular events in patients with AF (hazard ratio: 2.17, $P = .007$) [91]. After cardioversion, the ratio is an independent predictor of recurrent AF (HR: 1.88, $P = .03$) [92].

Epidemiologic studies report that VWF levels are increased in patients with nonvalvular AF as compared to those in sinus rhythm, irrespective of a history of stroke [93–96]. In the Atherosclerosis Risk in Communities Study, VWF was associated with AF independent of other CV risk factors [97]. In multivariable Cox models, the hazard ratio for incident AF was 1.17 (95% CI: 1.11–1.23) for each one-standard deviation increase in VWF. Conway et al. [98] and Krishnamoorthy et al. [99] reported that raised VWF levels in patients with AF are predictive of stroke and vascular events, and Roldan et al. [100] found that high VWF levels are an independent risk factor for adverse events in AF patients on anticoagulant therapy.

MODIFICATION OF VWF

Excessive VWF activity might be decreased by altering the synthesis or increasing the clearance of the molecule, but a more feasible approach is to modify its activity. Caplacizumab is an anti-VWF humanized single-variable-domain immunoglobulin (nanobody) that inhibits the interaction between VWF multimers and platelet glycoprotein 1b. A recent study showed that caplacizumab reduced the frequency of major thromboembolic events and death in patients with acquired thrombotic thrombocytopenic purpura (TTP) [101]. Another therapeutic approach is to increase VWF proteolysis by infusing ADAMTS13. There is evidence that ADAMTS13 has a destabilizing effect on thrombus growth and delays the occlusion of experimentally injured vessels [102]. A recombinant ADAMTS13 product has been developed for the treatment of TTP [103], but might also be considered for limiting the frequency and extent of thrombosis in patients with severe vascular disease. The safety and effectiveness of targeting VWF for the treatment of patients with thrombotic disorders requires further investigation.

SUMMARY

Von Willebrand factor (VWF) contributes to the pathogenesis of atherosclerosis, the development of arterial and venous thrombosis, and thromboembolic risks in people with atrial fibrillation. Experimental studies demonstrate that VWF mediates critical platelet-endothelial cell interactions involved in atherogenesis. Epidemiological investigations report that elevated levels of VWF in conjunction with other risk factors predict cardiovascular events; for example, in patients with angina, hospitalization for acute coronary syndromes and 1-year all-cause mortality. Furthermore, high VWF levels are associated with subclinical brain infarcts and ischemic strokes in children and adults.

People with genetic polymorphisms that increase the release of VWF from endothelial cells are at risk for venous thrombosis, and VTE is associated with elevated VWF levels in patients with cancer, hyperthyroidism, and chronic renal disease. In addition, VWF is increased in the left atrial blood of patients with atrial fibrillation and predicts stroke and vascular events in these individuals. Decreases in ADAMTS13 relative to VWF are associated with ischemic stroke, myocardial infarction, venous thrombosis, and cardiovascular events in patients with atrial fibrillation. Currently under development are ADAMTS13 concentrates and drugs that modify the binding of VWF to platelets, but whether these approaches will be safe and effective for the treatment of patients with thrombotic disorders requires further investigation.

Future research might examine the following:

- Conduct genomic studies of VWF and other hemostatic genes to determine whether specific polymorphisms are associated with atherothrombotic disease
- Identify additional genetic variants associated with elevated levels of VWF in familial venous thrombosis
- Investigate the effect of atrial myopathy on the activation of hemostasis and the development of atrial thrombi
- Determine whether quantitative or qualitative characteristics of VWF are associated with atrial fibrosis, strain, and flow in models of atrial fibrillation
- Evaluate pharmaceutical products that might safely reduce VWF levels in thrombotic disorders

References

[1] Chow TW, Hellums JD, Moake JL, Kroll MH. Shear stress-induced von Willebrand factor binding to platelet glycoprotein Ib initiates calcium influx associated with aggregation. Blood 1992;80:113–20.
[2] Lee KW, Lip GYH, Tayebjee M, Foster W, Blann AD. Circulating endothelial cells, von Willebrand factor, interleukin-6, and prognosis in patients with acute coronary syndromes. Blood 2005;105:526–32.

[3] Reitsma PH, Branger J, Van Den Blink B, Weijer S, Van Der Poll T, Meijers JC. Procoagulant protein levels are differentially increased during human endotoxemia. J Thromb Haemost 2003;1:1019–23.

[4] Fuster V, Badimon L, Badimon JJ, Chesebro JH. The pathogenesis of coronary artery disease and the acute coronary syndromes. N Engl J Med 1992;326:242–50.

[5] Wu MD, Atkinson TM, Lindner JR. Platelets and von Willebrand factor in atherogenesis. Blood 2017;129:1415–9.

[6] Andre P, Denis CV, Ware J, Saffaripour S, Hynes RO, Ruggieri ZM, Wagner DD. Platelets adhere to and translocate on von Willebrand factor presented by endothelium in stimulated veins. Blood 2000;96:3322–8.

[7] Ananyeva NM, Kouiavskaia DV, Shima M, Saenko EL. Intrinsic pathway of blood coagulation contributes to thrombogenicity of atherosclerotic plaque. Blood 2002;99:4475–85.

[8] Dayananda KM, Singh I, Mondai N, Neelamegham S. von Willebrand factor self-association on platelet GpIbα under hydrodynamic shear: effect on shear-induced platelet activation. Blood 2010;116:3990–8.

[9] Westein E, van der Meer AD, Kuipers MJ, Frimat JP, van den Berg A, Heemskerk JW. Atherosclerotic geometries exacerbate pathological thrombus formation poststenosis in a von Willebrand factor-dependent manner. Proc Natl Acad Sci U S A 2013;110:1357–62.

[10] Reininger AJ, Heijnen HFG, Schumann H, Specht HM, Schramm W, Ruggieri ZM. Mechanism of platelet adhesion to von Willebrand factor and microparticle formation under high shear stress. Blood 2006;107:3537–45.

[11] Shim CY, Liu YN, Atkinson T, Xie A, Foster T, Davidson BP, Treible M, Qi Y, Lopez JA, Munday A, Ruggeri Z, Lindner JR. Molecular imaging of platelet-endothelial interactions and endothelial von Willebrand factor in early and mid-stage atherosclerosis. Circ Cardiovasc Imaging 2015;8.

[12] Methia N, Andre P, Denis CV, Economopoulos M, Wagner DD. Localized reduction of atherosclerosis in von Willebrand factor-deficient mice. Blood 2001;98:1424–8.

[13] Kleinschnitz C, De Meyer SF, Schwarz T, Austinat M, Vanhoorelbeke K, Nieswandt B, Deckmyn H, Stoll G. Deficiency of von Willebrand factor protects mice from ischemic stroke. Blood 2009;113:3600–3.

[14] Khan MM, Motto DG, Lentz SR, Chauhan AK. ADAMTS13 reduces VWF-mediated acute inflammation following focal cerebral ischemia in mice. J Thromb Haemost 2012;10:1665–71.

[15] Theilmeier G, Michiels C, Spaepen E, Vreys I, Collen D, Vermylen J, Hoylaerts MF. Endothelial von Willebrand factor recruits platelets to atherosclerosis-prone sites in response to hypercholesterolemia. Blood 2002;99:4486–93.

[16] Fuster V, Lie JT, Badimon L, Rosemark JA, Badimon JJ, Bowie EJ. Spontaneous and diet-induced coronary atherosclerosis in normal swine and swine with von Willebrand disease. Arteriosclerosis 1985;5:67–73.

[17] Nichols TC, Bellinger DA, Reddick RL, Koch GG, Sigman JL, Erickson G, du Laney T, Johnson T, Read MS, Griggs TR. Von Willebrand factor does not influence atherogenesis in arteries subjected to altered shear stress. Arterioscler Thromb Vasc Biol 1998;18:323–30.

[18] Hoshiba Y, Hatakeyama K, Tanabe T, Asada Y, Goto S. Co-localization of von Willebrand factor with platelet thrombi, tissue factor and platelets with fibrin, and consistent presence of inflammatory cells in coronary thrombi obtained by an aspiration device from patients with acute myocardial infarction. J Thromb Haemost 2006;4:114–20.

[19] Peyvandi F, Hollestelle MJ, Palla R, Merlini PA, Feys HB, Vanhoorelbeke K, Lenting PJ, Mannucci PM. Active platelet-binding conformation of plasma von

Willebrand factor in young women with acute myocardial infarction. J Thromb Haemost 2010;8:1653–6.

[20] Gandhi C, Khan MM, Lentz SR, Chauhan AK. ADAMTS13 reduces vascular inflammation and the development of early atherosclerosis in mice. Blood 2012;119:2385–91.

[21] Gandhi C, Mottto DG, Jensen M, Lentz SR, Chauhan AK. ADAMTS13 deficiency exacerbates VWF-dependent acute myocardial ischemia/reperfusion injury in mice. Blood 2012;120:5224–30.

[22] De Meyer SF, Savchenko AS, Haas MS, Schatzberg D, Carroll MC, Schiviz A, Dietrich B, Rottensteiner H, Scheiflinger F, Wagner DD. Protective anti-inflammatory effect of ADAMTS13 on myocardial ischemia/reperfusion injury in mice. Blood 2012;120:5217–23.

[23] Bongers TN, de Bruijne ELE, Dippel DWJ, de Jong AJ, Deckers JW, Poldermans D, de Maat MPM, Leebeek FWG. Lower levels of ADAMTS13 are associated with cardiovascular disease in young patients. Atherosclerosis 2009;207:250–4.

[24] Kokame K, Sakata T, Kokubo Y, Miyata T. von Willebrand factor-to-ADAMTS13 ratio increases with age in a Japanese population. J Thromb Haemost 2011;9:1426–8.

[25] Timm A, Fahrenkrug J, Jørgensen HL, Sennels HP, Goetze JP. Diurnal variation of von Willebrand factor in plasma: the Bispebjerg study of diurnal variations. Eur J Haematol 2014;93:48–53.

[26] Tofler GH, Brezinski D, Schafer AI, Czeisler CA, Rutherford JD, Willich SN, Gleason RE, Williams GH, Muller JE. Concurrent morning increase in platelet aggregability and the risk of myocardial infarction and sudden cardiac death. N Engl J Med 1987;316:1514–8.

[27] Ghebre MA, Wannamethee SG, Rumley A, Whincup PH, Lowe GDO. Prospective study of seasonal patterns in hemostatic factors in older men and their relation to excess winter coronary heart disease deaths. J Thromb Haemost 2012;10:352–8.

[28] Albanez S, Ogiwara K, Michels A, Hopman W, Grabell J, James P, Lillicrap D. Aging and ABO blood type influence von Willebrand factor and factor VIII levels through interrelated mechanisms. J Thromb Haemost 2016;14:953–63.

[29] Rumley A, Lowe GD, Sweetnam PM, Yarnell JW, Ford RP. Factor VIII, von Willebrand factor and the risk of major ischaemic heart diease in the Caerphilly Heart Study. Br J Haematol 1999;105:110–6.

[30] Thompson SG, Kienast J, Pyke SDM, Haverkate F, van de Loo JCW for the European concerted Action on Thrombosis and Disabilities Angina Pectoris Study Group. Hemostatic factors and the risk of myocardial infarction or sudden death in patients with angina pectoris. N Engl J Med 1995;332:635–41.

[31] Ossei-Gerning N, Wilson IJ, Grant PJ. Sex differences in coagulation and fibrinolysis in subjects with coronary artery disease. Thromb Haemost 1998;79:736–40.

[32] Sonneveld MA, Cheng JM, Oemrawsingh RM, de Maat MP, Kardys I, Garcia-Garcia HM, van Geuns RJ, Regar E, Serruys PW, Boersma E, Akkerhuis KM, Leebeek FW. Von Willebrand factor in relation to coronary plaque characteristics and cardiovascular outcome. Results of the ATHEROREMO-IVUS study. Thromb Haemost 2015;113:577–84.

[33] Ristić GG, Subota V, Lepić T, Stanisavljević D, Glišić B, Ristić AD, Petronijević M, Stefanović DZ. Subclinical atherosclerosis in patients with rheumatoid arthritis and low cardiovascular risk: the role of von Willebrand factor activity. PLoS ONE 2015;10.

[34] Paramo JA, Beloqui O, Colina I, Diez J, Orbe J. Independent association of von Willebrand factor with surrogate markers of atherosclerosis in middle-aged asymptomatic subjects. J Thromb Haemost 2005;3(4):662.

[35] Gottesman RF, Cummiskey C, Chambless L, Wu KK, Aleksic N, Folsom AR, Sharrett AR. Hemostatic factors and subclinical brain infarction in a community-based sample: the ARIC study. Cerebrovasc Dis 2009;28:589–94.

[36] Catto AJ, Carter AM, Barrett JH, Bamford J, Rice PJ, Grant PJ. Von Willebrand factor and factor VIII:C in acute cerebrovascular disease. Relationship to stroke subtype and mortality. Thromb Haemost 1997;77:1104–8.

[37] Wannamethee SG, Whincup PH, Lennon L, Rumley A, Lowe GDO. Fibrin d-dimer, tissue-type plasminogen activator, von Willebrand factor, and risk of incident stroke in older men. Stroke 2012;43:1206–11.

[38] Pedersen A, Redfors P, Lundberg L, Gils A, Declerck PJ, Nilsson S, Jood K, Jern C. Haemostatic biomarkers are associated with long-term recurrent vascular events after ischemic stroke. Thromb Haemost 2016;116:537–43.

[39] Stoll M, Ruhle F, Witten A, Barysenka A, Arning A, Strauss CD, Nowak-Gottl U. Rare variants in the ADAMTS13 von Willebrand factor-binding domain contribute to pediatric stroke. Circ Cardiovasc Genet 2016;9:357–67.

[40] Andersson HM, Siegerink B, Luken BM, Crawley JTB, Algra A, Lane DA, Rosendaal FR. High VWF, low ADAMTS13, and oral contraceptives increase the risk of ischemic stroke and myocardial infarction in young women. Blood 2012;119:1555–60.

[41] Frankel DS, Meigs JB, Massaro JM, Wilson PWF, O'Donnell CJ, D'Agostino RB, Tofler GH. Von Willebrand factor, type 2 diabetes mellitus, and risk of cardiovascular disease. Circulation 2008;118:2533–9.

[42] Folsom AR, Wu KK, Rosamond WD, Sharrett AR, Chambless LE. Prospective study of hemostatic factors and incidence of coronary heart disease. Circulation 1997;96:1102–8.

[43] Morange PE, Bickel C, Nicaud V, Schnabel R, Rupprecht HJ, Peetz D, Lackner KJ, Cambien F, Blankenberg S and Tiret L for the AtheroGene Investigators. Haemostatic factors and the risk of cardiovascular death in patients with coronary artery disease: the AtheroGene study. Arterioscler Thromb Vasc Biol 2006;26:2793–9.

[44] Wannamethee SG, Whincup PH, Shaper AG, Rumley A, Lennon L, Lowe GDO. Circulating inflammatory and hemostatic biomarkers are associated with risk of myocardial infarction and coronary death, but not angina pectoris, in older men. J Thromb Haemost 2009;7:1605–11.

[45] Woodward M, Rumley A, Welsh P, MacMahon S, Lowe GDO. A comparison of the associations between seven hemostatic or inflammatory variables and coronary heart disease. J Thromb Haemost 2007;5:1795–800.

[46] Green D, Tian L, Greenland P, Liu K, Kibbe M, Tracy R, Shah S, Wilkins JT, Huffman MD, Liao Y, Lloyd Jones D, McDermott MM. Association of the von Willebrand factor-ADAMTS13 ratio with incident cardiovascular events in patients with peripheral arterial disease. Clin Appl Thromb Hemost 2017;23:807–13.

[47] Van Loon JE, Kavousi M, Leebeek FW, Felix JF, Hofman A, Witteman JHC, de Maat MP. Von Willebrand factor plasma levels, genetic variations and coronary heart disease in an older population. J Thromb Haemost 2012;10:1262–9.

[48] A1 S, Bucciarelli P, Federici AB, Mannucci PM, De Rosa V, Castaman G, Morfini M, Mazzucconi MG, Rocino A, Schiavoni M, Scaraggi FA, Reiber JH, Rosendaal FR. Patients with type 3 severe von Willebrand disease are not protected against atherosclerosis: results from a multicenter study in 47 patients. Circulation 2004;109:740–4.

[49] Zwiers M, Lefrandt JD, Mulder DJ, Smit AJ, Gans RO, Vliegenthart R, Brands-Nijenhuis AV, Kluin-Nelemans JC, Meijer K. Coronary artery calcification score and carotid intima media thickness in patients with von Willebrand disease. Haemophilia 2013;19:e186–8.

[50] Sanders YV, Eikenboom J, de Wee EM, van der Bom JG, Cnossen MH, Degenaar-Dujardin ME, Fijnvandraat K, Kamphuisen PW, Laros-van Gorkom BA, Meijer K, Mauser-Bunschoten EP, Leebeek FW, WiN Study Group. Reduced prevalence of arterial thrombosis in von Willebrand disease. J Thromb Haemost 2013;11:845–54.

[51] Seaman CD, Yabes J, Comer DM, Ragni MV. Does deficiency of von Willbrand factor protect agianst cardiovascular disease? Analysis of a national discharge register. J Thromb Haemost 2015;13:1999–2003.

[52] Smeets MWJ, Mourik MJ, Niessen HWM, Hordijk PL. Stasis promotes erythrocyte adhesion to von Willebrand factor. Arterioscler Thromb Vasc Biol 2017, https://doi.org/10.1161/ATVBAHA.117.309885.

[53] Girolami A, Tasinato V, Sambado L, Peroni E, Casonato A. Venous thormbosis in von Willebrand disease as observed in one centre and as reported in the literature. Blood Coagul Fibrinolysis 2015;26:54–8.

[54] Koster T, Blann AD, Briet E, Vandenbroucke JP, Rosendaal FR. Role of clotting factor VIII in effect of von Willebrand factor on occurrence of deep-vein thrombosis. Lancet 1995;345:152–6.

[55] Nossent AY, van Marion V, van Tilburg NH, Rosendaal FR, Bertina RM, van Mourik JA, Eikenboom HC. Von Willebrand factor and its propeptide: the influence of secretion and clearance on protein levels and the risk of venous thrombosis. J Thromb Haemost 2006;4:2556–62.

[56] Nossent AY, Robben JH, Deen PM, Vos HL, Rosendaal FR, Doggen CJ, Hansen JL, Sheikh SP, Bertina RM, Eikenboom JC. Functional variation in the arginine vasopressin 2 receptor as a modifier of human plasma von Willebrand factor levels. J Thromb Haemost 2010;8:1547–54.

[57] Desch K, Ozel AB, Halvorsen M, Michels AL, Swystun LL, Mokry L, Richards B, Germain M, Tregouet DA, Reitsma PH, Cl K, Goldstein DB, Lillicrap D, Ginsburg D. Exome sequencing studies identify mutations in STAB2 as a genetic risk for venous thromboembolic disease. Blood 2017;130:457 [abstract].

[58] Smith NL, Rice KM, Bovill EG, Cushman M, Bis JC, McKnight B, Lumley T, Glazer NL, van Hylckama Vlieg A, Tang W, Dehghan A, Strachan DP, O'Donnell CJ, Rotter JI, Heckbert SR, Psaty BM, Rosendaal FR. Genetic variation associated with plasma von Willebrand factor levels and the risk of incident venous thrombosis. Blood 2011;117:6007–11.

[59] Zhu Q, Yamakuchi M, Ture S, de la Luz Garcia-Hernandez M, Ko KA, Modjeski KL, MB LM, Johnson AD, O'Donnell CJ, Takai Y, Morrell CN, Lowenstein CJ. Syntaxin-binding protein STXBP5 inhibits endothelial exocytosis and promotes platelet secretion. J Clin Invest 2014;124:4503–16.

[60] Lotta LA, Tuana G, Yu J, Martinelli I, Wang M, Yu F, Passamonti SM, Pappalardo E, Valsecchi C, Scherer SE, Hale IV W, Muzny DM, Randi G, Rosendaal FR, Gibbs RA, Peyvandi F. Next-generation sequencing study finds an excess of rare, coding single-nucleotide variants of ADAMTS13 in patients with deep vein thrombosis. J Thromb Haemost 2013;11:1228–39.

[61] Llobet D, Tirado I, Vilalta N, Vallvé C, Oliver A, Vázquez-Santiago M, Mateo J, Millón J, Fontcuberta J, Souto JC. Low ADAMTS13 levels are associated with venous thrombosis risk in women. Thromb Res 2017;157:38–40.

[62] Payne AB, Miller CH, Hooper WC, Lally C, Austin HD. High factor VIII, von Willebrand factor, and fibrinogen levels and risk for venous thromboembolism in blacks and whites. Ethn Dis 2014;24:169–74.

[63] Ocak G, Vossen CY, Lijfering WM, Verduijn M, Dekker FW, Rosendaal FR, Cannegieter SC. Role of hemostatic factors on the risk of venous thrombosis in people with impaired kidney function. Circulation 2014;129:683–91.

[64] Pépin M, Kleinjan A, Hajage D, Büller HR, Di Nisio M, Kamphuisen PW, Salomon L, Veyradier A, Stepanian A, Mahé I. ADAMTS-13 and von Willebrand factor predict venous thromboembolism in patients with cancer. J Thromb Haemost 2016;14:306–15.

[65] Debeij J, van Zaane B, Dekkers OM, Doggen CJ, Smit JW, van Zanten AP, Brandjes DP, Büller HR, Gerdes VE, Rosendaal FR, Cannegieter SC. High levels of procoagulant factors mediate the association between free thyroxine and the risk of venous thrombosis: the MEGA study. J Thromb Haemost 2014;12:839–46.

[66] Phiri HT, Bridges DJ, Glover SJ, van Mourik JA, de Laat B, M'baya B, Taylor TE, Seydel KB, Molyneux ME, Faragher EB, Craig AG, Bunn JE. Elevated plasma von Willebrand factor and propeptide levels in Malawian children with malaria. PLoS ONE 2011;6.

[67] Bridges DJ, Bunn J, van Mourik JA, Grau G, Preston RJ, Molyneux M, Combes V, O'Donnell JS, de Laat B, Craig A. Rapid activation of endothelial cells enables *Plasmodium falciparum* adhesion to platelet-decorated von Willebrand factor strings. Blood 2010;115:1472–4.

[68] Naccarelli GV, Varker H, Lin J, Schulman KL. Increasing prevalence of atrial fibrillation and flutter in the United States. Am J Cardiol 2009;104:1534–9.

[69] Magnani JW, Rienstra M, Lin H, Sinner MF, Lubitz SA, McManus DD, Dupuis J, Ellinor PT, Benjamin EJ. Atrial fibrillation: current knowledge and future directions in epidemiology and genomics. Circulation 2011;124:1982–93.

[70] Go AS, Mozaffarian D, Roger VL, Benjamin EJ, Berry JD, Blaha MJ, Dai S, Ford ES, Fox CS, Franco S, Fullerton HJ, Gillespie C, Hailpern SM, Heit JA, Howard VJ, Huffman MD, Judd SE, Kissela BM, Kittner SJ, Lackland DT, Lichtman JH, Lisabeth LD, Mackey RH, Magid DJ, Marcus GM, Marelli A, Matchar DB, DK MG, Mohler III ER, Moy CS, Mussolino ME, Neumar RW, Nichol G, Pandey DK, Paynter NP, Reeves MJ, Sorlie PD, Stein J, Towfighi A, Turan TN, Virani SS, Wong ND, Woo D, Turner MB, American Heart Association Statistics Committee and Stroke Statistics Subcommittee. Heart disease and stroke statistics—2014 update: a report from the American Heart Association. Circulation 2014;129:e28–e292.

[71] Miyasaka Y, Barnes ME, Gersh BJ, Cha SS, Bailey KR, Abhayaratna WP, Seward JB, Tsang TS. Secular trends in incidence of atrial fibrillation in Olmsted County, Minnesota, 1980 to 2000, and implications on the projections for future prevalence. Circulation 2006;114:119–25.

[72] Goldman ME, Pearce LA, Hart RG, Zabalgoitia M, Asinger RW, Safford R, Halperin JL. Pathophysiologic correlates of thromboembolism in nonvalvular atrial fibrillation: I. Reduced flow velocity in the left atrial appendage (The Stroke Prevention in Atrial Fibrillation [SPAF-III] study). J Am Soc Echocardiogr 1999;12:1080–7.

[73] Handke M, Harloff A, Hetzel A, Olschewski M, Bode C, Geibel A. Left atrial appendage flow velocity as a quantitative surrogate parameter for thromboembolic risk: determinants and relationship to spontaneous echocontrast and thrombus formation—a transesophageal echocardiographic study in 500 patients with cerebral ischemia. J Am Soc Echocardiogr 2005;18:1366–72.

[74] Pollick C, Taylor D. Assessment of left atrial appendage function by transesophageal echocardiography. Implications for the development of thrombus. Circulation 1991;84:223–31.

[75] Asinger RW, Koehler J, Pearce LA, Zabalgoitia M, Blackshear JL, Fenster PE, Strauss R, Hess D, Pennock GD, Rothbart RM, Halperin JL. Pathophysiologic correlates of thromboembolism in nonvalvular atrial fibrillation: II. Dense spontaneous echocardiographic contrast (The Stroke Prevention in Atrial Fibrillation [SPAF-III] study). J Am Soc Echocardiogr 1999;12:1088–96.

[76] Goldberger JJ, Arora R, Green D, Greenland P, Lee DC, Lloyd-Jones DM, Markl M, Ng J, Shah SJ. Evaluating the atrial myopathy underlying atrial fibrillation: identifying the arrhythmogenic and thrombogenic substrate. Circulation 2015;132:278–91.

[77] Zotz RJ, Muller M, Genth-Zotz S, Darius H. Spontaneous echo contrast caused by platelet and leukocyte aggregates? Stroke 2001;32:1127–33.

[78] Lim HS, Willoughby SR, Schultz C, Gan C, Alasady M, Lau DH, Leong DP, Brooks AG, Young GD, Kistler PM, Kalman JM, Worthley MI, Sanders P. Effect of atrial fibrillation on atrial thrombogenesis in humans: impact of rate and rhythm. J Am Coll Cardiol 2013;61:852–60.

[79] Cervero J, Montes R, Espana F, Esmon CT, Hermida J. Limited ability to activate protein C confers left atrial endocardium a thrombogenic phenotype: a role in cardioembolic stroke? Stroke 2011;42:2622–4.

[80] Christersson C, Wallentin L, Andersson U, Alexander JH, Ansell J, De Caterina R, Gersh BJ, Granger CB, Hanna M, Horowitz JD, Huber K, Husted S, Hylek EM, Lopes RD, Siegbahn A. D dimer and risk of thromboembolic and bleeding events in patients with atrial fibrillation—observations from the ARISTOTLE trial. J Thromb Haemost 2014;12:1401–12.

[81] Siegbahn A, Oldgren J, Andersson U, Ezekowitz MD, Reilly PA, Connolly SJ, Yusuf S, Wallentin L, Eikelboom JW. D-dimer and factor VIIa in atrial fibrillation—prognostic values for cardiovascular events and effects of anticoagulation therapy. A RE-LY substudy. Thromb Haemost 2016;115:921–30.

[82] Wang TL, Hung CR, Chang H. Evolution of plasma D-dimer and fibrinogen in witnessed onset of paroxysmal atrial fibrillation. Cardiology 2004;102:115–8.

[83] Mahe I, Drouet L, Chassany O, Mazoyer E, Simoneau G, Knellwolf AL, Caulin C, Bergmann JF. D-dimer: a characteristic of the coagulation state of each patient with chronic atrial fibrillation. Thromb Res 2002;107:1–6.

[84] Feinberg WM, Macy E, Cornell ES, Nightingale SD, Pearce LA, Tracy RP, Bovill EG. Plasmin-alpha2-antiplasmin complex in patients with atrial fibrillation. Stroke Prevention in Atrial Fibrillation Invaestigators. Thromb Haemost 1999;82:100–3.

[85] Cai H, Li Z, Goette A, Mera F, Honeycutt C, Feterik K, Wilcox JN, Dudley Jr. SC, Harrison DG, Langberg JJ. Downregulation of endocardial nitric oxide synthase expression and nitric oxide production in atrial fibrillation: potential mechanisms for atrial thrombosis and stroke. Circulation 2002;106:2854–8.

[86] Akar JG, Jeske W, Wilber DJ. Acute onset human atrial fibrillation is associated with local cardiac platelet activation and endothelial dysfunction. J Am Coll Cardiol 2008;51:1790–3.

[87] Scridon A, Girerd N, Rugeri L, Nonin-Babary E, Chevalier P. Progressive endothelial damage revealed by multilevel von Willebrand factor plasma concentrations in atrial fibrillation patients. Europace 2013;15:1562–6.

[88] Ammash N, Konik EA, McBane RD, Chen D, Tange JI, Grill DE, Herges RM, McLeod TG, Friedman PA, Wysokinski WE. Left atrial blood stasis and Von Willebrand factor-ADAMTS13 homeostasis in atrial fibrillation. Arterioscler Thromb Vasc Biol 2011;31:2760–6.

[89] Girerd N, Scridon A, Bessiere F, Chauveau S, Geloen A, Boussel L, Morel E, Chevalier P. Periatrial epicardial fat is associated with markers of endothelial dysfunction in patients with atrial fibrillation. PLoS ONE 2013;8.

[90] Uemura T, Kaikita K, Yamabe H, Soejima K, Matsukawa M, Fuchigami S, Tanaka Y, Morihisa K, Enomoto K, Sumida H, Sugiyama S, Ogawa H. Changes in plasma von Willebrand factor and ADAMTS13 levels associated with left atrial remodeling in atrial fibrillation. Thromb Res 2009;124:28–32.

[91] Freynhofer MK, Gruber SC, Bruno V, Hochtl T, Farhan S, Zaller V, Wojta J, Huber K. Prognostic value of plasma von Willebrand factor and its cleaving protease ADAMTS13 in patients with atrial fibrillation. Int J Cardiol 2013;168:317–25.

[92] Freynhofer MK, Bruno V, Jarai R, Gruber S, Hochtl T, Brozovic I, Farhan S, Wojta J, Huber K. Levels of von Willebrand factor and ADAMTS13 determine clinical outcome after cardioversion for atrial fibrillation. Thromb Haemost 2011;105:435–43.

[93] Gustafsson C, Blomback M, Britton M, Hamsten A, Svensson J. Coagulation factors and the increased risk of stroke in nonvalvular atrial fibrillation. Stroke 1990;21:47–51.

[94] Li-Saw-Hee FL, Blann AD, Lip GY. A cross-sectional and diurnal study of thrombogenesis among patients with chronic atrial fibrillation. J Am Coll Cardiol 2000;35:1926–31.

[95] Hatzinikolaou-Kotsakou E, Kartasis Z, Tziakas D, Hotidis A, Stakos D, Tsatalas K, Bourikas G, Kotsakou ME, Hatseras DI. Atrial fibrillation and hypercoagulability: dependent on clinical factors or/and on genetic alterations? J Thromb Thrombolysis 2003;16:155–61.
[96] Freestone B, Chong AY, Lim HS, Blann A, Lip GY. Angiogenic factors in atrial fibrillation: a possible role in thrombogenesis? Ann Med 2005;37:365–72.
[97] Alonso A, Tang W, Agarwal SK, Soliman EZ, Chamberlain AM, Folsom AR. Hemostatic markers are associated with the risk and prognosis of atrial fibrillation: the ARIC study. Int J Cardiol 2012;155:217–22.
[98] Conway DS, Pearce LA, Chin BS, Hart RG, Lip GY. Prognostic value of plasma von Willebrand factor and soluble P-selectin as indices of endothelial damage and platelet activation in 994 patients with nonvalvular atrial fibrillation. Circulation 2003;107:3141–5.
[99] Krishnamoorthy S, Khoo CW, Lim HS, Lane DA, Pignatelli P, Basili S, Violi F, Lip GY. Prognostic role of plasma von Willebrand factor and soluble E-selectin levels for future cardiovascular events in a 'real-world' community cohort of patients with atrial fibrillation. Eur J Clin Invest 2013;43:1032–8.
[100] Roldan V, Marin F, Muina B, Torregrosa JM, Hernandez-Romero D, Valdes M, Vicente V, Lip GY. Plasma von Willebrand factor levels are an independent risk factor for adverse events including mortality and major bleeding in anticoagulated atrial fibrillation patients. J Am Coll Cardiol 2011;57:2496–504.
[101] Peyvandi F, Scully M, Kremer Hovinga JA, Knöbl P, Cataland S, De Beuf K, Callewaert F, De Winter H, Zeldin RK. Caplacizumab reduces the frequency of major thromboembolic events, exacerbations and death in patients with acquired thrombotic thrombocytopenic purpura. J Thromb Haemost 2017;15:1448–52.
[102] De Meyer SF, de Maeyer B, Deckmyn H, Vanhoorelbeke K. Von Willebrand factor: drug and drug target. Cardiovasc Hematol Disord Drug Targets 2009;9:9–20.
[103] Kopić A, Benamara K, Piskernik C, Plaimauer B, Horling F, Höbarth G, Ruthsatz T, Dietrich B, Muchitsch EM, Scheiflinger F, Turecek M, Höllriegl W. Preclinical assessment of a new recombinant ADAMTS-13 drug product (BAX930) for the treatment of thrombotic thrombocytopenic purpura. J Thromb Haemost 2016;14:1410–9.

Recommended Reading

[1] Rumley A, Lowe GD, Sweetnam PM, Yarnell JW, Ford RP. Factor VIII, von Willebrand factor and the risk of major ischaemic heart disease in the Caerphilly Heart Study. Br J Haematol 1999;105:110–6.
[2] Woodward M, Rumley A, Welsh P, MacMahon S, Lowe GDO. A comparison of the associations between seven hemostatic or inflammatory variables and coronary heart disease. J Thromb Haemost 2007;5:1795–800.
[3] Sonneveld MA, Cheng JM, Oemrawsingh RM, de Maat MP, Kardys I, Garcia-Garcia HM, van Geuns RJ, Regar E, Serruys PW, Boersma E, Akkerhuis KM, Leebeek FW. Von Willebrand factor in relation to coronary plaque characteristics and cardiovascular outcome. Results of the ATHEROREMO-IVUS study. Thromb Haemost 2015;113:577–84.
[4] Wannamethee SG, Whincup PH, Lennon L, Rumley A, Lowe GDO. Fibrin d-dimer, tissue-type plasminogen activator, von Willebrand factor, and risk of incident stroke in older men. Stroke 2012;43:1206–11.
[5] Nossent AY, van Marion V, van Tilburg NH, Rosendaal FR, Bertina RM, van Mourik JA, Eikenboom HC. Von Willebrand factor and its propeptide: the influence of secretion and clearance on protein levels and the risk of venous thrombosis. J Thromb Haemost 2006;4:2556–62.

[6] Alonso A, Tang W, Agarwal SK, Soliman EZ, Chamberlain AM, Folsom AR. Hemostatic markers are associated with the risk and prognosis of atrial fibrillation: the ARIC study. Int J Cardiol 2012;155:217–22.

[7] Goldberger JJ, Arora R, Green D, Greenland P, Lee DC, Lloyd-Jones DM, Markl M, Ng J, Shah SJ. Evaluating the atrial myopathy underlying atrial fibrillation: identifying the arrhythmogenic and thrombogenic substrate. Circulation 2015;132:278–91.

15

Factor VIII/Von Willebrand Factor: The Janus of Coagulation

The Roman God, Janus the Gatekeeper, controls beginnings and passages; analogously, the FVIII/VWF complex links the initiation with the amplification phases of coagulation. The unique value of this association has been noted by Pipe et al., who write "each protein is a separate gene product but the processes they regulate are coordinated and critical to maintenance of hemostasis" [1]. VWF initiates a series of events that require FVIII to complete the journey to a strong platelet-fibrin clot. Fig. 15.1 shows a depiction of Janus and an electronmicrograph of a portion of the FVIII/VWF complex.

The relationship between FVIII and VWF remained obscure for many decades. Early workers found FVIII activity near the void volume of chromatographic columns, raised rabbit antisera against this material, and observed immunoprecipitin lines with normal and hemophilic plasma, but not with plasma from patients with severe Von Willebrand disease (VWD) [2]. This was very puzzling until dialysis experiments revealed that the large protein in plasma could be dissociated into two discrete components [3]. The large-molecular-weight protein was absent in VWD and the smaller one, in hemophilia. The characteristics of the two proteins were displayed in Table 3.1. FVIII and VWF are held together by noncovalent bonds [4], and their association is mediated primarily by the FVIII C1 domain (Fig. 15.2) [5], with secondary binding sites the acidic a3 peptide located at the A3 terminus and a site in the C2 domain [6]. In addition, the interaction of FVIII with VWF requires full sulfation of Tyr1680 on FVIII [7]. The major binding site for FVIII on VWF is located within the first 272 amino acids of the mature VWF protein, in the D'D3 region [8], and the positively charged VWD-D' surface is thought to bind the acidic a3 peptide of FVIII [5,6]. The molar ratio of circulating FVIII (0.8 nM) to VWF monomer (35 nM) is 1:44, but when forming a complex, each VWF monomer binds one FVIII molecule [9].

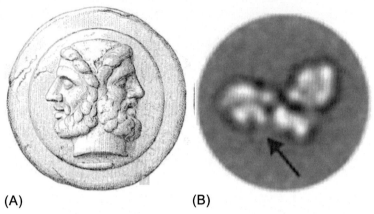

FIG. 15.1 (A) Roman coin depicting Janus. (B) Electronmicrograph of human FVIII in complex with the dimeric D'D3 domain of VWF. The *black arrow* shows the electron-dense D' handle of VWF interacting with the FVIII C domain. *Modified from Yee A, Oleskie AN, Dosey AM, Kretz CA, Gildersleeve RD, Dutta S, Su M, Ginsburg D, Skiniotis G. Visualization of an N-terminal fragment of von Willebrand factor in complex with factor VIII. Blood 2015;126:939–42, and reproduced with permission from the American Society of Hematology.*

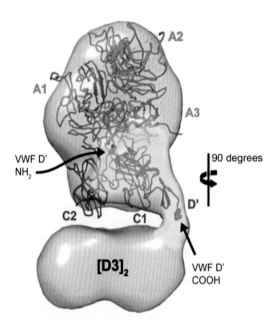

FIG. 15.2 The D9 domain interfaces primarily with the FVIII C1 domain. 3D reconstruction of the murine [D9D3]2-FVIII ternary complex (determined from 962 projections) reveals the D9 density (suggested by the *yellow shell*) extending from the [D3]2 core and enmeshed with the region corresponding to the FVIII C domains. The structure of FVIII14 (Protein Data Bank: 2R7E) fits the EM density in a unique orientation (A1, *olive*; A2, *green*; A3, *cyan*; C1, *blue*; C2, *purple*). The suboptimal agreement between the EM envelope and the position of the C domains suggests that they likely undergo conformational changes upon interaction with VWF D9. *Modified from Yee A, Oleskie AN, Dosey AM, Kretz CA, Gildersleeve RD, Dutta S, Su M, Ginsburg D, Skiniotis G. Visualization of an N-terminal fragment of von Willebrand factor in complex with factor VIII. Blood 2015;126:939–42, and reproduced with permission from the American Society of Hematology.*

The extent that FVIII binds to VWF varies among individuals and between populations; in African-Americans, common VWF coding variants decrease FVIII binding, and population studies report significant between group differences in FVIII:VWF ratios (range, 1.12–1.20; ($P = .02$) [10]. During pregnancy, both VWF and FVIII increase, but the capacity of VWF to bind FVIII declines so that the FVIII:VWF ratio falls from 1.54 to 1.13 [11]. Lastly, the decreased FVIII and FVIII:VWF ratio and impaired binding of FVIII in patients with type 2N VWD is consistent with the concept that VWF is the major modifier of circulating FVIII levels.

ORIGIN AND CLEARANCE OF THE FVIII/VWF COMPLEX

As noted in Chapter 3, mature FVIII is tightly packed into Weibel-Palade bodies and becomes noncovalently bound to high-molecular-weight VWF, and the two proteins are secreted together [12]. FVIII mRNA is expressed in lymphatic postcapillary endothelial venules and hepatic fenestrated sinusoidal endothelial cells, while VWF mRNA is expressed in capillary and postcapillary endothelial venules [13]. FVIII transduced into umbilical vein endothelial cells also enters the Weibel-Palade bodies and can be released along with the VWF [14]. Blood outgrowth endothelial cells that express FVIII have an altered morphology and short VWF tubules; released VWF strings retain bound FVIII but fail to recruit platelets [15]. Lymphatic endothelial cells secrete FVIII without VWF [8], and platelet α-granules store and release FVIII independent of VWF [16]. In addition, the desmopressin-induced release of FVIII in patients with type 2N VWD suggests that even impaired binding to VWF does not alter FVIII storage [17,18].

Secreted FVIII without its partner, VWF, has a very short half-life. But if patients with VWD are transfused, even with hemophilic plasma, prolonged increases in FVIII levels are observed, suggesting that stored and secreted FVIII readily binds to plasma VWF [19]. The VWF protects FVIII from cleavage by FXa [20], and VWF along with FIXa, decrease FVIII inactivation by activated protein C [21].

In polarized human umbilical vein endothelial cells, VWF is constitutively secreted basolaterally; in contrast, stimulated and continuous basal release occurs apically from the Weibel-Palade bodies [22]. Agonists induce the co-release of both FVIII and VWF into the blood stream and the two proteins circulate as a stable complex. Studies show that the complex undergoes phagocytosis by macrophages in vascular beds exposed to high shear stress [23]. The receptor on the macrophage for the FVIII/VWF complex is low-density lipoprotein receptor-related protein-1 (LRP-1) [24,25], and binding is facilitated by cell surface heparan sulfate proteoglycans [26]. The binding sites on FVIII are the C1 domain for LRP-1 [27] and the A2 domain for heparan sulfate proteoglycans. A second ligand for FVIII is the low-density lipoprotein receptor (LDLR) [28]. FVIII, bound to these receptors and free of VWF, is endocytosed and degraded by lysosomes.

PARTICIPATION IN HEMOSTASIS

Processes that activate or strip endothelial cells from the vessel wall trigger blood coagulation. Stimulated endothelial cells form secretory bodies that fuse with the cell membrane, releasing uncoiled VWF strings that adhere to subendothelial collagen type VI via their A1 domain; FVIII remains bound to these strings. Although the FVIII/VWF complex released from transduced blood outgrowth endothelial cells is unable to recruit platelets [29], the complex secreted from glomerular microvascular and umbilical vein endothelial cells readily binds to platelets [12]. Under shear stress, the binding of VWF to platelet GpIbα receptors promotes the influx of Ca^{2+}, inducing platelet aggregation [30]. VWF in the FVIII/VWF complex mediates platelet adherence to the damaged vessel wall and platelet aggregation, as well as conveys FVIII to the site of injury.

The traces of thrombin that form on the platelet surface cleave FVIII at residue 1689, dissociating it from VWF [31], and convert the single-chain molecule into a heterotrimeric activation product [32]. This permits the C2 domain on the FVIIIa light chain to bind to soluble fibrin bound to the platelet $\alpha_{IIb}\beta_3$ receptor; the free FVIIIa can also bind to phosphotidylserine exposed on superactivated platelets [33]. The adjacent C1 domain of FVIIIa also contributes to platelet binding [34,35]. In addition, FVIIIa can bind to the surface of endothelial cells; the binding site on the FVIIIa is the region between Ala2318 and Tyr2332 of the C2 domain [36]. Once it is membrane bound, FVIIIa is primed to form a complex with factors IXa and X. The affinity of activated FIX (FIXa) for FVIIIa is increased 2000-fold when FVIIIa is membrane bound; the site on FVIIIa that binds FIXa is in the A3-domain, between residues1803 and 1818 [37,38], and there is evidence for a second site in the A2-domain [39]. When bound to FVIIIa, the catalytic activity of FIXa toward FX is increased approximately 10^6-fold. FX binds to the FVIIIa light chain (A3C1C2 domains) [40] and undergoes hydrolytic cleavage by the hydroxyl side chain of Ser365 of FIXa [41]. Activation of FX depends on the binding of FVIIIa to the membranes of either platelets or endothelial cells, as well as the binding of FIXa to FVIIIa, and the binding of FX to the factor VIIIa/IXa complex.

ROLE IN BLEEDING DISORDERS

The FVIII/VWF complex is hobbled in VWD by mutations that affect the assembly, binding, and clearance of VWF. Type 1 mutations are mainly located in the C and D domains, and reduce the synthesis, increase the clearance, or promote the intracellular retention of the molecule. Type 2 mutations are associated with defective multimerization or enhanced proteolysis by ADAMTS13 (type 2A), heightened binding to platelet glycoprotein Ibα (type 2B), decreased collagen binding (type 2M), or impaired binding of FVIII (type 2N). Finally, type 3 is often due to mutations involving the propeptide, resulting in the failure of VWF synthesis or secretion.

FVIII in the complex can be altered by more than 2500 mutations, but intron 22 inversions in the *F8* gene account for the majority of those that are clinically relevant. Other mutations affect critical cleavage sites, the interaction of FVIIIa with FIXa, or decrease the catalytic capacity of the tenase complex. Still other mutations increase the rate of dissociation of the A2 domain from thrombin-activated FVIII, or are associated with misfolding of the molecule, impairing its secretion. Finally, the formation of the FVIII/VWF complex is reduced by mutations that affect the binding site for VWF.

ROLE IN THROMBOTIC DISORDERS

Elevated levels of the FVIII/VWF complex are associated with thrombotic disease. In particular, the risk for **venous thromboembolism** (VTE) increases by 10% for each 10 U/dL increment in FVIII; for recurrent episodes of VTE, the risk increases by 24% [42]. Studies of patients with VTE show that those with the highest FVIII levels have a threefold higher VTE recurrence rate than those with the lowest levels, and all-cause mortality increases in a dose-response fashion with increasing percentiles of FVIII levels [43]. Conditions associated with high levels of FVIII, such as cancer and Cushing syndrome, contribute to the development of VTE. And supertherapeutic levels that occasionally occur during the treatment of hemophilia appear to provoke VTE.

Hereditability of high FVIII levels has been demonstrated in families having one or more members with elevated FVIII and a history of VTE [44]. Increased FVIII is associated with specific polymorphisms of *F8*, *VWF*, and other genes; the extent of glycation of FVIII and VWF (ABO group); and factors that promote the release of VWF from endothelial cells [45]. FVIII and VWF levels correlate within families, and the familial influence is more prominent for VWF than for FVIII [46]. Significant associations with venous thrombosis are observed for rs1039084 in the syntaxin-binding protein 5 gene ($P = .005$), but not with the scavenger receptor class A or stabilin 2 variants [47]. Also associated with VTE are genetic polymorphisms of the *LRP-1* gene that increase FVIII levels by decreasing its clearance [48], and the extent of sialylation that affects the susceptibility of FVIII and VWF to proteolysis by ADAMTS13 [49]. High levels of VWF propeptide, consistent with increased release of VWF from endothelial cells, are reported in patients with elevated levels of FVIII and a history of VTE [50].

On the other hand, **arterial thrombosis** seems more closely linked to elevated VWF than FVIII. VWF released from the Weibel-Palade bodies of activated endothelial cells dramatically increases the concentration of platelets along the vessel wall through an interaction with GP1bα [51], and mediates platelet adhesion in areas of high shear stress—the branching points where atheroma typically form [52]. Microfluidic chambers that mimic the flow conditions in stenotic arteries display enhanced platelet aggregation that is critically dependent on blood-borne VWF [53]. Other in vitro studies show that platelets subject to high shear forces develop membrane tethers that adhere to VWF and generate procoagulant microparticles [54].

Epidemiologic studies consistently show associations between elevated VWF levels, coronary atherosclerosis and myocardial infarction, and carotid intimal-medial thickness and stroke. A 4.4-fold increase in arterial thrombosis was reported in middle-aged men and women with low levels of ADAMTS13, and a 7.7-fold increase if low ADAMTS13 was combined with high VWF [55]. Other evidence supporting an association of VWF with arterial thrombotic events is the observation that the diurnal and seasonal concentrations of VWF are highest when coronary deaths are most prevalent (noon and winter) [56–58]. Also, populations known to have increased VWF levels, such as the elderly and people with non-O blood group, have higher incidences of cardiovascular events than younger persons and those with blood group O [59].

VWF levels are increased in patients with nonvalvular atrial fibrillation (AF) as compared to those in sinus rhythm, irrespective of a history of stroke [60]. In the Atherosclerosis Risk in Communities Study, the hazard ratio for incident AF associated with a 1-standard deviation increase in VWF was significantly increased [61]. Raised VWF levels in patients with AF are predictive of stroke and vascular events, and are independent risk factors for adverse events in AF patients on anticoagulant therapy [62–64].

Genetic evidence that elevated VWF levels are a cause arterial thrombosis has been inconsistent. Three single-nucleotide polymorphisms in the *VWF* gene were associated with VWF antigen concentrations but not with coronary heart disease (CHD) [65]. On the other hand, single-nucleotide polymorphisms in the *ADAMTS13* gene affecting ADAMTS13 levels were associated with ischemic stroke susceptibility in children [66]. Also, young women with high VWF, low ADAMTS13, and oral contraceptive use were noted to have an increased risk of ischemic stroke and myocardial infarction [67]. Furthermore, high levels of VWF predicted the appearance of cardiovascular disease in people with type 2 diabetes and insulin resistance, but not in those who were insulin sensitive [68].

A role for FVIII in arterial thrombosis is suggested by a prospective case-control study that found high concentrations of FVIII and enhanced thrombin generation in diabetic patients prior to the development of macrovascular complications [69]. In addition, carriers of the LDLR-rs688 allele have high levels of FVIII as well as a significantly greater risk for coronary artery disease (odds ratio: 1.48), and the risk is independent of lipid levels [70]. A retrospective study of 584 first-degree relatives of 177 patients with elevated FVIII and a first MI or peripheral arterial thrombosis observed that those with high FVIII, as compared to relatives with normal levels, had an annual incidence of a first arterial thrombotic event that was threefold increased [71]. In particular, the risks for a first MI (odds ratio: 4.3) and first peripheral arterial thrombosis (odds ratio: 8.6) were significantly increased ($P = .046$).

A genetic influence on FVIII levels and arterial thrombosis is supported by a study reporting that two haplotypes in the *F8* gene are significantly associated with ischemic events [72]. Kamphuisen et al. [73] have calculated that an elevated level of FVIII accounts for 4% of arterial thrombotic events. Finally, although men with FVIII deficiency have a nonsignificant decrease in mortality due to arterial thrombosis [74], the female carriers of hemophilia have a reduction in overall mortality and deaths from ischemic heart disease [75]. Taken together, the epidemiological and genetic evidence are consistent with the concept that atherosclerosis is a polygenic disorder, and its development and progression are likely influenced by a variety of factors, including the FVIII/VWF complex.

THE FVIII/VWF COMPLEX IN THE TREATMENT OF BLEEDING DISORDERS

Therapeutic materials containing the FVIII/VWF complex, such as cryoprecipitate, and commercial plasma-derived concentrates such as Alphanate-SD and Humate-P, can be used for the treatment of either hemophilia or VWD. Another plasma-derived product, Wilate, is approved only for the treatment of VWD. Most recombinant FVIII products also contain VWF; in fact, Advate and Afstyla are specifically formulated to take advantage of the stabilizing effect of VWF on FVIII, but they are not used to treat VWD. On the other hand, the recombinant

VWF concentrate, Vonvendi, contains only trace amounts of FVIII and is not indicated for patients with hemophilia.

Baseline levels of FVIII are measured in patients with hemophilia, and ristocetin cofactor and FVIII in those with VWD [76]. These assays are repeated after the loading dose of concentrate to assure that target levels are achieved. A trough level is measured immediately preceding the next dose to determine whether hemostatic levels are being maintained; if inadequate, doses are increased or the concentrate given more frequently. Peak and trough measurements are obtained daily until wounds are healed and the risk of bleeding has subsided. To reduce the risk of thrombosis, trough levels of FVIII should not exceed 150 IU/dL.

AREAS FOR FUTURE INVESTIGATION

Severe hemophilia is characterized by the development of hemarthroses. FVIII is required for the formation of the tenase complex (FVIII and factors IXa and X bound to a phospholipid membrane), but why is deficiency of this complex associated with a predilection to bleeding from synovial membranes? Is the integrity of the synovium maintained by the tenase complex, continuous deposition of fibrin, or breached by excessive fibrinolysis due to failure to generate thrombin activatable fibrinolysis inhibitor (TAFI)?

Why do some carriers with FVIII levels >5% have recurrent episodes of bleeding? Could uncontrolled fibrinolysis due to impaired thrombin generation and decreased formation of TAFI account for this excessive blood loss?

FVIII and VWF concentrates are being modified to prolong their half-lives, but there are opportunities to do more to increase the usefulness of these therapeutic products. First, new methods for modifying concentrates should be investigated that decrease their uptake by LRP1 and CLEM4M receptors on macrophages and endothelial cells. Second, it might be possible to develop depot or sustained release factor preparations that could be administered subcutaneously. Third, efforts should be expended to reduce the immunogenicity of FVIII concentrates. But the most immediate need is to reduce the cost of currently available concentrates. A recent study estimated the economic burden of blood disorders in the 28 countries of the European Union, Iceland, Norway, and Switzerland [77]. The total cost was €23 billion in 2012, of which €11 billion was for nonmalignant disorders (hemorrhagic diseases, anemia, etc.). It is likely that clotting factor concentrates were major contributors to these healthcare costs. Surely, executives can have their bonuses, shareholders their profits, and patients with bleeding disorders effective therapies, without putting exorbitant price tags on products. Lowering prices would increase product accessibility in the developed and developing world, and would prevent a great deal of morbidity and mortality.

Inhibitor formation occurs in up to a third of hemophiliacs exposed to currently available concentrates. The sensitivity and the specificity of methods for assaying FVIII allo- and autoantibodies need to be improved. The incidence of antibody formation could be established by conducting randomized trials comparing modified and unmodified recombinant FVIII concentrates in previously untreated patients (PUPs). Research should be undertaken to identify agents that might inhibit antibody formation when coadministered with the concentrates. Better ways to induce tolerance are needed; for example, by shifting the preponderance of immunologic activity from T-effector cells to T-regulatory cells, or by suppressing Th17 cells. New bypassing agents should be developed whose activity can be monitored and reversed, if necessary.

Specific criteria for VWD need to be formulated so that the true incidence of the disorder can be established, and an assay developed that measures the functional activities inherent in the molecule. A delivery system that enables desmopressin to be effective when given by mouth would be much appreciated by patients. It should also be possible to synthesize analogs of interleukin-11 that stimulate VWF release but are orally available and have fewer adverse reactions. Also needed are prospective studies of prophylactic therapy in patients with severe bleeding phenotypes to identify appropriate doses and dose intervals of long-acting VWF concentrates. A priority for gene therapy is the identification of vectors that do not elicit harmful immunologic reactions, and the initiation of therapeutic trials using these vectors in patients with severe type I and type 3 VWD.

In patients with thrombotic disorders, it should be possible to determine whether increased levels of FVIII/VWF are due to polymorphisms in the *F8* and *VWF* genes or to other factors, and to determine whether quantitative or qualitative characteristics of VWF are associated with atrial fibrosis, strain, and flow in models of atrial fibrillation. And finally, pharmaceutical products such as caplacizumab should be evaluated for their ability to safely reduce FVIII/VWF levels in patients at risk for thrombosis.

SUMMARY

The FVIII/VWF complex links the initiation and amplification phases of blood coagulation. The binding sites for the two proteins are located within the C1 domain of FVIII and the D'D3 domain of VWF. They become associated in the Weibel-Palade bodies of endothelial cells and are released together when these cells are stimulated. After circulating for several hours (FVIII $T/2$ is 12 h and VWF $T/2$ is 15 h), the complex binds to receptors on macrophages and endothelial cells, and is degraded. VWF is the principal modulator of circulating FVIII.

Shear stress uncoils VWF, exposing sites that bind platelets and induce platelet aggregation. FVIII binds to the membranes of activated platelets and is converted to FVIIIa by trace amounts of thrombin generated through the tissue factor-FVIIa pathway. FVIIIa in the tenase complex juxtaposes FIXa with FX, enabling the formation of FXa and the generation of sufficient thrombin for hemostasis. Mutations that decrease the activity of VWF or FVIII result in VWD or hemophilia A, respectively, and those associated with higher than normal levels of the proteins promote thrombosis. Currently, deficiencies of the proteins are managed with plasma-derived or recombinant concentrates, but they are so expensive as to be unaffordable for as many as 70% of people with bleeding disorders. In addition, there is no specific therapy for patients with thrombotic disorders associated with supernormal levels of either FVIII or VWF. This chapter concludes with suggestions for potential areas of future investigation.

References

[1] Pipe SW, Montgomery RR, Pratt KP, Lenting PJ, Lillicrap D. Life in the shadow of a dominant partner: the FVIII-VWF association and its clinical implications for hemophilia A. Blood 2016;128:2007–16.

[2] Meyer D, Lavergne J-M, Larrieu M-J, Josso F. Cross-reacting material in congenital factor VIII deficiencies (haemophilia A and von Willebrand's disease). Thromb Res 1972;1:183–96.

[3] Van Mourik JA, Bouma BN, LaBruyere WT, de Graaf S, Mochtar IA. Factor VIII, a series of homologous oligomers and a complex of two proteins. Thromb Res 1974;4:155–64.

[4] Poon M-C, Ratnoff OD. Evidence that functional subunits of antihemophilic factor (factor VIII) are linked by noncovalent bonds. Blood 1976;48:87–94.

[5] Yee A, Oleskie AN, Dosey AM, Kretz CA, Gildersleeve RD, Dutta S, Su M, Ginsburg D, Skiniotis G. Visualization of an N-terminal fragment of von Willebrand factor in complex with factor VIII. Blood 2015;126:939–42.

[6] Chiu P-L, Bou-Assaf GM, Chhabra ES, Chambers MG, Peters RT, Kulman JD, Walz T. Mapping the interaction between factor VIII and von Willebrand factor by electron microscopy and mass spectrometry. Blood 2015;126:935–8.

[7] Leyte A, van Schijndel HB, Niehrs C, Huttner WB, Verbeet MP, Mertens K, van Mourik JA. Sulfation of Tyr1680 of human blood coagulation factor VIII is essential for the interaction of factor VIII with von Willebrand factor. J Biol Chem 1991;266:740–6.

[8] Foster PA, Fulcher CA, Mari T, Titani K, Zimmerman TS. A major factor VIII binding domain resides within the amino-terminal 272 amino acid residues of von Willebrand factor. J Biol Chem 1987;262:8443–6.

[9] Vlot AJ, Koppelman SJ, van den Berg MH, Bouma BN, Sixma JJ. The affinity and stoichiometry of binding of human factor VIII to von Willebrand factor. Blood 1995;85:3150–7.

[10] Johnsen JM, Auer PL, Morrison AC, Jiao S, Wei P, Haessler J, Fox K, McGee SR, Smith JD, Carlson CS, Smith N, Boerwinkle E, Kooperberg C, Nickerson DA, Rich SS, Green D, Peters U, Cushman M, Reiner AP, NHLBI Exome Sequencing Project. Common and rare

von Willebrand factor (VWF) coding variants, VWF levels, and factor VIII levels in African Americans: the NHLBI Exome Sequencing Project. Blood 2013;122:590–7.

[11] Drury-Stewart DN, Lannert KW, Chung DW, Teramura GT, Zimring JC, Konkle BA, Gammill HS, Johnsen JM. Complex changes in von Willebrand factor-associated parameters are acquired during uncomplicated pregnancy. PLoS ONE 2014;9.

[12] Turner NA, Moake JL. Factor VIII is synthesized in human endothelial cells, packaged in Weibel-Palade bodies and secreted bound to ULVWF strings. PLoS ONE 2015;10.

[13] Pan J, Dinh TT, Rajaraman A, Lee M, Scholz A, Czupalla CJ, Kiefel H, Zhu L, Xia L, Morser J, Jiang H, Santambrogio L, Butcher EC. Patterns of expression of factor VIII and von Willebrand factor by endothelial cell subsets in vivo. Blood 2016;128:104–9.

[14] Rosenberg JB, Greengard JS, Montgomery RR. Genetic induction of a releasable pool of factor VIII in human endothelial cells. Arterioscler Thromb Vasc Biol 2000;20:2689–95.

[15] Bouwens EAM, Mourik MJ, van den Biggelaar M, Eikenboom JCJ, Voorberg J, Valentijn KM, Mertens K. Factor VIII alters tubular organization and functional properties of von Willebrand factor stored in Weibel-Palade bodies. Blood 2011;118:5947–56.

[16] Yarovoi H, Nurden AT, Montgomery RR, Nurden P, Poncz M. Intracellular interaction of von Willebrand factor and factor VIII depends on cellular context: lessons from platelet-expressed factor VIII. Blood 2005;105:4674–6.

[17] Mazurier C, Gaucher C, Jorieux S, Goudemand M. Biological effect of desmopressin in eight patients with type 2N ('Normandy') von Willebrand disease. Collaborative Group. Br J Haematol 1994;88:849–54.

[18] van den Biggelaar M, Bouwens EA, Voorberg J, Mertens K. Storage of factor VIII variants with impaired von Willebrand factor binding in Weibel-Palade bodies in endothelial cells. PLoS ONE 2011;6.

[19] Cornu P, Larrieu MJ, Caen JP, Bernard J. Transfusion studies in von Willebrand's disease: effect on bleeding time and factor VIII. Br J Haematol 1963;9:189–202.

[20] Koedam JA, Hamer RJ, Beeser-Visser NH, Bouma BN, Sixma JJ. The effect of von Willebrand factor on activation of factor VIII by factor Xa. Eur J Biochem 1990;189:229–34.

[21] Rick ME, Esmon NL, Krizek DM. Factor IX and von Willebrand factor modify the inactivation of factor VIII by activated protein C. J Lab Clin Med 1990;115:415–21.

[22] Lopes da Silva M, Cutler DF. Von Willebrand factor multimerization and the polarity of secretory pathways in endothelial cells. Blood 2016;128:277–85.

[23] Rastegarlari G, Pegon JN, Casari C, Odouard S, Navarrete AM, Saint-Lu N, van Vlijmen BJ, Legendre P, Christophe OD, Denis CV, Lenting PJ. Macrophage LRP1 contributes to the clearance of von Willebrand factor. Blood 2012;119:2126–34.

[24] Saenko EL, Yakhyaev AV, Mikhailenko I, Strickland DK, Sarafanov AG. Role of the low density lipoprotein-related protein receptor in mediation of factor VIII catabolism. J Biol Chem 1999;274:37685–92.

[25] Lenting PJ, Neels JG, van den Berg BM, Clijsters PP, Meijerman DW, Pannekoek H, van Mourik JA, Mertens K, van Zonneveld AJ. The light chain of factor VIII comprises a binding site for low density lipoprotein receptor-related protein. J Biol Chem 1999;274:23734–9.

[26] Sarafanov AG, Ananyeva NM, Shima M, Saenko EL. Cell surface heparan sulfate proteoglycans participate in factor VIII catabolism mediated by low density lipoprotein receptor-related protein. J Biol Chem 2001;276:11970–9.

[27] Bloem E, van den Biggelaar M, Wroblewska A, Voorberg J, Faber JH, Kjalke M, Stennicke HR, Mertens K, Meijer AB. Factor VIII C1 domain spikes 2092-2093 and 2158-2159 comprise regions that modulate cofactor function and cellular uptake. J Biol Chem 2013;288:29670–9.

[28] Bovenschen N, Mertens K, Hu L, Havekes LM, van Vijmen BJM. LDL receptor cooperates with LDL receptor-related protein in regulating plasma levels of coagulation factor VIII in vivo. Blood 2005;106:906–12.

[29] Wang JW, Bouwens EA, Pintao MC, Voorberg J, Safdar H, Valentijn KM, de Boer HC, Mertens K, Reitsma PH, Eikenboom J. Analysis of the storage and secretion of von Willebrand factor in blood outgrowth endothelial cells derived from patients with von Willebrand disease. Blood 2013;121:2762–72.

[30] Chow TW, Hellums JD, Moake JL, Kroll MH. Shear stress-induced von Willebrand factor binding to platelet glycoprotein Ib initiates calcium influx associated with aggregation. Blood 1992;80:113–20.

[31] Lollar P, Hill-Eubanks DC, Parker CG. Association of the factor VIII light chain with von Willebrand factor. J Biol Chem 1988;263:10451–5.

[32] Fay PJ, Anderson MT, Chavin SI, Marder VJ. The size of human factor VIII heterodimers and the effects produced by thrombin. Biochim Biophys Acta 1986;871:268–78.

[33] Gilbert GE, Novakovic VA, Shi J, Rasmussen J, Pipe SW. Platelet binding sites for factor VIII in relation to fibrin and phosphatidylserine. Blood 2015;126:1237–44.

[34] Hsu T-C, Pratt KP, Thompson AR. The factor VIII C1 domain contributes to platelet binding. Blood 2008;111:200–8.

[35] Meems H, Meijer AB, Cullinan DB, Mertens K, Gilbert GE. Factor VIII C1 domain residues Lys 2092 and Phe 2093 contribute to membrane binding and cofactor activity. Blood 2009;114:3938–46.

[36] Brinkman H-JM, Mertens K, van Mourik JA. Phospholipid-binding domain of factor VIII is involved in endothelial cell-mediated activation of factor X by factor IXa. Arterioscler Thromb Vasc Biol 2002;22:511–6.

[37] Lenting PJ, van de Loo JW, Donath MJ, et al. The sequence Glu1811-Lys1818 of human blood coagulation factor VIII comprises a binding site for activated factor IX. J Biol Chem 1996;271:1935–40.

[38] Bovenschen N, Boertjes RC, van Stempvoort G, et al. Low density lipoprotein receptor-related protein and factor IXa share structural requirements for binding to the A3 domain of coagulation factor VIII. J Biol Chem 2003;278:9370–7.

[39] Bajaj SP, Schmidt AE, Mathur A, et al. Factor IXa:factor VIIIa interaction. Helix 330-338 of factor IXa interacts with residues 558-565 and spatially adjacent regions of the a2 subunit of factor VIIIa. J Biol Chem 2001;276:16302–9.

[40] Takeyama M, Wakabayashi H, Fay PJ. Factor VIII light chain contains a binding site for factor X that contributes to the catalytic efficiency of factor Xase. Biochemistry 2012;51:820–8.

[41] Bajaj SP, Thompson AR. [Chapter 7]. Molecular and structural biology of factor IX. In: Colman RW, Clowes AW, Goldhaber SZ, Marder VJ, George JN, editors. Hemostasis and thrombosis. 5th ed. Philadelphia: Lippincott Williams & Wilkins; 2006. p. 140.

[42] Kraaijenhagen RA, in't Anker PS, MMW K, Reitsma PH, Prins MH, van den Ende A, Buller HR. High plasma concentration of factor VIIIc is a major risk factor for venous thromboembolism. Thromb Haemost 2000;83:5–9.

[43] Yap ES, Timp JF, Flinterman LE, van Hylckama Vlieg A, Rosendaal FR, Cannegieter SC, Lijfering WM. Elevated levels of factor VIII and subsequent risk of all-cause mortality: results from the MEGA follow-up study. J Thromb Haemost 2015;13:1833–42.

[44] Schambeck CM, Hinney K, Haubitz I, Taleghani BM, Wahler D, Keller F. Familial clustering of high factor VIII levels in patients with venous thromboembolism. Arterioscler Thromb Vasc Biol 2001;21:289–92.

[45] Jenkins PV, Rawley O, Smith OP, O'Donnell JS. Elevated factor VIII levels and risk of venous thrombosis. Br J Haematol 2012;157:653–63.

[46] Kamphuisen PW, Houwing-Duistermaat JJ, van Houwelingen HC, Eikenboom JCJ, Bertina RM, Rosendaal FR. Familial clustering of factor VIII and von Willebrand factor levels. Thromb Haemost 1998;79:323–7.

[47] Smith NL, Rice KM, Bovill EG, Cushman M, Bis JC, McKnight B, Lumley T, Glazer NL, van Hylckama Vlieg A, Tang W, Dehghan A, Strachan DP, O'Donnell CJ, Rotter JI,

Heckbert SR, Psaty BM, Rosendaal FR. Genetic variation associated with plasma von Willebrand factor levels and the risk of incident venous thrombosis. Blood 2011;117:6007–11.

[48] Vormittag R, Bencur P, Ay C, Tengler T, Vukovich T, Quehenberger P, Mannhalter C, Pabinger I. Low-density lipoprotein receptor-related protein 1 polymorphism 663 C > T affects clotting factor VIII activity and increases the risk of venous thromboembolism. J Thromb Haemost 2007;5:497–502.

[49] McGrath RT, McKinnon TAJ, Byrne B, O'Kennedy R, Terraube V, McRae E, Preston RJS, Laffan MA, O'Donnell JS. Expression of terminal α2-6-linked sialic acid on von Willebrand factor specifically enhances proteolysis by ADAMTS13. Blood 2010;115:2666–73.

[50] Nossent AY, Van Marion V, Van Tilburg NH, Rosendaal FR, Bertina RM, Van Mourik JA, Eikenboom HC. Von Willebrand factor and its propeptide: the influence of secretion and clearance on protein levels and the risk of venous thrombosis. J Thromb Haemost 2006;4:2556–62.

[51] Andre P, Denis CV, Ware J, Saffaripour S, Hynes RO, Ruggieri ZM, Wagner DD. Platelets adhere to and translocate on von Willebrand factor presented by endothelium in stimulated veins. Blood 2000;96:3322–8.

[52] Ananyeva NM, Kouiavskaia DV, Shima M, Saenko EL. Intrinsic pathway of blood coagulation contributes to thrombogenicity of atherosclerotic plaque. Blood 2002;99:4475–85.

[53] Westein E, van der Meer AD, Kuipers MJ, Frimat JP, van den Berg A, Heemskerk JW. Atherosclerotic geometries exacerbate pathological thrombus formation poststenosis in a von Willebrand factor-dependent manner. Proc Natl Acad Sci U S A 2013;110:1357–62.

[54] Reininger AJ, Heijnen HFG, Schumann H, Specht HM, Schramm W, Ruggieri ZM. Mechanism of platelet adhesion to von Willebrand factor and microparticle formation under high shear stress. Blood 2006;107:3537–45.

[55] Bongers TN, de Bruijne ELE, Dippel DWJ, de Jong AJ, Deckers JW, Poldermans D, de Maat MPM, Leebeek FWG. Lower levels of ADAMTS13 are associated with cardiovascular disease in young patients. Atherosclerosis 2009;207:250–4.

[56] Timm A, Fahrenkrug J, Jørgensen HL, Sennels HP, Goetze JP. Diurnal variation of von Willebrand factor in plasma: the Bispebjerg study of diurnal variations. Eur J Haematol 2014;93:48–53.

[57] Tofler GH, Brezinski D, Schafer AI, Czeisler CA, Rutherford JD, Willich SN, Gleason RE, Williams GH, Muller JE. Concurrent morning increase in platelet aggregability and the risk of myocardial infarction and sudden cardiac death. N Engl J Med 1987;316:1514–8.

[58] Ghebre MA, Wannamethee SG, Rumley A, Whincup PH, Lowe GDO. Prospective study of seasonal patterns in hemostatic factors in older men and their relation to excess winter coronary heart disease deaths. J Thromb Haemost 2012;10:352–8.

[59] Albanez S, Ogiwara K, Michels A, Hopman W, Grabell J, James P, Lillicrap D. Aging and ABO blood type influence von Willebrand factor and factor VIII levels through interrelated mechanisms. J Thromb Haemost 2016;14:953–63.

[60] Gustafsson C, Blomback M, Britton M, Hamsten A, Svensson J. Coagulation factors and the increased risk of stroke in nonvalvular atrial fibrillation. Stroke 1990;21:47–51.

[61] Alonso A, Tang W, Agarwal SK, Soliman EZ, Chamberlain AM, Folsom AR. Hemostatic markers are associated with the risk and prognosis of atrial fibrillation: the ARIC study. Int J Cardiol 2012;155:217–22.

[62] Conway DS, Pearce LA, Chin BS, Hart RG, Lip GY. Prognostic value of plasma von Willebrand factor and soluble P-selectin as indices of endothelial damage and platelet activation in 994 patients with nonvalvular atrial fibrillation. Circulation 2003;107:3141–5.

[63] Krishnamoorthy S, Khoo CW, Lim HS, Lane DA, Pignatelli P, Basili S, Violi F, Lip GY. Prognostic role of plasma von Willebrand factor and soluble E-selectin levels for future cardiovascular events in a 'real-world' community cohort of patients with atrial fibrillation. Eur J Clin Invest 2013;43:1032–8.

[64] Roldan V, Marin F, Muina B, Torregrosa JM, Hernandez-Romero D, Valdes M, Vicente V, Lip GY. Plasma von Willebrand factor levels are an independent risk factor for adverse events including mortality and major bleeding in anticoagulated atrial fibrillation patients. J Am Coll Cardiol 2011;57:2496–504.

[65] Van Loon JE, Kavousi M, Leebeek FW, Felix JF, Hofman A, Witteman JHC, de Maat MP. Von Willebrand factor plasma levels, genetic variations and coronary heart disease in an older population. J Thromb Haemost 2012;10:1262–9.

[66] Stoll M, Ruhle F, Witten A, Barysenka A, Arning A, Strauss CD, Nowak-Gottl U. Rare variants in the ADAMTS13 von Willebrand factor-binding domain contribute to pediatric stroke. Circ Cardiovasc Genet 2016;9:357–67.

[67] Andersson HM, Siegerink B, Luken BM, Crawley JTB, Algra A, Lane DA, Rosendaal FR. High VWF, low ADAMTS13, and oral contraceptives increase the risk of ischemic stroke and myocardial infarction in young women. Blood 2012;119:1555–60.

[68] Frankel DS, Meigs JB, Massaro JM, Wilson PWF, O'Donnell CJ, D'Agostino RB, Tofler GH. Von Willebrand factor, type 2 diabetes mellitus, and risk of cardiovascular disease. Circulation 2008;118:2533–9.

[69] Kim HK, Kim JE, Park SH, Kim YI, Nam-Goong IS, Kim ES. High coagulation factor levels and low protein C levels contribute to enhanced thrombin generation in patients with diabetes who do not have macrovascular complications. J Diabetes Complications 2014;28:365–9.

[70] Martinelli N, Girelli D, Lunghi B, Pinotti M, Marchetti G, Malerba G, Pignatti PF, Corrocher R, Olivieri O, Bernardi F. Polymorphisms at LDLR locus may be associated with coronary artery disease through modulation of coagulation factor VIII activity and independently from lipid profile. Blood 2010;116:5688–97.

[71] Bank I, Libourel EJ, Middeldorp S, Hamulyak K, van Pampus ECM, Koopman MMW, Prins MH, van der Meer J, Buller HR. Elevated levels of FVIII:C within families are associated with an increased risk for venous and arterial thrombosis. J Thromb Haemost 2005;3:79–84.

[72] Smith NL, Bis JC, Biagiotti S, Rice K, Lumley T, Kooperberg C, Wiggins KL, Heckbert SR, Psaty BM. Variation in 24 hemostatic genes and associations with non-fatal myocardial infarction and ischemic stroke. J Thromb Haemost 2008;6:45–53.

[73] Kamphuisen PW, Eikenboom CJ, Bertina RM. Elevated factor VIII levels and the risk of thrombosis. Arterioscler Thromb Vasc Biol 2001;21:731–8.

[74] Biere-Rafi S, Zwiers M, Peters M, van der Meer J, Rosendaal FR, Büller HR, Kamphuisen PW. The effect of haemophilia and von Willebrand disease on arterial thrombosis: a systematic review. Neth J Med 2010;68:207–14.

[75] Sramek A, Kriek M, Rosendaal FR. Decreased mortality of ischaemic heart disease among carriers of haemophilia. Lancet 2003;362:351–4.

[76] Laffan MA, Lester W, O'Donnell JS, Will A, Tait RC, Goodeve A, Millar CM, Keeling DM. The diagnosis and management of von Willebrand disease: a United Kingdom Haemophilia Centre Doctors Organization guideline approved by the British Committee for Standards in Haematology. Br J Haematol 2014;167:453–65.

[77] Luengo-Fernandez R, Burns R, Leal J. Economic burden of non-malignant blood disorders across Europe: a population-based cost study. Lancet Haematol 2016;3:e371–78.

Recommended Reading

[1] Pipe SW, Montgomery RR, Pratt KP, Lenting PJ, Lillicrap D. Life in the shadow of a dominant partner: the FVIII-VWF association and its clinical implications for hemophilia A. Blood 2016;128:2007–16.

[2] Haberichter SL. VWF and FVIII: the origins of a great friendship. Blood 2009;113:2813–4.

[3] Turner NA, Moake JL. Factor VIII is synthesized in human endothelial cells, packaged in Weibel-Palade bodies and secreted bound to ULVWF strings. PLoS ONE 2015;10.

[4] Gilbert GE, Novakovic VA, Shi J, Rasmussen J, Pipe SW. Platelet binding sites for factor VIII in relation to fibrin and phosphatidylserine. Blood 2015;126:1237–44.

[5] Laffan MA, Lester W, O'Donnell JS, et al. The diagnosis and management of von Willebrand disease: a United Kingdom Haemophilia Centre Doctors Organization guideline approved by the British Committee for Standards in Haematology. Br J Haematol 2014;167:453–65.

Index

Note: Page numbers followed by *f* indicate figures, and *t* indicate tables.

Printed in the United States
By Bookmasters